CLIMATE CHANGE AND THE OCEANIC CARBON CYCLE

Variables and Consequences

CLIMATE CHANGE AND THE OCEANIC CARBON CYCLE

Variables and Consequences

Edited by

Isabel Ferrera, PhD

Apple Academic Press Inc.	Apple Academic Press Inc.
3333 Mistwell Crescent	9 Spinnaker Way
Oakville, ON L6L 0A2	Waretown, NJ 08758
Canada	USA

First issued in paperback 2021

Exclusive worldwide distribution by CRC Press, a member of Taylor & Francis Group

ISBN 13: 978-1-77-463669-5 (pbk)
ISBN 13: 978-1-77-188536-2 (hbk)

Library and Archives Canada Cataloguing in Publication

Climate change and the oceanic carbon cycle : variables and consequences/edited by Isabel Ferrera, PhD.

Includes bibliographical references and index.
Issued in print and electronic formats.
ISBN 978-1-77188-536-2 (hardcover).--ISBN 978-1-315-20749-0 (PDF)

1. Chemical oceanography. 2. Carbon cycle (Biogeochemistry). 3. Climatic changes. I. Ferrera, Isabel, editor
GC117.C37C65 2017 551.46'6 C2016-907923-6 C2016-907924-4

CIP data on file with US Library of Congress

Apple Academic Press also publishes its books in a variety of electronic formats. Some content that appears in print may not be available in electronic format. For information about Apple Academic Press products, visit our website at **www.appleacademicpress.com** and the CRC Press website at **www.crcpress.com**

About the Editor

ISABEL FERRERA, PhD

Dr. Isabel Ferrera holds a PhD from the Autonomous University of Barcelona since 2004. After a long postdoctoral stay in the USA, she joined the Marine Sciences Institute in Barcelona where she carries out research on the ecology of marine bacteria. In recent years she has specialized in the study of photoheterotrophic bacteria and on how their diversity and activity influence biogeochemical cycling in the ocean. She is author of more than 30 publications and has extensive experience in teaching in the field of environmental microbiology.

Contents

Acknowledgment and How to Cite .... *ix*

List of Contributors .... *xi*

Introduction .... *xvii*

Part I: Understanding the Importance of Ocean Biogeochemistry

1. Grand Challenges in Marine Biogeochemistry 3

Eric P. Achterberg

Part II: Quantifying Oceanic Carbon Variables

2. A Statistical Gap-Filling Method to Interpolate Global Monthly Surface Ocean Carbon Dioxide Data 17

Steve D. Jones, Corinne Le Quéré, Christian Rödenbeck, Andrew C. Manning, and Are Olsen

3. The Seasonal Sea-Ice Zone in the Glacial Southern Ocean as a Carbon Sink 63

Andrea Abelmann, Rainer Gersonde, Gregor Knorr, Xu Zhang, Bernhard Chapligin, Edith Maier, Oliver Esper, Hans Friedrichsen, Gerrit Lohmann, Hanno Meyer, and Ralf Tiedemann

4. On the Influence of Interseasonal Sea Surface Temperature on Surface Water pCO_2 at 49.0°N/16.5°W and 56.5°N/52.6°W in the North Atlantic Ocean 97

Nsikak U. Benson, Oladele O. Osibanjo, Francis E. Asuquo, and Winifred U. Anake

5. Carbon Export by Small Particles in the Norwegian Sea 107

Giorgio Dall'Olmo and Kjell Arne Mork

Part III: Phytoplankton and Oceanic Carbon Cycle

6. Ubiquitous Healthy Diatoms in the Deep Sea Confirm Deep Carbon Injection by the Biological Pump 125

S. Agusti, J. I. González-Gordillo, D. Vaqué, M. Estrada, M. I. Cerezo, G. Salazar, J. M. Gasol, and C. M. Duarte

7. **Carbon Export Efficiency and Phytoplankton Community Composition in the Atlantic Sector of the Arctic Ocean 149**
Frédéric A. C. Le Moigne, Alex J. Poulton, Stephanie A. Henson, Chris J. Daniels, Glaucia M. Fragoso, Elaine Mitchell, Sophie Richier, Benjamin C. Russell, Helen E. K. Smith, Geraint A. Tarling, Jeremy R. Young, and Mike Zubkov

Part IV: Ocean Acidification

8. **Ocean Warming–Acidification Synergism Undermines Dissolved Organic Matter Assembly 191**
Chi-Shuo Chen, Jesse M. Anaya, Eric Y-T Chen, Erik Farr, and Wei-Chun Chin

9. **Ocean Acidification with (De)Eutrophication Will Alter Future Phytoplankton Growth and Succession 207**
Kevin J. Flynn, Darren R. Clark, Aditee Mitra, Heiner Fabian, Per J. Hansen, Patricia M. Glibert, Glen L. Wheeler, Diane K. Stoecker, Jerry C. Blackford, and Colin Brownlee

10. ***Coccolithophore* Calcification Response to Past Ocean Acidification and Climate Change 219**
Sarah A. O'Dea, Samantha J. Gibbs, Paul R. Bown, Jeremy R. Young, Alex J. Poulton, Cherry Newsam, and Paul A. Wilson

11. **Near-Shore Antarctic pH Variability has Implications for the Design of Ocean Acidification Experiments 239**
Lydia Kapsenberg, Amanda L. Kelley, Emily C. Shaw, Todd R. Martz, and Gretchen E. Hofmann

Author Notes 267

Index 273

Acknowledgment and How to Cite

The editor and publisher thank each of the authors who contributed to this book. The chapters in this book were previously published elsewhere. To cite the work contained in this book and to view the individual permissions, please refer to the citation at the beginning of each chapter. Each chapter was carefully selected by the editor; the result is a book that looks at climate change and the oceanic cycle from a variety of perspectives. The chapters included are broken into four sections.

List of Contributors

Andrea Abelmann
Alfred Wegener Institute Helmholtz Centre for Polar and Marine Research, Am Alten Hafen 26, Bremerhaven 27568, Germany

Eric P. Achterberg
Chemical Oceanography Group, Marine Biogeochemistry Division, GEOMAR Helmholtz Centre for Ocean Research Kiel, Kiel, Germany

S. Agusti
Red Sea Research Center, King Abdullah University of Science and Technology (KAUST), Thuwal 23955-6900, Kingdom of Saudi Arabia and Department of Global Change Research, IMEDEA (CSIC-UIB), Miquel Marqués 21, Esporles 07190, Spain

Winifred U. Anake
Department of Chemistry, School of Natural and Applied Sciences, Covenant University, Ota, Ogun State, Nigeria.

Jesse M. Anaya
School of Natural Sciences, University of California Merced, Merced, California, United States of America

Francis E. Asuquo
Institute of Oceanography, University of Calabar, P. M. B. 1115, Calabar, Cross River State, Nigeria.

Nsikak U. Benson
Department of Chemistry, School of Natural and Applied Sciences, Covenant University, Ota, Ogun State, Nigeria.

Jerry C. Blackford
Plymouth Marine Laboratory, Prospect Place, Plymouth PL1 3DH, UK

Paul R. Bown
Department of Earth Sciences, University College London, Gower Street, London WC1E 6BT, UK

Colin Brownlee
Marine Biological Association, Citadel Hill, Plymouth PL1 2PB, UK

M. I. Cerezo
Department of Global Change Research, IMEDEA (CSIC-UIB), Miquel Marqués 21, Esporles 07190, Spain

Bernhard Chapligin
Alfred Wegener Institute Helmholtz Centre for Polar and Marine Research, Am Alten Hafen 26, Bremerhaven 27568, Germany

Chi-Shuo Chen
School of Engineering, University of California Merced, Merced, California, United States of America

Eric Y-T Chen
School of Engineering, University of California Merced, Merced, California, United States of America

Wei-Chun Chin
School of Engineering, University of California Merced, Merced, California, United States of America

Darren R. Clark
Plymouth Marine Laboratory, Prospect Place, Plymouth PL1 3DH, UK

Giorgio Dall'Olmo
Plymouth Marine Laboratory, Plymouth, UK and National Centre for Earth Observation, Plymouth, UK

Chris J. Daniels
University of Southampton, European Way, Southampton, UK

C. M. Duarte
Red Sea Research Center, King Abdullah University of Science and Technology (KAUST), Thuwal 23955-6900, Kingdom of Saudi Arabia and Department of Global Change Research, IMEDEA (CSIC-UIB), Miquel Marqués 21, Esporles 07190, Spain

Oliver Esper
Alfred Wegener Institute Helmholtz Centre for Polar and Marine Research, Am Alten Hafen 26, Bremerhaven 27568, Germany

M. Estrada
Institut de Ciències del Mar, CSIC, Passeig Marítim de la Barceloneta 37-49, Barcelona, Catalunya E 08003, Spain

Heiner Fabian
Centre for Sustainable Aquatic Research, Swansea University, Swansea SA2 8PP, UK

Erik Farr
School of Natural Sciences, University of California Merced, Merced, California, United States of America

Kevin J. Flynn
Centre for Sustainable Aquatic Research, Swansea University, Swansea SA2 8PP, UK

Glaucia M. Fragoso
University of Southampton, European Way, Southampton, UK

Hans Friedrichsen
Free University of Berlin, Berlin 14195, Germany

J. M. Gasol
Institut de Ciències del Mar, CSIC, Passeig Marítim de la Barceloneta 37-49, Barcelona, Catalunya E 08003, Spain

Rainer Gersonde
Alfred Wegener Institute Helmholtz Centre for Polar and Marine Research, Am Alten Hafen 26, Bremerhaven 27568, Germany

Samantha J. Gibbs
Department of Ocean and Earth Science, National Oceanography Centre, Southampton, University of Southampton, Waterfront Campus, European Way, Southampton SO14 3ZH, UK

Patricia M. Glibert
University of Maryland Center for Environmental Science, Horn Point Laboratory, PO Box 775, Cambridge, MD 21613, USA

J. I. González-Gordillo
Department of Biology, Campus de Excelencia Internacional del Mar (CEIMAR), Universidad de Cádiz, Puerto Real (Cádiz) 11510, Spain

Per J. Hansen
Marine Biological Section, University of Copenhagen, Strandpromenaden 5, 3000 Helsingør, Denmark

Stephanie A. Henson
National Oceanography Centre, European Way, Southampton, UK

Gretchen E. Hofmann
Department of Ecology Evolution and Marine Biology, University of California Santa Barbara, Santa Barbara, California, 93106, United States of America

Steve D. Jones
Tyndall Centre for Climate Change Research, University of East Anglia, Norwich, UK. Now at College of Life and Environmental Sciences, University of Exeter, Exeter, UK

Lydia Kapsenberg
Department of Ecology Evolution and Marine Biology, University of California Santa Barbara, Santa Barbara, California, 93106, United States of America

Amanda L. Kelley
Department of Ecology Evolution and Marine Biology, University of California Santa Barbara, Santa Barbara, California, 93106, United States of America

Gregor Knorr
Alfred Wegener Institute Helmholtz Centre for Polar and Marine Research, Am Alten Hafen 26, Bremerhaven 27568, Germany and Cardiff School of Earth and Ocean Sciences, Cardiff, Wales CF10 3AT, UK

Frédéric A. C. Le Moigne
National Oceanography Centre, European Way, Southampton, UK and GEOMAR, Helmholtz Centre for Ocean Research, Kiel, Germany

Corinne Le Quéré
Tyndall Centre for Climate Change Research, University of East Anglia, Norwich, UK

Gerrit Lohmann
Alfred Wegener Institute Helmholtz Centre for Polar and Marine Research, Am Alten Hafen 26, Bremerhaven 27568, Germany

Edith Maier
Alfred Wegener Institute Helmholtz Centre for Polar and Marine Research, Am Alten Hafen 26, Bremerhaven 27568, Germany

Andrew C. Manning
Centre for Ocean and Atmospheric Sciences, School of Environmental Sciences, University of East Anglia, Norwich, UK

Todd R. Martz
Scripps Institution of Oceanography, University of California San Diego, La Jolla, California, 92093, United States of America

Hanno Meyer
Alfred Wegener Institute Helmholtz Centre for Polar and Marine Research, Am Alten Hafen 26, Bremerhaven 27568, Germany

Elaine Mitchell
Scottish Association for Marine Sciences, Oban, Argyll, UK

Aditee Mitra
Centre for Sustainable Aquatic Research, Swansea University, Swansea SA2 8PP, UK

Kjell Arne Mork
Institute for Marine Research, Bergen, Norway and Centre for Climate Dynamics, Bergen, Norway

Cherry Newsam
Department of Ocean and Earth Science, National Oceanography Centre, Southampton, University of Southampton, Waterfront Campus, European Way, Southampton SO14 3ZH, UK, Department of Earth Sciences, University College London, Gower Street, London WC1E 6BT, UK and Department of Earth Sciences, University College London, Gower Street, London WC1E 6BT, UK

Sarah A. O'Dea
Department of Ocean and Earth Science, National Oceanography Centre, Southampton, University of Southampton, Waterfront Campus, European Way, Southampton SO14 3ZH, UK

Are Olsen
Geophysical Institute, University of Bergen and Bjerknes Centre for Climate Research, Bergen, Norway

Oladele O. Osibanjo
Department of Chemistry, University of Ibadan, Ibadan, Oyo State, Nigeria.

Alex J. Poulton
National Oceanography Centre, European Way, Southampton, UK

Sophie Richier
University of Southampton, European Way, Southampton, UK

Christian Rödenbeck
Max Planck Institute for Biogeochemistry, Jena, Germany

Benjamin C. Russell
University of Southampton, European Way, Southampton, UK

G. Salazar
Institut de Ciències del Mar, CSIC, Passeig Marítim de la Barceloneta 37-49, Barcelona, Catalunya E 08003, Spain

Emily C. Shaw
Biophysical Remote Sensing Group, School of Geography, Planning and Environmental Management, The University of Queensland, Brisbane, Queensland, 4072, Australia

Helen E. K. Smith
University of Southampton, European Way, Southampton, UK

Diane K. Stoecker
University of Maryland Center for Environmental Science, Horn Point Laboratory, PO Box 775, Cambridge, MD 21613, USA

Geraint A. Tarling
British Antarctic Survey, Natural Environment Research Council, High Cross, Cambridge, UK

Ralf Tiedemann
Alfred Wegener Institute Helmholtz Centre for Polar and Marine Research, Am Alten Hafen 26, Bremerhaven 27568, Germany

D. Vaqué
Institut de Ciències del Mar, CSIC, Passeig Marítim de la Barceloneta 37-49, Barcelona, Catalunya E 08003, Spain

Glen L. Wheeler
Plymouth Marine Laboratory, Prospect Place, Plymouth PL1 3DH, UK and Marine Biological Association, Citadel Hill, Plymouth PL1 2PB, UK

Paul A. Wilson
Department of Ocean and Earth Science, National Oceanography Centre, Southampton, University of Southampton, Waterfront Campus, European Way, Southampton SO14 3ZH, UK

Jeremy R. Young
Department of Earth Sciences, University College London, Gower Street, London WC1E 6BT, UK

Xu Zhang
Alfred Wegener Institute Helmholtz Centre for Polar and Marine Research, Am Alten Hafen 26, Bremerhaven 27568, Germany

Mike Zubkov
National Oceanography Centre, European Way, Southampton, UK

Introduction

Oceans cover about 70 percent of the Earth's surface and contain roughly 97 percent of the Earth's water supply, representing the largest ecosystem in our planet. Microorganisms are abundant in marine waters and, due to their vast numbers, they play critical roles in the planet's functioning, being responsible for the flow and fate of carbon and nutrients. Likewise, their activity can influence the properties of the atmosphere and thus, climate at planetary scales. Marine ecosystems are, though, being increasingly threatened by growing human pressures, including climate change. Understanding the consequences that climate change may have is crucial to predict the future of our oceans. Rising temperatures and ocean acidification may profoundly alter the mode of matter and energy transformation in marine ecosystems, which could have irreversible consequences for our planet on ecological timescales. For that reason, the scientific community has engaged in the grand challenge of studying the variables and consequences of oceanic carbon cycling in the context of climate change, which has emerged as a relevant field of science. The articles included in this compendium provide an overview of the topic highlighting the importance of marine plankton in carbon processing as well as the effects of rising CO_2 and temperature in their functioning.

—Isabel Ferrera

The ocean plays a central role in our earth's climate system and also provides a range of important ecosystem services, including food, energy, transport, and nutrient cycling. Marine biogeochemistry focuses on the study of complex biological, chemical, and physical processes involved in the cycling of key chemical elements within the ocean, and between the ocean and the seafloor, land and atmosphere. The ocean is increasingly perturbed by human induced alterations to our planet, including anthropogenic emissions of nitrogen, phosphorus, carbon and trace elements, and climate change. The establishment of a detailed understanding of biogeochemical processes, including their rates, is essential to the identification and assessment of climatic and chemical feedbacks associated with changes in the chemical and physical environment that are mediated through ocean biology, chemistry and physics. Chapter 1, by Achterberg, details important research areas in marine biogeochemistry involving the cycling of organic and inorganic forms of carbon, nitrogen and phosphorus, the cycling and biological roles of essential trace elements, and the fate and climatic impact of marine produced trace gases.

In Chapter 2, Jones and colleagues have developed a statistical gap-filling method adapted to the specific coverage and properties of observed fugacity of surface ocean CO_2 (fCO_2). We have used this method to interpolate the Surface Ocean CO_2 Atlas (SOCAT) v2 database on a 2.5°×2.5° global grid (south of 70°N) for 1985–2011 at monthly resolution. The method combines a spatial interpolation based on a "radius of influence" to determine nearby similar fCO_2 values with temporal harmonic and cubic spline curve-fitting, and also fits long-term trends and seasonal cycles. Interannual variability is established using deviations of observations from the fitted trends and seasonal cycles. An uncertainty is computed for all interpolated values based on the spatial and temporal range of the interpolation. Tests of the method using model data show that it performs as well as or better than previous regional interpolation methods, but in addition it provides a near-global and interannual coverage.

Reduced surface–deep ocean exchange and enhanced nutrient consumption by phytoplankton in the Southern Ocean have been linked to lower glacial atmospheric CO_2. However, identification of the biological and physical conditions involved and the related processes remains incomplete. In Chapter 3, Abelmann and colleagues specify Southern Ocean surface–subsurface contrasts using a new tool, the combined oxygen and silicon isotope measurement of diatom and radiolarian opal, in combination with numerical simulations. The data do not indicate a permanent glacial halocline related to melt water from icebergs. Corroborated by numerical simulations, the authors find that glacial surface stratification was variable and linked to seasonal sea-ice changes. During glacial spring–summer, the mixed layer was relatively shallow, while deeper mixing occurred during fall–winter, allowing for surface-ocean refueling with nutrients from the deep reservoir, which was potentially richer in nutrients than today. This generated specific carbon and opal export regimes turning the glacial seasonal sea-ice zone into a carbon sink.

In Chapter 4, Benson and colleagues employed the sea surface temperature (SST) and partial pressure of carbon dioxide (pCO_2) derived from hourly in situ measurements at Northwest (56.5°N, 52.6°W) and Northeast (49.0°N, 16.5°W) subpolar sites of the Atlantic Ocean from 2003–2005 to investigate the seasonal pCO_2–SST relationship. The results indicate weak to moderately strong significant negative relationships ($r = -0.04$ to -0.89, $p<0.0001$) and ($r = -0.56$ to -0.97, $p<0.0001$) between SST and pCO_2 for the Northeast and Northwest observed data respectively. At the Northwestern site, the variation in surface water pCO_2 might be partly controlled by the seasonal change in SST as well as biological activities and other physical processes. The variability in pCO_2 distribution at the Northeastern oceanographic site were attributed principally to mixing and stratification processes during the autumn and spring seasons, while the pCO_2–SST interrelationship obtained during summertime suggested that pCO_2 variability could have been induced mainly by thermodynamic effects.

Despite its fundamental role in controlling the Earth's climate, present estimates of global organic carbon export to the deep sea are affected by relatively large uncertainties. These uncertainties are due to lack of observations as well as disagreement among methods and assumptions used to estimate carbon export. Complementary observations are thus needed to reduce these uncertainties. In Chapter 5, Dall'Olmo and Mork show that optical backscattering measured by Bio-Argo floats can detect a seasonal carbon export flux in the Norwegian Sea. This export was most likely due to small particles (i.e., 0.2–20 μm), was comparable to published export values, and contributed to long-term carbon sequestration. The findings highlight the importance of small particles and of physical mixing in the biological carbon pump and support the use of autonomous platforms as tools to improve our mechanistic understanding of the ocean carbon cycle.

The role of the ocean as a sink for CO_2 is partially dependent on the downward transport of phytoplankton cells packaged within fast-sinking particles. However, whether such fast-sinking mechanisms deliver fresh organic carbon down to the deep bathypelagic sea and whether this mechanism is prevalent across the ocean requires confirmation. In Chapter 6, Agusti and colleagues report the ubiquitous presence of healthy photosynthetic cells, dominated by diatoms, down to 4,000 m in the deep dark ocean. Decay experiments with surface phytoplankton suggested that the large proportion (18%) of healthy photosynthetic cells observed, on average, in the dark ocean, requires transport times from a few days to a few weeks, corresponding to sinking rates (124–732 $m\,d^{-1}$) comparable to those of fast-sinking aggregates and faecal pellets. These results confirm the expectation that fast-sinking mechanisms inject fresh organic carbon into the deep sea and that this is a prevalent process operating across the global oligotrophic ocean.

Arctic primary production is sensitive to reductions in sea ice cover, and will likely increase into the future. Whether this increased primary production (PP) will translate into increased export of particulate organic carbon (POC) is currently unclear. In Chapter 7, Le Moigne and colleagues report on the POC export efficiency during summer 2012 in the Atlantic sector of the Arctic Ocean. The authors coupled 234-thorium based estimates of the export flux of POC to onboard incubation-based estimates of PP. Export efficiency (defined as the fraction of PP that is exported below 100 m depth: ThE-ratio) showed large variability (0.09±0.19–1.3±0.3). The highest ThE-ratio (1.3±0.3) was recorded in a mono-specific bloom of *Phaeocystis pouchetii* located in the ice edge. Blooming diatom dominated areas also had high ThE-ratios (0.1±0.1–0.5±0.2), while mixed and/or prebloom communities showed lower ThE-ratios (0.10±0.03–0.19±0.05). Furthermore, using oxygen saturation, bacterial abundance, bacterial production, and zooplankton oxygen demand, the authors also investigated spatial variability in the degree to which this sinking material may be remineralized in the upper mesopelagic (<300 m). The results suggest that blooming diatoms and *P. pouchetii* can export a significant fraction of their biomass below the surface layer (100 m) in the open Arctic Ocean. Also, the paper shows evidence that the material sinking from a *P. pouchetii* bloom may be remineralized (>100 m) at a similar rate as the material sinking from diatom blooms in the upper mesopelagic, contrary to previous findings.

Understanding the influence of synergisms on natural processes is a critical step toward determining the full-extent of anthropogenic stressors. As carbon emissions continue unabated, two major stressors—warming and acidification—threaten marine systems on several scales. In Chapter 8, Chen and colleagues report that a moderate temperature increase (from 30°C to 32°C) is sufficient to slow— even hinder—the ability of dissolved organic matter, a major carbon pool, to self-assemble to form marine microgels, which contribute to the particulate organic matter pool. Moreover, acidification lowers the temperature threshold at which we observe our results. These findings carry implications for the marine carbon cycle, as self-assembled marine microgels generate an estimated global seawater budget of ~1016 g C. The authors used laser scattering spectroscopy to test the influence of temperature and pH on spontaneous marine gel assembly. The results of independent experiments revealed that at a particular point, both pH and temperature block microgel formation (32°C, pH 8.2), and disperse existing gels (35°C). The authors then tested the hypothesis that temperature and pH have a synergistic influence on marine gel dispersion. They found that the dispersion temperature decreases concurrently with pH: from 32°C at pH 8.2, to 28°C at pH 7.5. If the laboratory observations can be extrapolated to complex marine environments, the results suggest that a warming–acidification synergism can decrease carbon and nutrient fluxes, disturbing marine trophic and trace element cycles, at rates faster than projected.

Human activity causes ocean acidification (OA) though the dissolution of anthropogenically generated CO_2 into seawater, and eutrophication through the addition of inorganic nutrients. Eutrophication increases the phytoplankton biomass that can be supported during a bloom, and the resultant uptake of dissolved inorganic carbon during photosynthesis increases water-column pH (bloom-induced basification). This increased pH can adversely affect plankton growth. With OA, basification commences at a lower pH. Using experimental analyses of the growth of three contrasting phytoplankton under different pH scenarios, coupled with mathematical models describing growth and death as functions of pH and nutrient status, Flynn and colleagues show in Chapter 9 how different conditions of pH modify the scope for competitive interactions between phytoplankton species. The authors then use the models previously configured against experimental data to explore how the commencement of bloom-induced basification at lower pH with OA, and operating against a background of changing patterns in nutrient loads, may modify phytoplankton growth and competition. The conclude that OA and changed nutrient supply into shelf seas with eutrophication or de-eutrophication (the latter owing to pollution control) has clear scope to alter phytoplankton succession, thus affecting future trophic dynamics and impacting both biogeochemical cycling and fisheries.

Anthropogenic carbon dioxide emissions are forcing rapid ocean chemistry changes and causing ocean acidification (OA), which is of particular significance for calcifying organisms, including planktonic coccolithophores. Detailed analysis of coccolithophore skeletons enables comparison of calcite production in modern and fossil cells in order to investigate biomineralization response of ancient coccolithophores to climate change. In Chapter 10 O'Dea and colleagues show that the two dominant coccolithophore taxa across the Paleocene–Eocene Thermal Maximum (PETM) OA global warming event (~56 million years ago) exhibited morphological response to environmental change and both showed reduced calcification rates. However, only *Coccolithus pelagicus* exhibits a transient thinning of coccoliths, immediately before the PETM, that may have been OA-induced. Changing coccolith thickness may affect calcite production more significantly in the dominant modern species *Emiliania huxleyi*, but, overall, these PETM records indicate that the environmental factors that govern taxonomic composition and growth rate will most strongly influence coccolithophore calcification response to anthropogenic change.

Understanding how declining seawater pH caused by anthropogenic carbon emissions, or ocean acidification, impacts Southern Ocean biota is limited by a paucity of pH time-series. In Chapter 11, Kapsenberg presents the first high-frequency in-situ pH time-series in near-shore Antarctica from spring to winter under annual sea ice. Observations from autonomous pH sensors revealed a seasonal increase of 0.3 pH units. The summer season was marked by an increase in temporal pH variability relative to spring and early winter, matching coastal pH variability observed at lower latitudes. Using our data, simulations of ocean acidification show a future period of deleterious wintertime pH levels potentially expanding to 7–11 months annually by 2100. Given the presence of (sub)seasonal pH variability, Antarctica marine species have an existing physiological tolerance of temporal pH change that may influence adaptation to future acidification. Yet, pH-induced ecosystem changes remain difficult to characterize in the absence of sufficient physiological data on present-day tolerances. It is therefore essential to incorporate natural and projected temporal pH variability in the design of experiments intended to study ocean acidification biology.

PART I

UNDERSTANDING THE IMPORTANCE OF OCEAN BIOGEOCHEMISTRY

CHAPTER 1

Grand Challenges in Marine Biogeochemistry

ERIC P. ACHTERBERG

1.1 INTRODUCTION

The ocean plays a central role in our earth's climate system and also provides a range of important ecosystem services, including food, energy, transport, and nutrient cycling. Marine biogeochemistry focuses on the study of complex biological, chemical, and physical processes involved in the cycling of key chemical elements within the ocean, and between the ocean and the seafloor, land and atmosphere. The ocean is increasingly perturbed by human induced alterations to our planet, including anthropogenic emissions of nitrogen, phosphorus, carbon and trace elements, and climate change. The establishment of a detailed understanding of biogeochemical processes, including their rates, is essential to the identification and assessment of climatic and chemical feedbacks associated

Grand Challenges in Marine Biogeochemistry. Frontiers in Marine Science ***1***,*7 (2014). doi: 10.3389/fmars.2014.00007.*

with changes in the chemical and physical environment that are mediated through ocean biology, chemistry and physics. Important research areas in marine biogeochemistry involve the cycling of organic and inorganic forms of carbon, nitrogen and phosphorus, the cycling and biological roles of essential trace elements, and the fate and climatic impact of marine produced trace gases.

1.2 GREENHOUSE GASES

The concentrations of atmospheric greenhouse gases (GHGs; primarily CO_2, CH_4 and N_2O) are currently ca. 40% higher for CO_2, 150% for CH_4 and 19% for N_2O compared with pre-industrial levels (Myhre et al., 2013), and surpass levels seen over the past 650,000 years or more. Furthermore, there is no paleo analogue available for the present rate of increase in the atmospheric GHG concentrations. The anthropogenic inputs of the GHGs interact with their natural cycles in the oceans and troposphere, resulting in climate feedbacks and impacts on the environment. The ocean takes up about one quarter of the anthropogenic CO_2 emissions (ca. 10 Pg C per year) (Canadell et al., 2007), and as a consequence the accumulation in the atmosphere is reduced (Le Quere et al., 2009). Whilst our understanding of the oceanic carbonate system is well developed at a fundamental level, there are still important questions about the regional, seasonal, and multi-annual variability of ocean uptake of CO_2, and the biological and physical processes that determine this uptake and their variability. On-going observational studies with improved methodologies, including novel autonomous carbonate chemistry sensors (e.g., Martz et al., 2010; Rérolle et al., 2012), will provide answers to these questions over the coming years. Furthermore, we can expect an improved knowledge of the role of biological processes in oceanic CO_2 uptake from a combined use of ocean color remote sensing tools, in-situ observations and modeling activities, (e.g., Friedrich and Oschlies, 2009).

An important step-change in our understanding and quantification of the physical air-sea transfer of CO_2 and other GHGs is required, as the current modeling of this key process is underdeveloped and primarily driven through dependencies on windspeed only. New measurement approaches for GHGs are becoming available, such as eddy-covariance (Miller et al., 2010) which, with improved physical parameterization will lead to new insights and improved quantification of physical air-sea exchange of GHGs. Surface active components (surfactants) of the sea surface microlayer influence air-sea exchange through acting as a physicochemical barrier and through modifications of the sea surface hydrodynamics (Cunliffe et al., 2013). A major uncertainty in the determination of the air-sea exchange lies in the unknown role of biologically produced surfactants in modifying gas transfer, which can be suppressed by 5–50% as a result of natural phytoplankton exudates (Frew et al., 1990). Work is required in different biogeochemical regimes and over seasonal cycles and varying wind strengths to compare the effects of different surfactant levels and compositions on air-sea exchange of GHGs.

Our understanding of the global cycles of N_2O and CH_4 is less developed compared with CO_2, which is a concern as these GHGs have significantly stronger effects on radiative forcing than CO_2. Changing oceanic conditions, in terms of warming, water column stratification, nutrient status, acidification, and de-oxygenation, will potentially result in changes in oceanic emissions of these GHGs (e.g., Nevison et al., 2003; Berndt et al., 2014). However, our understanding of the production and removal of these gases is only just emerging, as are global maps of their emissions which currently only concern a 20 year old effort on N_2O by Nevison et al. (1995), and no climatology for CH_4 yet. Increased research efforts are therefore underway and for example promoted by Scientific Committee on Ocean Research (SCOR) working group 143 (Working toward a global network of ocean time series measurements of N_2O and CH_4).

1.3 CHANGING OCEANS

The polar regions are strongly influenced by climate change and will be subject to intensive research activities in future years. The Arctic region faces dramatic changes including rapid sea ice loss, alterations to CO_2 exchange, acidification, increased freshwater inputs with associated carbon supply through terrestrial run off (e.g., Stroeve et al., 2007; Manizza et al., 2013). Whilst our understanding of the productivity in the contemporary Arctic Ocean is underdeveloped, for example, in relation to primary productivity in sea ice systems (Arrigo et al., 2012), future changes to primary and bacterial productivity will be challenging to predict. Upcoming research activities will need to provide detailed insights into the functioning of biogeochemical processes and ecosystems in the Arctic Ocean and thereby facilitate projections under future climate conditions.

The Southern Ocean plays a central role in the earth climate system as a global hub in overturning circulation (Marinov et al., 2006), and a key region for ocean CO_2 uptake (Le Quere et al., 2009). This ocean region is facing important changes with reported warming of waters up to ca. 1100 m (Levitus et al., 2005), and shifts in wind patterns which involve strengthening of mid-latitude westerlies related to an increasingly positive Southern Annular Mode Index (Marshall et al., 2004). The changing wind patterns are postulated to enhance upwelling of relatively warm circumpolar deep waters (CDW) and flooding of Antarctic continental shelves (Jacobs, 2006), resulting in glacier melt (e.g., Pine Glacier) (Holland et al., 2008; Jacobs et al., 2011; Dutrieux et al., 2014) and associated sea level rise. The more pronounced upwelling of CDW near the Antarctic continent is also thought to enhance primary productivity through increased Fe supply to high nitrate low chlorophyll (HNLC) waters (Planquette et al., 2013). The shifts in wind patterns and associated changes in mixing, up- and downwelling patterns may change the Southern Ocean CO_2 sink, thereby providing unknown climatic feedbacks. The implications of changes to the Southern Ocean carbon cycle are very important to the earth climate system, but the current uncertainties are significant. Major future research efforts are foreseen in the high latitude oceans. The Arctic and Southern Ocean are currently undersampled with significant challenges of data collection using vessels due to inhospitable conditions in autumn and winter periods, when important transfers of CO_2 to deep waters occur. The more widespread deployment of autonomous platforms (e.g., gliders/Argo floats) with biogeochemical sensors is envisioned within the next years and would transform our observational capabilities (Johnson et al., 2009).

Climate change is considered to result in the expansion of the low nitrate low chlorophyll (LNLC) regions (Steinacher et al., 2010), which currently occupy ca. 60% of the world's ocean and are characterized by chlorophyll a concentrations <0.07 mg.m^{-3}, but nevertheless play an important role in the global carbon cycle (Lomas et al., 2010; Steinacher et al., 2010). Deposition of aerosols to these regions forms a key supply of nutrients (e.g., N, P) and also trace elements (e.g., Fe, Co, Mn, Zn) to surface ocean microbial communities. Whereas the atmospheric supply of Fe to HNLC regions has been a focus of attention in recent years (Jickells et al., 2005), this focus is shifting toward the LNLC regions. Indications are that the supplies to these regions are incorrectly represented in models, with the pulsed aerosol supplies providing more pronounced effects on dinitrogen fixation and heterotrophy than the continuous supply currently assumed (Guieu et al., under review). Therefore, improved long-term observations of aerosol inputs to LNLC regions, in combination with measurements of surface water chemistry and microbial community diversity and microbial rates are required. This will allow improved parameterizations of ocean biogeochemical models for LNLC regions and facilitate projections of climate change related alterations to aerosol inputs on ocean productivity, nitrogen, and carbon cycles.

Shelf seas have been receiving increased attention in recent years, partly due to the recognition of their importance for ecosystem services. They comprise only about 5% of the global ocean, yet support 15–20% of global primary productivity (Simpson and Sharples, 2012) and form an important interface between the land and the ocean. On a global scale, ca. 30% of the air-to-sea CO_2 flux occurs in shelf seas (Chen and Borges, 2009), and the shelves are estimated to contribute to >40% of particulate carbon sequestration (Muller-Karger et al., 2005; Regnier et al., 2013). The exact mechanisms responsible for these processes are not fully clear, which hampers our ability to project the effects of future climate change, and other anthropogenic pressures, on the ability of these systems to absorb CO_2 (Regnier et al., 2013). The pressures on the coastal seas include overfishing, rising water temperatures, pollution inputs, acidification, eutrophication, and deoxygenation (Rabalais et al., 2009). Therefore, further efforts are required to constrain the carbon fluxes in the contemporary shelf seas with a view to allow predictions of fluxes under future climate conditions.

1.4 MULTIPLE STRESSORS IMPACTING BIOGEOCHEMICAL PROCESSES

The influence of single forcing factors on biogeochemical cycles and marine ecosystems has been an active area of research over the last decades. This work includes the influence of macronutrients (Timmermans et al., 2004), cobalt (Saito et al., 2002) and iron (Nielsdóttir et al., 2009) additions on the functioning and structure of microbial communities, effects of temperature (Eppley, 1972), light (Falkowski and Raven, 2007) and CO_2 (Hein and Sand-Jensen, 1997) on primary productivity, and effects of iron (Schlosser et al., 2014) and CO_2 (Hutchins et al., 2007) on dinitrogen fixation. The future ocean is predicted to face a multitude of changes, including warming and increased water column stratification, reductions in ice cover, enhanced reactive nitrogen inputs, changes in atmospheric dust deposition, spread and intensification of oxygen minimum zones and ocean acidification (Gruber, 2011; Guieu et al., under review). Combined effects of two or more of these future changes on ocean biogeochemical cycles and ecosystems are challenging to predict as additive, synergistic and antagonistic effects may occur in addition to transitions in oceanic microbial communities. Research on the effects of multiple stressors on marine biogeochemical cycles and ecosystems is still in an initial phase, but will increase in volume despite significant logistical and intellectual challenges involved in the experimental and interpretation stages. A number of approaches to achieve this are currently employed, including laboratory experiments, mesocosm studies, and also oceanic observational studies conducted across strong biogeochemical and physical gradients. Current research efforts are focused on the impacts and feedbacks of combinations of forcing factors including high pCO_2, nutrient limitation, Fe availability, aerosol dust addition, oxygen, and temperature. An important outstanding question involves the hypothesized decrease in strength of the biological carbon pump due to reduced $CaCO_3$ ballasting and decreased size of phytoplankton cells in a more acidic, warmer, stratified and increasingly oligotrophic ocean. Furthermore, the combined effects of ocean acidification, de-oxygenation, water column stratification and (micro-) nutrient supply changes on dinitrogen fixation, nitrification, denitrification, and

N_2O emissions forms a key research area due to the central role of the nitrogen cycle in global ocean productivity.

1.5 BASIN SCALE OBSERVATIONS OF TRACE ELEMENTS AND ISOTOPES: GEOTRACES

Major international observational programmes including GEOSECS, JGOFS, and WOCE mapped the global oceanic distributions of the macronutrients N, P and Si and provided understanding of the processes involved in their biogeochemical cycling. The GEOTRACES programme (Henderson et al., 2007) is underway since 2004, with currently about 30 completed cruises, and as main aim to develop an understanding of the distribution and cycling of trace elements and isotopes to complement our knowledge of the macronutrients. This programme has been made possible by improved trace metal clean sampling methods and analytical techniques, and the emergence of high quality reference seawaters for trace metals (Cutter and Bruland, 2012; Anderson et al., 2014). The publications of this programme are now emerging, showing remarkable oceanic distributions of elements including Zn (Wyatt et al., 2014) and Fe (Nishioka et al., 2013) which indicate sources (including hydrothermal vents, benthic supply and atmospheric inputs), transport pathways and relationships with macronutrient cycles. The programme has also allowed us to understand the observed latitudinal migration of dinitrogen fixation in the tropical Atlantic, which is forced by the seasonal shifts of the Intertropical Convergence Zone and associated supply of the essential micronutrient Fe by wet deposition (Schlosser et al., 2014). Wet and dry aerosol inputs form an important but poorly constrained source of trace metals and nutrients to the surface ocean (Baker et al., 2007), with many outstanding questions regarding solubility of elements following aerosol deposition in the surface waters, (e.g., Baker and Croot, 2010). Aerosol collection on the GEOTRACES cruises, in combination with surface water measurements, will provide unique and important new information on the role of aerosols in marine biogeochemical cycles and ecosystems.

The emerging datasets from the GEOTRACES programme for trace elements and their isotopes will provide pivotal information for modelers to improve their models of the oceanic carbon cycle. Present models typically have at best a rather simple representation of the key micronutrient Fe, limiting the accuracy of the models' response to drivers involving changes in strengths of micronutrient supplies and removal mechanisms. Important future changes in micronutrient cycles through processes such as increased continental aridity, ocean acidification, or decreasing ocean oxygen levels can only be modeled through improvements of our representation of these processes in biogeochemical models. A key upcoming challenge will be to improve our understanding of the scavenging and stabilisation processes of trace elements (notably Fe) in the ocean. The emerging full ocean depth datasets of trace elements and indicators of elemental scavenging [radionuclides ^{234}Th and ^{238}U (Honeyman et al., 1988)], in combination with modeling approaches, will likely allow us to understand and quantify the scavenging rates in the global oceans.

1.6 CONCLUSION

Progress in marine biogeochemistry is driven by innovations in observational capabilities, which yield an improved assessment of climatic and chemical feedbacks (e.g., GEOTRACES). There is currently a drive toward increased use of biological and chemical sensors on autonomous platforms, providing large high frequency temporal and spatial data sets which will require multi-disciplinary data convolution approaches, directly linked to modeling efforts. The revolution in our data-acquisition approaches will require training of the new generation of marine biogeochemists in the statistical and numerical skills needed to handle complex chemical, biological, and physical datasets.

REFERENCES

1. Anderson, R. F., Mawji, E., Cutter, G. A., Measures, C. I., and Jeandel, C. (2014). GEOTRACES: changing the way we explore ocean chemistry. Oceanography 27, 50–61. doi: 10.5670/oceanog.2014.07

2. Arrigo, K. R., Perovich, D. K., Pickart, R. S., Brown, Z. W., Van Dijken, G. L., Lowry, K. E., et al. (2012). Massive phytoplankton blooms under arctic sea ice. Science 336, 1408. doi: 10.1126/science.1215065
3. Baker, A. R., and Croot, P. L. (2010). Atmospheric and marine controls on aerosol iron solubility in seawater. Mar. Chem. 120, 4–13. doi: 10.1016/j.marchem.2008.09.003
4. Baker, A. R., Weston, K., Kelly, S. D., Voss, M., Streu, P., and Cape, J. N. (2007). Dry and wet deposition of nutrients from the tropical Atlantic atmosphere: links to primary productivity and nitrogen fixation. Deepsea Res. I Oceanogr. Res. Papers 54, 1704–1720. doi: 10.1016/j.dsr.2007.07.001
5. Berndt, C., Feseker, T., Treude, T., Krastel, S., Liebetrau, V., Niemann, H., et al. (2014). Temporal constraints on hydrate-controlled methane seepage off svalbard. Science 343, 284–287. doi: 10.1126/science.1246298
6. Canadell, J. G., Le Quéré, C., Raupach, M. R., Field, C. B., Buitenhuis, E. T., Ciais, P., et al. (2007). Contributions to accelerating atmospheric CO2 growth from economic activity, carbon intensity, and efficiency of natural sinks. Proc. Natl. Acad. Sci. U.S.A. 104, 18866–18870. doi: 10.1073/pnas.0702737104
7. Chen, C.-T. A., and Borges, A. V. (2009). Reconciling opposing views on carbon cycling in the coastal ocean: continental shelves as sinks and near-shore ecosystems as sources of atmospheric CO2. Deep Sea Res. II Top. Stud. Oceanogr. 56, 578–590. doi: 10.1016/j.dsr2.2009.01.001
8. Cunliffe, M., Engel, A., Frka, S., Gašparović, B., Guitart, C., Murrell, J. C., et al. (2013). Sea surface microlayers: a unified physicochemical and biological perspective of the air–ocean interface. Prog. Oceanogr. 109, 104–116. doi: 10.1016/j.pocean.2012.08.004
9. Cutter, G. A., and Bruland, K. W. (2012). Rapid and noncontaminating sampling system for trace elements in global ocean surveys. Limnol. Oceanogr. Methods 10, 425–436. doi: 10.4319/lom.2012.10.425
10. Dutrieux, P., De Rydt, J., Jenkins, A., Holland, P. R., Ha, H. K., Lee, S. H., et al. (2014). Strong sensitivity of pine island ice-shelf melting to climatic variability. Science 343, 174–178. doi: 10.1126/science.1244341
11. Eppley, R. W. (1972). Temperature and phytoplankton growth in the sea. Fishery Bull. 70, 1063–1085.
12. Falkowski, P. G., and Raven, J. A. (2007). Aquatic Photosynthesis. Princeton, NJ: Princeton University Press.
13. Frew, N. M., Goldman, J. C., Dennett, M. R., and Johnson, A. S. (1990). Impact of phytoplankton-generated surfactants on air-sea gas exchange. J. Geophys. Res. Oceans 95, 3337–3352. doi: 10.1029/JC095iC03p03337
14. Friedrich, T., and Oschlies, A. (2009). Neural network-based estimates of North Atlantic surface pCO2 from satellite data: a methodological study. J. Geophys. Res. Oceans 114, C03020. doi: 10.1029/2007JC004646
15. Gruber, N. (2011). Warming up, turning sour, losing breath: ocean biogeochemistry under global change. Phil. Trans. R. Soc. A 369, 1980–1996. doi: 10.1098/rsta.2011.0003
16. Hein, M., and Sand-Jensen, K. (1997). CO2 increases oceanic primary production. Nature 388, 526–527. doi: 10.1038/41457

17. Henderson, G. M., Anderson, R. F., Adkins, J., Andersson, P., Boyle, E. A., Cutter, G., et al. (2007). GEOTRACES - an international study of the global marine biogeochemical cycles of trace elements and their isotopes. Chemie Der Erde-Geochem. 67, 85–131. doi: 10.1016/j.chemer.2007.02.001
18. Holland, P. R., Jenkins, A., and Holland, D. M. (2008). The response of ice shelf basal melting to variations in ocean temperature. J. Clim. 21, 2558–2572. doi: 10.1175/2007JCLI1909.1
19. Honeyman, B. D., Balistrieri, L. S., and Murray, J. W. (1988). Oceanic trace metal scavenging: the importance of particle concentration. Deep Sea Res. 35, 227–246. doi: 10.1016/0198-0149(88)90038-6
20. Hutchins, D. A., Fu, F. X., Zhang, Y., Warner, M. E., Feng, Y., Portune, K., et al. (2007). CO2 control of Trichodesmium N-2 fixation, photosynthesis, growth rates, and elemental ratios: implications for past, present, and future ocean biogeochemistry. Limnol. Oceanogr. 52, 1293–1304. doi: 10.4319/lo.2007.52.4.1293
21. Jacobs, S. S. (2006). Observations of change in the Southern Ocean. Philos. Trans. R. Soc. Lond. A 364, 1657–1681. doi: 10.1098/rsta.2006.1794
22. Jacobs, S. S., Jenkins, A., Giulivi, C. F., and Dutrieux, P. (2011). Stronger ocean circulation and increased melting under Pine Island Glacier ice shelf. Nat. Geosci. 4, 519–523. doi: 10.1038/ngeo1188
23. Jickells, T. D., An, Z. S., Andersen, K. K., Baker, A. R., Bergametti, G., Brooks, N., et al. (2005). Global iron connections between desert dust, ocean biogeochemistry, and climate. Science 308, 67–71. doi: 10.1126/science.1105959
24. Johnson, K. S., Berelson, W. M., Boss, E. S., Chase, Z., Claustre, H., Emerson, S. R., et al. (2009). Observing biogeochemical cycles at global scales with profiling floats and gliders: prospects for a global array. Oceanography 22, 216–225. doi: 10.5670/oceanog.2009.81
25. Le Quere, C., Raupach, M. R., Canadell, J. G., Marland, G., Bopp, L., Ciais, P., et al. (2009). Trends in the sources and sinks of carbon dioxide. Nat. Geosci. 2, 831–836. doi: 10.1038/ngeo689
26. Levitus, S., Antonov, J., and Boyer, T. (2005). Warming of the world ocean, 1955–2003. Geophys. Res. Lett. 32, L02604. doi: 10.1029/2004GL021592
27. Lomas, M. W., Steinberg, D. K., Dickey, T., Carlson, C. A., Nelson, N. B., Condon, R. H., et al. (2010). Increased ocean carbon export in the Sargasso Sea linked to climate variability is countered by its enhanced mesopelagic attenuation. Biogeosciences 7, 57–70. doi: 10.5194/bg-7-57-2010
28. Manizza, M., Follows, M. J., Dutkiewicz, S., Menemenlis, D., Hill, C. N., and Key, R. M. (2013). Changes in the Arctic Ocean CO2 sink (1996–2007): a regional model analysis. Global Biogeochem. Cycles 27, 1108–1118. doi: 10.1002/2012GB004491
29. Marinov, I., Gnanadesikan, A., Toggweiler, J. R., and Sarmiento, J. L. (2006). The Southern Ocean biogeochemical divide. Nature 441, 964–967. doi: 10.1038/nature04883
30. Marshall, G. J., Stott, P. A., Turner, J., Connolley, W. M., King, J. C., and Lachlan-Cope, T. A. (2004). Causes of exceptional atmospheric circulation changes in the Southern Hemisphere. Geophys. Res. Lett. 31, L14205. doi: 10.1029/2004GL019952

31. Martz, T. R., Connery, J. G., and Johnson, K. S. (2010). Testing the Honeywell Durafet® for seawater pH applications. Limnol. Oceanogr. Methods 8, 172–184. doi: 10.4319/lom.2010.8.172
32. Miller, S. D., Marandino, C., and Saltzman, E. S. (2010). Ship-based measurement of air-sea CO2 exchange by eddy covariance. J. Geophys. Res. Atmos. 115, D02304. doi: 10.1029/2009JD012193
33. Muller-Karger, F. E., Varela, R., Thunell, R., Luerssen, R., Hu, C., and Walsh, J. J. (2005). The importance of continental margins in the global carbon cycle. Geophys. Res. Lett. 32, L01602. doi: 10.1029/2004GL021346
34. Myhre, G., Shindell, D., Bréon, F. M., Collins, W., Fuglestvedt, J., Huang, J., et al. (2013). "Anthropogenic and natural radioative forcing," in Climate Change 2013: The Physical Science Basis. Contribution of Working Group I to the Fifth Assessment Report of the Intergovernmental Panel on Climate Change, eds T. F. Stocker, D. Qin, G.-K. Plattner, M. Tignor, S. K. Allen, J. Boschung, et al. (NewYork, NY: Cambridge University Press), 659–740.
35. Nevison, C., Butler, J. H., and Elkins, J. W. (2003). Global distribution of N2O and the ΔN2O-AOU yield in the subsurface ocean. Global Biogeochem. Cycles 17, 1119. doi: 10.1029/2003GB002068
36. Nevison, C. D., Weiss, R. F., and Erickson, D. J. (1995). Global oceanic emissions of nitrous oxide. J. Geophys. Res. Oceans 100, 15809–15820. doi: 10.1029/95JC00684
37. Nielsdóttir, M. C., Moore, C. M., Sanders, R., Hinz, D. J., and Achterberg, E. P. (2009). Iron limitation of the postbloom phytoplankton communities in the Iceland Basin. Global Biogeochem. Cycles 23, GB3001. doi: 10.1029/2008GB003410
38. Nishioka, J., Obata, H., and Tsumune, D. (2013). Evidence of an extensive spread of hydrothermal dissolved iron in the Indian Ocean. Earth Planet. Sci. Lett. 361, 26–33. doi: 10.1016/j.epsl.2012.11.040
39. Planquette, H., Sherrell, R. M., Stammerjohn, S., and Field, M. P. (2013). Particulate iron delivery to the water column of the Amundsen Sea, Antarctica. Mar. Chem. 153, 15–30. doi: 10.1016/j.marchem.2013.04.006
40. Rabalais, N. N., Turner, R. E., Díaz, R. J., and Justic, D. (2009). Global change and eutrophication of coastal waters. ICES J. Mar. Sci. 66, 1528–1537. doi: 10.1093/icesjms/fsp047
41. Regnier, P., Friedlingstein, P., Ciais, P., Mackenzie, F. T., Gruber, N., Janssens, I. A., et al. (2013). Anthropogenic perturbation of the carbon fluxes from land to ocean. Nat. Geosci. 6, 597–607. doi: 10.1038/ngeo1830
42. Rérolle, V. M. C., Floquet, C. F. A., Mowlem, M. C., Connelly, D. P., Achterberg, E. P., and Bellerby, R. R. G. J. (2012). Seawater-pH measurements for ocean-acidification observations. Trends Anal. Chem. 40, 146–157. doi: 10.1016/j.trac.2012.07.016
43. Saito, M. A., Chisholm, S. W., Moffett, J. W., and Waterbury, J. (2002). Cobalt limitation and uptake in the marine cyanobacterium Prochlorococcus. Limnol. Oceanogr. 47, 1629–1636. doi: 10.4319/lo.2002.47.6.1629
44. Schlosser, C., Klar, J. K., Wake, B. D., Snow, J. T., Honey, D. J., Woodward, E. M. S., et al. (2014). Seasonal ITCZ migration dynamically controls the location of the (sub)tropical Atlantic biogeochemical divide. Proc. Natl. Acad. Sci. U.S.A. 111, 1438–1442. doi: 10.1073/pnas.1318670111

45. Simpson, J. H., and Sharples, J. (2012). Introduction to the Physical and Biological Oceanography of Shelf Seas. Cambridge: Cambridge University Press. doi: 10.1017/CBO9781139034098
46. Steinacher, M., Joos, F., Frölicher, T. L., Bopp, L., Cadule, P., Cocco, V., et al. (2010). Projected 21st century decrease in marine productivity: a multi-model analysis. Biogeosciences 7, 979–1005. doi: 10.5194/bg-7-979-2010
47. Stroeve, J., Holland, M. M., Meier, W., Scambos, T., and Serreze, M. (2007). Arctic sea ice decline: faster than forecast. Geophys. Res. Lett. 34, L09501. doi: 10.1029/2007GL029703
48. Timmermans, K. R., Wagt, B., and De Baar, H. J. W. (2004). Growth rates, half saturation constants, and silicate, nitrate, and phosphate depletion in relation to iron availability of four large open-ocean diatoms from the Southern Ocean. Limnol. Oceanogr. 49, 2141–2151. doi: 10.4319/lo.2004.49.6.2141
49. Wyatt, N. J., Milne, A., Woodward, E. M. S., Rees, A. P., Browning, T. J., Bouman, H. A., et al. (2014). Biogeochemical cycling of dissolved zinc along the GEOTRACES South Atlantic transect GA10 at 40°S. Global Biogeochem. Cycles 28, 2013GB004637. doi: 10.1002/2013GB004637

PART II

QUANTIFYING OCEANIC CARBON VARIABLES

CHAPTER 2

A Statistical Gap-Filling Method to Interpolate Global Monthly Surface Ocean Carbon Dioxide Data

STEVE D. JONES, CORINNE LE QUÉRÉ, CHRISTIAN RÖDENBECK, ANDREW C. MANNING, AND ARE OLSEN

2.1 INTRODUCTION

The world's oceans absorb approximately 25% of the total anthropogenic emissions of carbon dioxide (CO_2) released into the atmosphere every year [MikalofFletcher et al., 2006; Le Quéré et al., 2009]. Understanding oceanic fluxes of CO_2 is critical to explain present and project future perturbations of the global carbon cycle caused by human activities. The air-sea fluxes are driven primarily by the difference in the concentration of CO_2 between the atmosphere and the ocean surface. The concentration of CO_2 in surface water is commonly expressed as either the partial pressure (pCO_2) or fugacity (fCO_2) of carbon dioxide. Over 10 million surface ocean fCO_2 measurements have been collected since 1968 [Takahashi and

Sutherland, 2013; Pfeil et al., 2013; Bakker et al., 2014]. The majority of these measurements have been obtained in the northern hemisphere (Figure 1a) during the past 20 years (Figure 1b), which restricts in-depth analysis of global patterns and long-term trends.

The relatively limited distribution of surface ocean CO_2 measurements has restricted most mapping efforts to calculating climatological products of the seasonal cycle [Takahashi et al., 2002] or examining long-term trends [e.g., Takahashi et al., 2003, 2009; Fay and McKinley, 2013], with little or no emphasis on variability at other temporal scales. Until recently, only regional studies have focused on CO_2 variability on subannual time scales [e.g., Bates et al., 1998; Sarma, 2003; Shim et al., 2007; Olsen et al., 2008; Litt et al., 2010] and on interannual variability [Bates et al., 1996; Gruber et al., 2002; Cosca et al., 2003; Wong et al., 2010]. Some spatial and temporal interpolation efforts have been published using a variety of techniques based on harmonic curve fitting [e.g., Schuster et al., 2009] or on empirical relationships between CO_2 and proxy variables such as sea surface temperature, salinity, chlorophyll and mixed layer depth [Boutin et al., 1999; Lefèvre and Taylor, 2002; Cosca et al., 2003; Ono et al., 2004; Olsen et al., 2004; Lohrenz and Cai, 2006; Park et al., 2006; Jamet et al., 2007; Watson et al., 2009; Park et al., 2010; Telszewski et al., 2009]. The geographic and temporal scope of most of these studies has been limited to the relatively observation-rich regions. Furthermore, the use of proxy variables creates additional uncertainties due to the assumption that the relationships between CO_2 and these proxy variables are constant in time [Boutin et al., 1999; Lefèvre and Taylor, 2002; Cosca et al., 2003; Jamet et al., 2007; Park et al., 2010].

The recent release of two global databases of surface ocean CO_2 observations compiled by the Lamont-Doherty Earth Observatory (pCO_2) [Takahashi et al., 2009] and by the Surface Ocean CO_2 Atlas (SOCAT) project (fCO_2) [Pfeil et al., 2013; Bakker et al., 2014] has provided opportunities for a more detailed global analysis of surface ocean CO_2 over multiple time scales. Interpolated products of surface ocean CO_2 observations covering multiple years are valuable to characterize trends and variability. They can provide insights into the response of oceanic CO_2 to climate change and variability and the driving processes [Le Quéré et al., 2015], provide the prior estimates necessary for atmospheric inverse methods [e.g., Gurney et al., 2002], and help validate ocean biogeochemical models [e.g., Le Quéré et al., 2009].

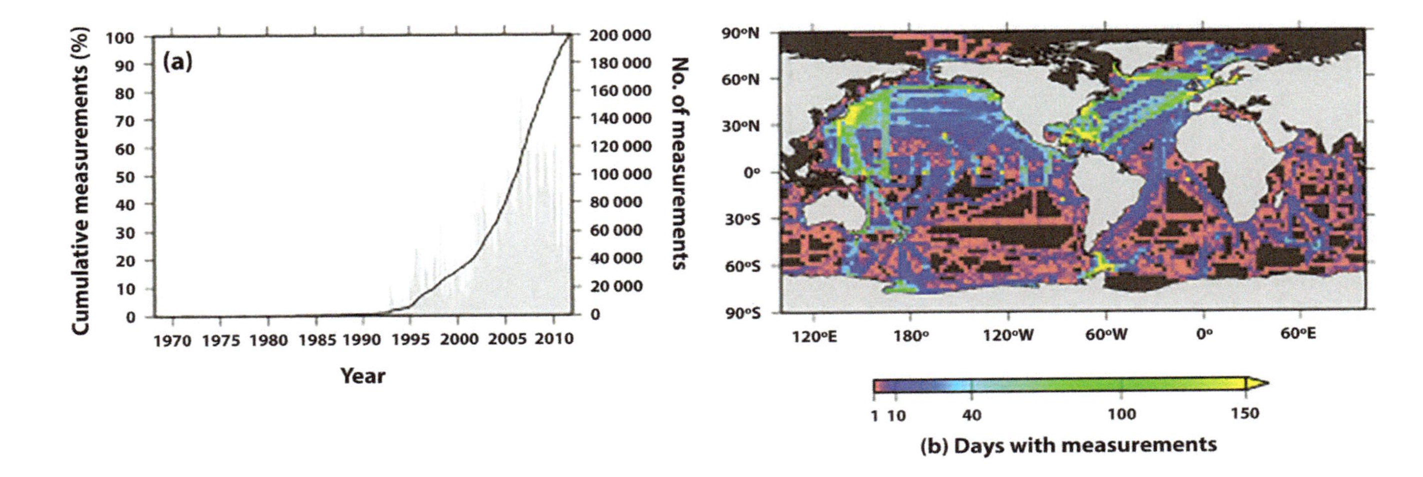

FIGURE 1: Density of the SOCAT v2 fCO_2 data coverage. (a) The total number of measurements (gray bars; right axis) and the cumulative percentage of all measurements (black line; left axis) from 1968 to 2011. (b) The number of total days between 1985 and 2011 with fCO_2 measurements in each 2.5°×2.5° grid cell.

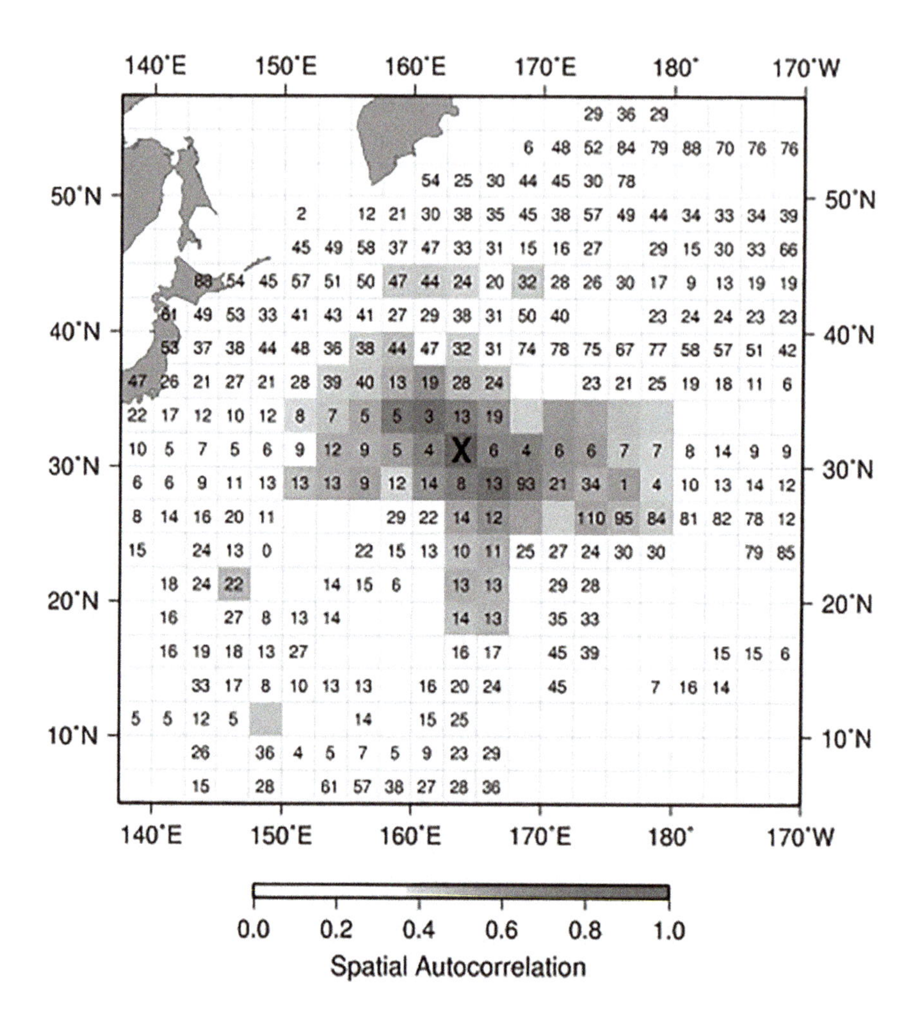

FIGURE 2: Example of the two metrics of spatial variability of fCO_2 used in the gap-filling method calculated for a target cell centered on 163.73°E 31.25°N (marked with X). The red scale indicates the strength of the spatial autocorrelation for cells within the decorrelation length. Numbers in each grid cell indicate the mean absolute difference (in μatm) between each cell's fCO_2 measurements and those of the target cell. Red cells without a number indicate that a cruise passed through both grid cells (allowing the spatial correlation to be calculated) but measurements were not taken within 7 days of each other and so were excluded from the difference calculation.

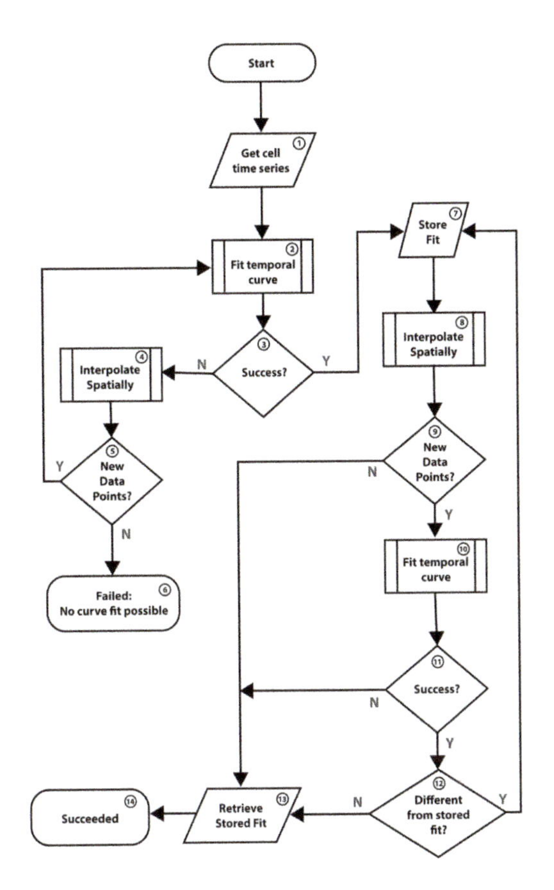

FIGURE 3: Flow diagram of the steps used to gap-fill fCO_2 data in a single grid cell. A curve is fitted to the cell's time series (2). If the fit cannot be made, or the fitted curve does not meet the criteria in Table 1, spatial interpolation is performed (4) and the curve fitting is repeated (2). If there are no new data points (5), then the interpolation fails (6) and the grid cell is processed in the next iteration of the gap-filling. If the curve fit succeeds, the fitted curve is stored (7) while more checks are performed. Another spatial interpolation is made (8). If this adds more data points (9), the subsequent curve fit is successful (10) and the new curve is significantly different from the stored fit (12), then the new fit is stored (7) and the process repeated. If the extra interpolation does not change the curve, the process reverts to the original stored fit (13) and the interpolation is marked as successful (14).

A number of methods are currently being developed to globally interpolate surface CO_2 observations using a range of techniques, including neural networks [Sasse et al., 2013; Landschützer et al., 2014; Zeng et al., 2014], diagnostic inverse models [Rödenbeck et al., 2013], biogeochemical models [Valsala and Maksyutov, 2010], and multilinear regressions [Park et al., 2010]. Many of these methods are extensions of previous regional scale interpolations [e.g., Schuster et al., 2009; Telszewski et al., 2009].

This paper presents a statistical gap-filling method to interpolate surface ocean fCO_2 in space and time for the entire global ocean south of 70°N. The method interpolates fCO_2 observations using a combination of spatial autocorrelations of fCO_2 observations within a "radius of influence" as used in the World Ocean Atlas [Jones et al., 2012; Cressman, 1959; Barnes, 1964; Levitus, 1982], harmonic curve fitting as used in GLOBALVIEW [Keeling et al., 1989; Masarie and Tans, 1995], and cubic spline fitting as used, for example, in Bacastow et al. [1985]. Our approach does not rely on proxy data, but uses only fCO_2 observations. The method includes an assessment of the uncertainty for every interpolated value, both by considering the amount of interpolation performed, and by carrying out a verification using model output, which provides information on the limitations of the interpolation.

2.2 METHODS

Our method uses the SOCAT v2 database [Bakker et al., 2014], which contains 10.1 million individual surface fCO_2 observations obtained between 1968 and 2011. We focused on the 1985–2011 time period, which encompasses 99.7% of the observations.

2.2.1 DATA PREPARATION

2.2.1.1 GRIDDING

The SOCAT v2 observations were binned into 2.5°×2.5° grid cells with daily temporal resolution. Leap years were converted to 365 day duration by dividing the year into even lengths of inline image calendar days. The complete data set was analyzed to remove any outliers that would adversely affect the subsequent curve and cubic spline fitting routines: in each grid cell, observations falling outside three standard deviations of the mean daily fCO_2 value were discarded in an iterative process, repeated until no further outliers were detected. 604 days' outliers (0.5% of the gridded observations) were eliminated in this manner. Setting the threshold to two standard deviations would have removed 12.2% of observations, severely impacting the performance of the method.

2.2.1.2 SPATIAL VARIABILITY OF FCO_2

The radius of influence over which observations were interpolated was dependent on the spatial variability of the fCO_2 observations, which was measured using two metrics: a spatial autocorrelation analysis to quantify the spatial extent over which fCO_2 observations are related, and the mean difference in fCO_2 between all neighboring grid cells to assign a magnitude of difference between observations.

For the spatial autocorrelation analysis, autocorrelation functions (ACFs) were calculated for each grid cell following the technique used in Jones et al. [2012]. A spatial ACF was calculated for each cruise in the SOCAT database using the Moran's I method [Moran, 1950], and the resulting e-folding length (i.e., the distance at which the autocorrelation coefficient drops below 1/e) assigned to each cell through which the cruise passed. This was used as a first guess of the decorrelation length for that cell, i.e. the distance beyond which fCO_2 observations were deemed unrelated. The cell's decorrelation length was refined by calculating the ACF for only those observations within a radius of five times the first guess ACF. The e-folding length of this refined ACF was used as the final spatial decorrelation length for the cell. The mean e-folding length was used where multiple cruises passed through a given cell. The decorrelation length for a given cell varied with the compass direction of the ship's heading as it passed through the cell, particularly in strong ocean currents. To provide the greatest accuracy, four directional spatial decorrelation lengths were calculated for each cell, one for each direction (i.e., north-south, east-west, northeast-southwest, and northwest-southeast). The mean decorrelation length for each direction was calculated using only those cruises traveling in the relevant direction. A "directionless" decorrelation length was calculated where there were insufficient ACF data to construct directional decorrelation lengths, using all cruises regardless of their direction of travel.

The difference in fCO_2 between grid cells was calculated for each pair of grid cells in turn. Whenever observations were made in both grid cells within 7 days, the absolute difference between them was recorded. The mean of these differences over the entire time period was recorded as the difference in fCO_2 between the two grid cells. This process was repeated for every pair of grid cells in the ocean.

The combination of these two metrics provided a dual assessment of the spatial variability of fCO_2. For a given grid cell, it was possible to estimate both the spatial autocorrelation with other grid cells (Figure 2, red shading) and the magnitude of the difference in fCO_2 with other grid cells (Figure 2, numbers). Both of these metrics of spatial variability were used when spatially interpolating the observations (see section 2.2.3).

2.2.2 GAP FILLING

The method developed here combined temporal [Masarie and Tans, 1995] and spatial [Cressman, 1959; Barnes, 1964] interpolation techniques. No interpolation was attempted poleward of 70°N as there were too few observations available.

The interpolation technique developed here comprised a series of distinct stages applied iteratively on each 2.5°×2.5° grid cell (hereafter referred to as the target cell). Here we provide an overview of the interpolation technique and detail the individual steps afterward.

2.2.2.1 OVERVIEW OF THE GAP-FILLING METHOD

Figure 3 provides an overview of the complete process used to gap-fill the fCO_2 data, applied in parallel to each cell. For each target cell, its time series of observations was retrieved (Figure 3, box 1) and a temporal curve fit was attempted (Figure 3, box 2; section 2.2.2). If the curve fit was not successful, the spatial interpolations were performed (Figure 3, box 4; section 2.2.3) and the temporal curve fit attempted again. This was repeated until either a successful curve fit was achieved or no new observations could be interpolated into the target cell's time series (Figure 3, box 6).

Even if a curve fit successfully passed the criteria for a valid fit (see section 2.2.2), it was possible that additional spatial interpolation could further constrain the fCO_2 curve fit for the target cell. This was because a time series with only a few data points may not have captured the full characteristics of the temporal variation of fCO_2. The gap-filling method accounted for this by performing one more iteration of the spatial interpolation to produce an extra curve fit (Figure 3, boxes 8–10). The two curves were then correlated. If the correlation coefficient $r^2 \geq 0.99$ (an empirically determined limit), the curves were deemed to be identical for the purposes of the interpolation, and the original curve fit was used as the final output of the interpolation for the target cell (Figure 3, box 13). A correlation coefficient of $r^2 < 0.99$ indicated that the curve fit benefitted from the extra interpolated observations. In this case the interpolation was repeated (Figure 3, box 12) until either the correlation coefficients of two consecutive curve fits reached $r^2 \geq 0.99$ (Figure 3, box 13) or no more observations could be added via spatial interpolation (Figure 3, box 9). The corresponding steps are further detailed in the following sections.

2.2.2.2 STEP I: CURVE FITTING

Step I of the gap-filling method fits a curve to the time series of each target cell (Figure 3, boxes 2 and 10) of the form:

$$f(t) = a_o + a_1 t + \sum_{k=1}^{n} [b_{2k-1} \sin(2\pi kt) + b_{2k} \cos(2\pi kt)] \quad (1)$$

where t is the time in days since 1 January 1985, a_0 the y-axis intercept, a1 the linear trend, and n the maximum number of harmonics used to represent the seasonal cycle. n is initially set to four to allow the fitted curve to encompass deviations from a purely sinusoidal progression of the seasonal cycle that may be caused by biological activity and temperature changes [e.g., Lüger et al., 2004; Körtzinger et al., 2008]. Equation (1) is a simplified version of that used by Masarie and Tans [1995] for atmospheric CO_2 mole fraction, which includes a polynomial term to account for changes in the long-term trend. We omitted the polynomial term here because there were insufficient observations to fit varying long-term trends over the time period examined.

With no constraints beyond the fCO_2 observations, the curve fitting algorithm frequently produced unrealistic fits due to the relative lack of observations in any given cell. Each curve fit was therefore assessed against a number of criteria to ensure that it produced a realistic result, as listed in Table 1. The criteria ensured that the curve was based on data covering an extended time period with observations representing a large proportion of the calendar year; that the fitted curve was representative of the range of fCO_2 observations and exhibited a plausible seasonal cycle; and that the trend of the fitted curve was within known reasonable limits. If the fit failed to meet all criteria, the number of harmonics, n, was reduced by 1 and the curve fit repeated until a good fit was achieved. If no good fit was achieved after n was reduced to 1, the curve fitting was deemed to have failed. A flowchart showing the progression of the curve fit is presented in Figure 4.

2.2.2.3 STEP II: SPATIAL INTERPOLATION

Step II of the gap-filling method was applied to target cells where a curve fit could not be found using the target cell's own observations alone. In this case, observations from nearby cells were added to the target cell's time series (Figure 3, boxes 4 and 8) and the curve fit was attempted again. Candidate cells for this spatial interpolation were chosen based on the spatial autocorrelation of the SOCAT database (see section 2.1.2). Only cells within the decorrelation length (i.e., whose spatial autocorrelation coefficient was >= 1/e) were included. These candidate cells were then sorted in order of those whose observations had the smallest difference to the target cell's fCO_2 (see section 2.1.2 and Figure 2). Any cell that had no concurrent observations with the target cell was excluded from the interpolation. In the example shown in Figure 2, only candidate cells with both red shading (within the decorrelation length) and a number (concurrent observations) were used in the spatial interpolation.

TABLE 1: Criteria Used to Determine Whether or Not a Curve Fitted to a Time Series of fCO_2 Measurements is Plausible

Criterion Name	Description	Justification
Total time range	The timespan covered by the earliest and latest measurements in the time series must be at least 5 years.	Short timespans of measurements are unlikely to reflect the long-term characteristics of pCO_2.
Standard deviation	The standard deviation of the available measurements must not exceed 75 μatm.	Curve fits applied to time series with only extreme low and high measurements are frequently unrealistic.
Populated months	Measurements must be available in at least eight of the twelve calendar months at some point in the time series.	Unless at least three of the four annual seasons are represented in the time series, the fitted curve is unlikely to represent a realistic seasonal cycle.
Curve ratio	The amplitude of the fitted curve must be between 50% and 150% of the range of values represented by the measurements.	A fitted curve whose amplitude is too small or too large does not represent an accurate fit to the measurements.
	The upper and lower limits of the curve must not exceed the limits of the measurements by more than 75 μatm.	
Seasonal peaks	Plankton blooms can produce a secondary peak in an otherwise sinusoidal seasonal cycle. Only one such peak should exist in the fitted curve. The size of the secondary peak must not exceed 33% of the total magnitude of the seasonal cycle.	Fits of multiple harmonics can produce an over-fitted curve with multiple peaks in the seasonal cycle. This is unrepresentative of the known annual cycles of pCO_2 concentrations.
Linear trend	The fitted linear trend (a1 in equation (1)) must be in the range $-2.5 \leq a1 \leq 4.75$ μatm yr^{-1}.	Linear trends outside these limits are unlikely to be realistic.

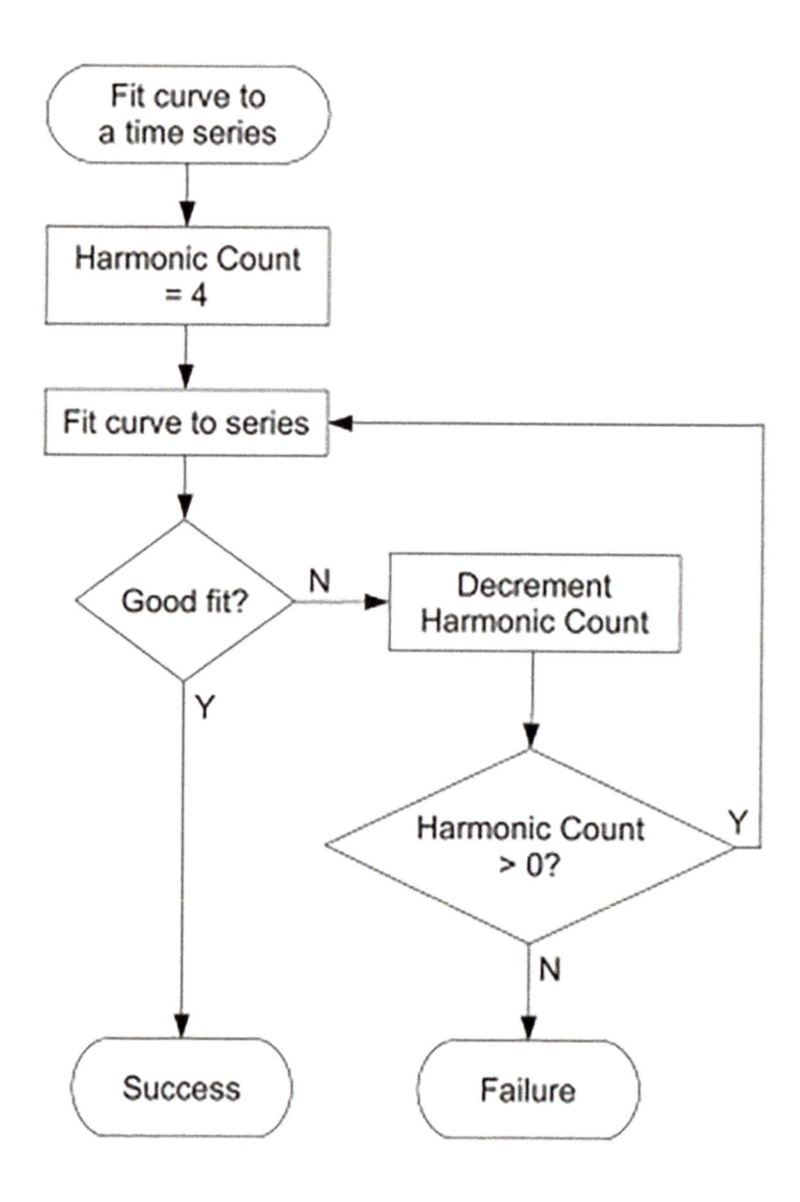

FIGURE 4: Flow diagram of the steps used to fit a curve to a single grid cell's time series (boxes 2 and 10 in Figure 3). Initially, a curve is fitted of the form in equation (1), with four harmonics. If a curve is fitted and it meets the criteria for a good fit (Table 1) then the fit is considered successful. If not, the number of harmonics is reduced by one and the fit tried again. If a successful fit cannot be made with three, two or a single harmonic, the fit is deemed to have failed.

The observations from the first candidate cell were merged with the time series of the target cell. For days where the target cell's time series contained an observation, no interpolated value was used. For the remaining time steps, the observations from the interpolated cells were added and given a weighting according to the spatial autocorrelation value (i.e., between inline image and 1). Original observations from the target cell were given a weighting of 1. The merged time series was then used in the next iteration of the temporal curve fitting algorithm.

If the new curve fit was still not successful according to the criteria in Table 1, the second candidate cell was added to the time series. If this cell had any observations on the same days as the previously interpolated cell, they were combined as a weighted mean according to the autocorrelation coefficient between the two grid cells. Again, original observations from the target cell remained unchanged. The curve fit was then attempted a third time. This was repeated until either a successful curve fit was achieved or no more candidate cells were available.

After one iteration 3,807 grid cells (54%) had successful curve fits. After nine iterations 4,736 cells (67%) had successful curve fits and no further curve fits could be achieved. Those cells that could not be interpolated either lacked sufficient observations for a curve to be fitted, or the local spatial variability was too high resulting in poor curve fits that were rejected based on the criteria in Table 1.

2.2.2.3.1 Uncertainty of Interpolated Observations

Original fCO_2 observations in the target cell's time series were given an uncertainty of ± 2.5 μatm as the default uncertainty for direct surface ocean CO_2 observations [Takahashi and Sutherland, 2013]. Estimated uncertainties in observations copied from candidate cells gap-filled fCO_2 values were calculated as the root mean squared sum of ± 2.5 μatm plus the variation in fCO_2 from the target cell's observations (i.e., the numbers on the map in Figure 2) to account for the spatial interpolation.

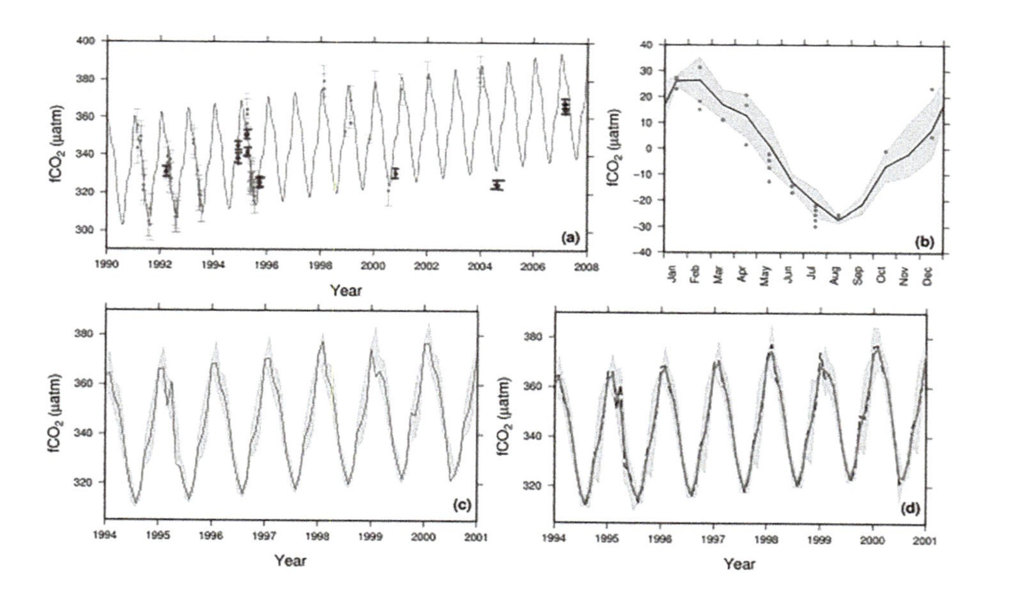

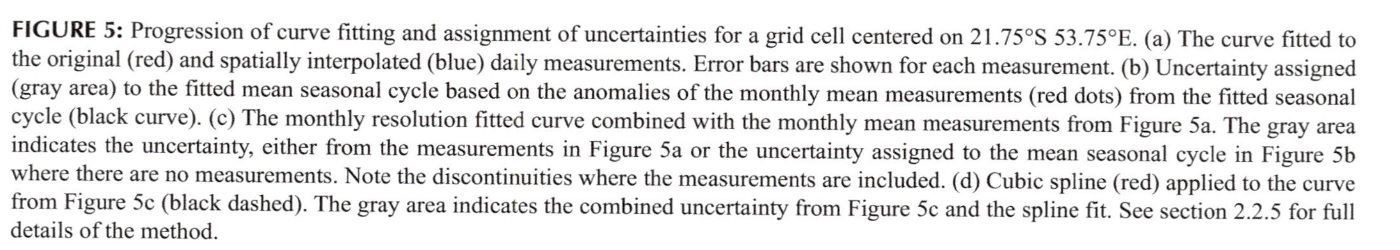

FIGURE 5: Progression of curve fitting and assignment of uncertainties for a grid cell centered on 21.75°S 53.75°E. (a) The curve fitted to the original (red) and spatially interpolated (blue) daily measurements. Error bars are shown for each measurement. (b) Uncertainty assigned (gray area) to the fitted mean seasonal cycle based on the anomalies of the monthly mean measurements (red dots) from the fitted seasonal cycle (black curve). (c) The monthly resolution fitted curve combined with the monthly mean measurements from Figure 5a. The gray area indicates the uncertainty, either from the measurements in Figure 5a or the uncertainty assigned to the mean seasonal cycle in Figure 5b where there are no measurements. Note the discontinuities where the measurements are included. (d) Cubic spline (red) applied to the curve from Figure 5c (black dashed). The gray area indicates the combined uncertainty from Figure 5c and the spline fit. See section 2.2.5 for full details of the method.

2.2.2.4 STEP III: CONVERSION TO MONTHLY RESOLUTION AND CALCULATION OF UNCERTAINTIES

Once all possible curve fits had been completed, each cell was converted to monthly resolution and uncertainties were calculated for the complete time series. Each monthly time series was constructed using the curve parameters established from the iterations of Steps I and II (Figure 5a). In months where original or interpolated observations were present, the weighted mean of those observations (weighted by the autocorrelation coefficient between the target cell and the cell from which the observations were interpolated) was inserted into the monthly time series, replacing the fitted curve value. Original observations from the cell were given a weighting of 1, while interpolated observations were weighted according to the spatial autocorrelation coefficient between the target and interpolated grid cell (red shading in Figure 2). The uncertainty for these observations was calculated as the root mean squared (RMS) uncertainty of the individual observations as in section 2.2.3.1. Uncertainties for the fitted curve where no observations were available were calculated for each month in the seasonal cycle in turn as follows. Any observations taken in January of any year in the time series were collected together. For each of these observations, the distance between the observation and the fitted curve was added to the uncertainty of the observation itself using a root mean squared sum (Figure 5b). This represents the uncertainty of the curve fit in relation to that observation. The root mean square of the uncertainties for all January measurements was used as the uncertainty for the curve fit. This was repeated for all other calendar months (Figure 5b, shaded area). If there were any months with no observations, a linear interpolation was performed between neighboring months to fill in the missing uncertainty. The observations (with their uncertainties) from the target cell and the spatially interpolated values were overlaid on the fitted curve to provide a complete time series (Figure 5c).

2.2.2.5 STEP IV: SPLINE FITTING

The time series generated from the combination of fitted curve and interpolated observations occasionally resulted in sharp and unrealistic discontinuities (Figure 5c). To eliminate these, each time series was smoothed by fitting a cubic spline function ("smooth spline") [Chambers and Hastie, 1991], with a smoothing parameter (0.3) chosen to compromise between smoothing out the discontinuities and maintaining the variability from the mean seasonal cycle that the observations represented (Figure 5d). Uncertainties for the spline fit were calculated as the uncertainty of the original time series plus the difference between the series and the fitted spline. The deviations of the spline fit (Figure 5d) from the fitted long-term trend and seasonal cycle (Figure 5a) was used to determine the interannual variability in each grid cell where observations were present or had been spatially interpolated.

2.2.2.6 STEP V: COMPLETING THE GAP FILLING

The grid cells for which no valid curve fits could be found were filled by spatial interpolation of the complete time series (original observations, interpolated observations and the fitted curve) from neighboring cells where curve fits were successfully generated. The time series were only interpolated from directly neighboring cells to reduce the uncertainty and likely errors in the interpolated values. If there was a large area of grid cells to be filled, the area was filled using several iterations with the outer edges (i.e., those with neighboring completed cells) interpolated first and progressing toward the center. Uncertainties for the interpolated values in these grid cells were calculated as for the spatial interpolation of individual observations described in section 2.2.3.1.

2.3 METHOD EVALUATION

2.3.1 RECONSTRUCTING MODEL OUTPUT

We assessed the performance of the gap-filling method by subsampling model output and applying our gap-filling method to these pseudo-data to recreate a complete pCO_2 field. We used pCO_2 output from a simulation of the PlankTOM5 model (updated from Buitenhuis et al. [2010]). pCO_2 is very similar to fCO_2 (typical differences are on the order of 1 μatm), so it is an effective measure to use for the method evaluation. PlankTOM5 is an ocean biogeochemical model forced with NCEP daily reanalysis meteorological data [Kalnay et al., 1996]. We use the ORCA2-LIM version which has a spatial model grid of 2° zonally and 0.5° to 2° meridionally [Madec and Imbard, 1996] and 15 time steps per day.

The PlankTOM5 output was regridded at 1°×1° spatial and daily temporal resolution. From this we reconstructed the individual cruise tracks in the SOCAT database that took place in between 1985 and 2011. The date and location of each observation in each SOCAT cruise was matched with the corresponding value in the regridded PlankTOM5 output to produce an analogous model "cruise." Where more than one observation was taken within a PlankTOM5 grid cell on the same day, only one value was recorded to prevent unrealistically strong spatial autocorrelations over short distances. The model "cruises" were then used to calculate the spatial autocorrelation characteristics of the sampled PlankTOM5 output as they were for the original SOCAT database (section 2.1.2). The PlankTOM5 decorrelation lengths were typically longer than those found in the SOCAT observations by a mean of 250±500 km due to PlankTOM5's 1° resolution (Figure 6). The observation locations were typically reported at much higher resolution (up to 0.001°), which means that much finer scales of variability could be detected in the actual observations. The pattern of decorrelation lengths was broadly similar for both the observations and PlankTOM5. The greatest differences were found in regions of high spatial variability, particularly the Indian Ocean and Eastern Equatorial Pacific and coastal regions, all of which were influenced by the model resolution. The coastal Southern Ocean was the only region with seemingly systematic differences in decorrelation lengths.

The input to the gap-filling method was constructed by sampling the 1°×1° PlankTOM5 output at the same spatial and temporal density as the SOCAT database (Figure 1), and then regridding this to the target resolution of 2.5°×2.5° in the same manner as the observations (section 2.1.1). The resulting sampled PlankTOM5 data were interpolated by applying the gap-filling method.

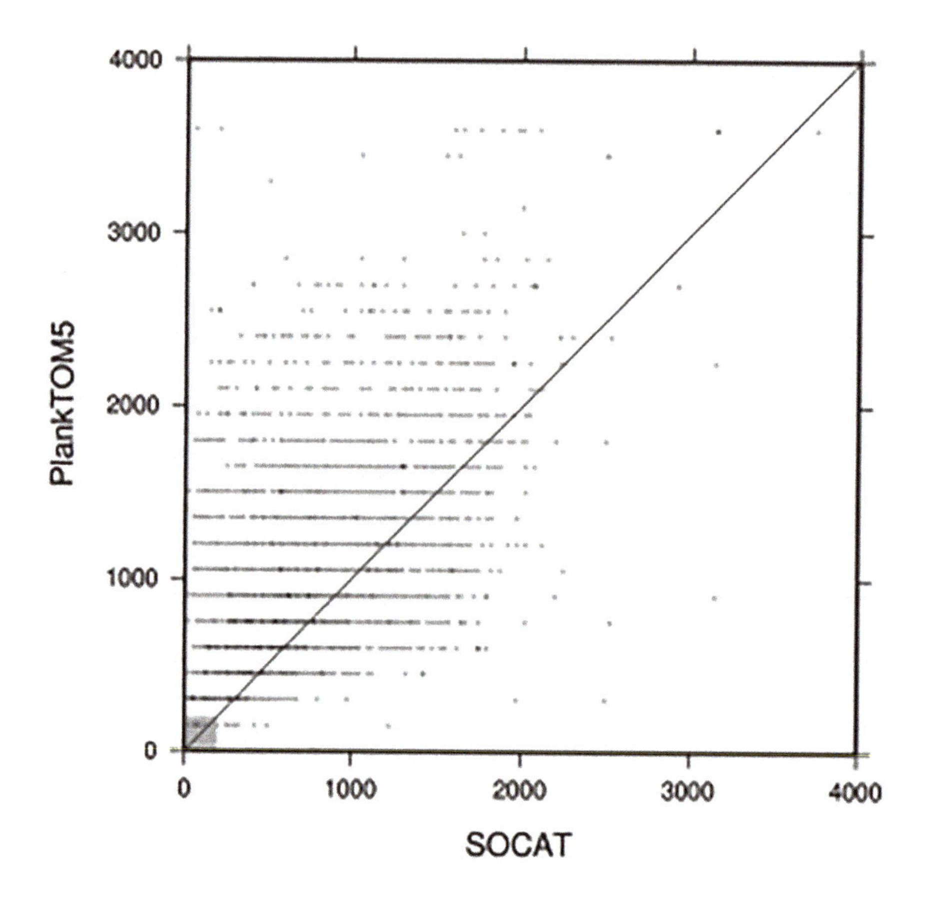

FIGURE 6: Decorrelation lengths (in km) calculated for each 2.5°×2.5° grid cell from the PlankTOM5 model output compared to the equivalent decorrelation lengths calculated from the SOCAT measurements. Darker shading indicates that multiple grid cells had the same comparative decorrelation lengths. The gray square represents the maximum decorrelation length between two neighboring grid cells (196.5 km). Cells with decorrelation lengths within this square cannot be spatially interpolated as they have no relationship to their neighbors.

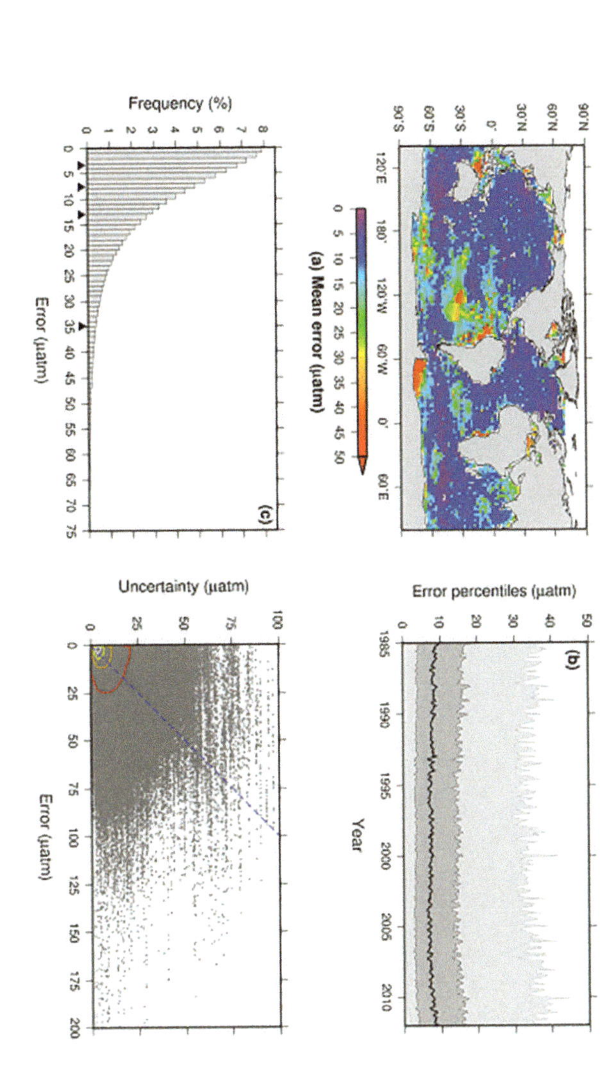

FIGURE 7: Root mean square errors between the original PlankTOM5 model data and the interpolated reconstruction using the gap-filling method applied with model data sampled at the same times and locations as the SOCAT observations (in µatm). (a) The mean error in each grid cell for the 1985–2011 time period. (b) The median error for each month (thick black line), with 25%/75% (dark gray) and 5%/95% (light gray) percentiles. (c) Histogram of errors for every data point in the reconstructed data set. Black triangles indicate the 25%, 50%, 75% and 95% quantiles. 0.3% of errors are larger than 75 µatm. (d) Errors in the gap-filled reconstruction versus the uncertainties calculated for the corresponding data points. Contours indicate regions encompassing 25% (white), 50% (yellow), 75% (orange) and 95% (red) of values in the plot. The blue dashed line indicates the 1:1 relationship. Both axes are truncated to better show the most significant values.

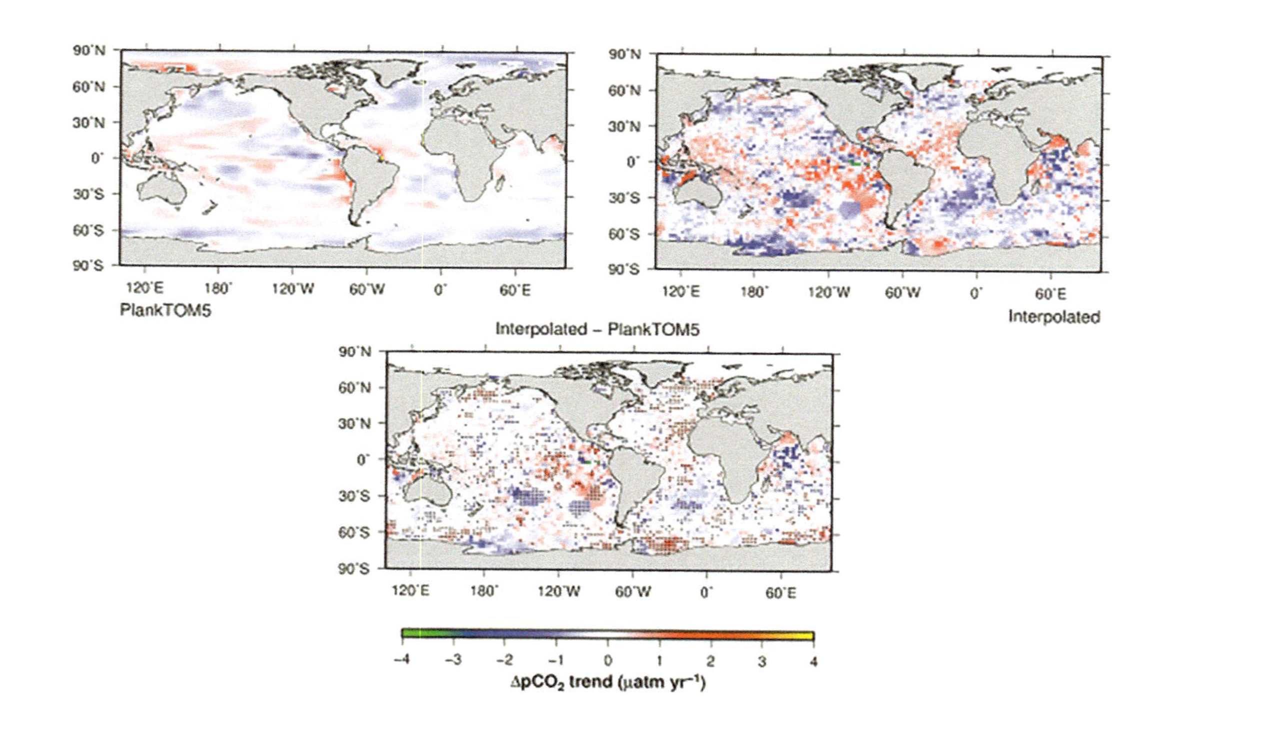

FIGURE 8: Trends of ocean minus atmosphere pCO_2 (ΔpCO_2) between 1985 and 2011 in μatm yr^{-1} for (top left) the PlankTOM5 model and (top right) the interpolated reconstruction. Trends are calculated as the difference between the 1985–1989 mean and the 2007–2011 mean, and are relative to the trend in atmospheric CO_2. The bottom map shows the difference between the two (interpolated minus PlankTOM5). Black dots indicate cells where the difference is larger than the uncertainty in the interpolated trend.

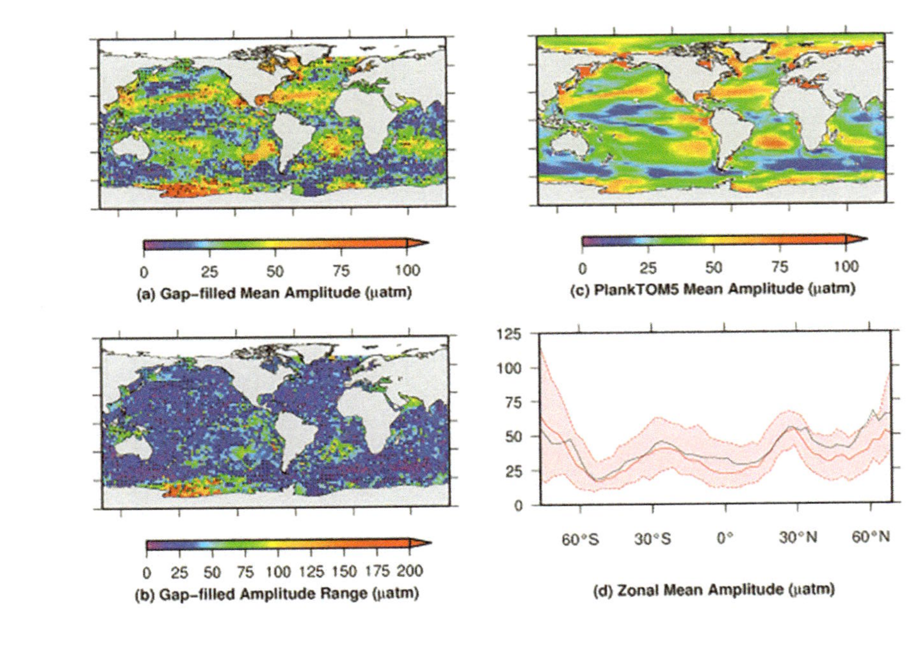

FIGURE 9: Comparison of the mean seasonal amplitude of pCO_2 in the PlankTOM5 model and the gap-filled reconstruction. (a) The amplitude from the gap-filled reconstruction. (b) The range of possible amplitudes in each cell based on the limits of the uncertainties of the fCO_2 values. This is typically twice as large as a normal $\pm$ uncertainty, which is not representative of the true uncertainty in the amplitude. (c) The amplitude from the PlankTOM5 model. (d) The zonal mean amplitude from PlankTOM5 (black) and the gap-filled reconstruction (red). The red envelope indicates the range limit of gap-filled amplitude. In Figures 9a and 9b, black dots indicate cells where the gap-filled data are significantly different from PlankTOM5, i.e., the PlankTOM5 value is not within the possible range of amplitudes in the gap-filled data.

We assessed the performance of our interpolation method by calculating the difference (error) between the original PlankTOM5 output and the result of the gap-filling performed on the sampled PlankTOM5 output at every month in each grid cell. Figure 7 shows the root mean square errors for each grid cell (Figure 7a) and each time step (Figure 7b). The largest errors were typically concentrated around those areas with few or no observations, namely the South Pacific, South Atlantic and Southern Ocean. These were caused by the lack of observations rather than differences in the decorrelation lengths that influence the method's operation. Areas of high fCO_2 variability such as the Eastern Equatorial Pacific and coastal regions also had relatively large errors (Figure 7a). The magnitude of the differences was stable through time and not significantly influenced by the density of observations in any given year (Figure 7b). 25% of the gap filled values had an error of $\leq$3.31 μatm. 50% had errors $\leq$7.62 μatm, and 75% were within 12.99 μatm (Figure 7c). In relative terms, 25% of gap-filled values were within 0.93% of the correct value; 50% were within 2.1%, and 75% within 4%.

This model evaluation also allowed us to assess the efficacy of the uncertainties calculated alongside the gap-filled pCO_2; they should be similar to the errors if they were truly representative of the limitations of the method. The mean error for the gap-filled model output was 11.3±12.4 μatm, while the mean calculated uncertainty was 10.5±9.5 μatm. 50% of the calculated uncertainties were within ±5.15 μatm of the interpolation error, and 75% were within ±10.41 μatm. Figure 7d shows a scatter plot comparing the errors in the reconstruction to the assigned uncertainties. These results indicate that the majority of uncertainty estimates calculated by our method are representative of the likely real errors in the estimate fCO_2 values.

Trends in the original PlankTOM5 output and the gap-filled interpolation were calculated as the difference between the 1985–1989 mean and the 2007–2011 mean divided by the 27 year period of the interpolation. For the interpolation, trends were also calculated using the extreme upper and lower limits of the gap-filled fCO_2 values bounded by their uncertainty, to give the maximum and minimum trend across the possible range of values. The range of these trends was used as the uncertainty range of the calculated trend. The trend in each grid cell was subtracted from the atmospheric CO_2 trend (calculated in the same manner, to account for the nonlinearity of the atmospheric trend [Dlugokencky and Tans, 2014]). We examined the difference between the two sets of trends (Figure 8).

The global mean gap-filled trends during 1985–2011 were marginally higher than the original PlankTOM5 output (means of −0.07 μatm yr^{-1} relative to the atmospheric trend versus −0.03 μatm yr^{-1} respectively). The gap-filled trends showed greater spatial variability than the trends of the PlankTOM5 output (0.55 μatm yr^{-1} and 0.25 μatm yr^{-1} respectively), but the broad spatial patterns of positive and negative trends were similar. As with the overall errors, the largest differences in trends occurred in those regions with the fewest observations, namely the southern hemisphere oceans. Trends were also difficult to reconstruct in the Eastern Equatorial Pacific because of high interannual variability from the El Niño Southern Oscillation (ENSO), and the sparse sampling of the observations in a region of complex currents and water masses. The seasonal cycle in the original PlankTOM5 output and the gap-filled data were also in good agreement in nearly all regions (Figure 9). The global mean amplitude of the seasonal cycle is 36.8 μatm in PlankTOM5 and 33.0 μatm in the gap-filled reconstruction, with an RMS error of 17.9 μatm. The correlation of amplitudes averaged at each latitude band (the zonal correlation) gave $r^2=0.73$.

The accuracy of the PlankTOM5 reconstruction was further tested by examining the relationship between pCO_2 and sea surface temperature (SST), and between pCO_2 and chlorophyll a (Chla). The detrended and deseasonalized time series of pCO_2 and SST (Chla) from the original model data were correlated in each 2.5°×2.5° grid cell (Figures 10a and 11a). The same correlation was performed using the reconstructed pCO_2 field (Figures 10b and 11b), and the differences examined (Figures 10c and 11c). 75% of the reconstructed correlations had an r value that is within 0.29 (0.23 for Chla) of the original correlations, and 95% were within 0.59 (0.49 for Chla) (Figures 10c and 11c). The largest differences were in the Southern Ocean, particularly with SST where the pattern of correlations along the Antarctic Circumpolar Current was not captured.

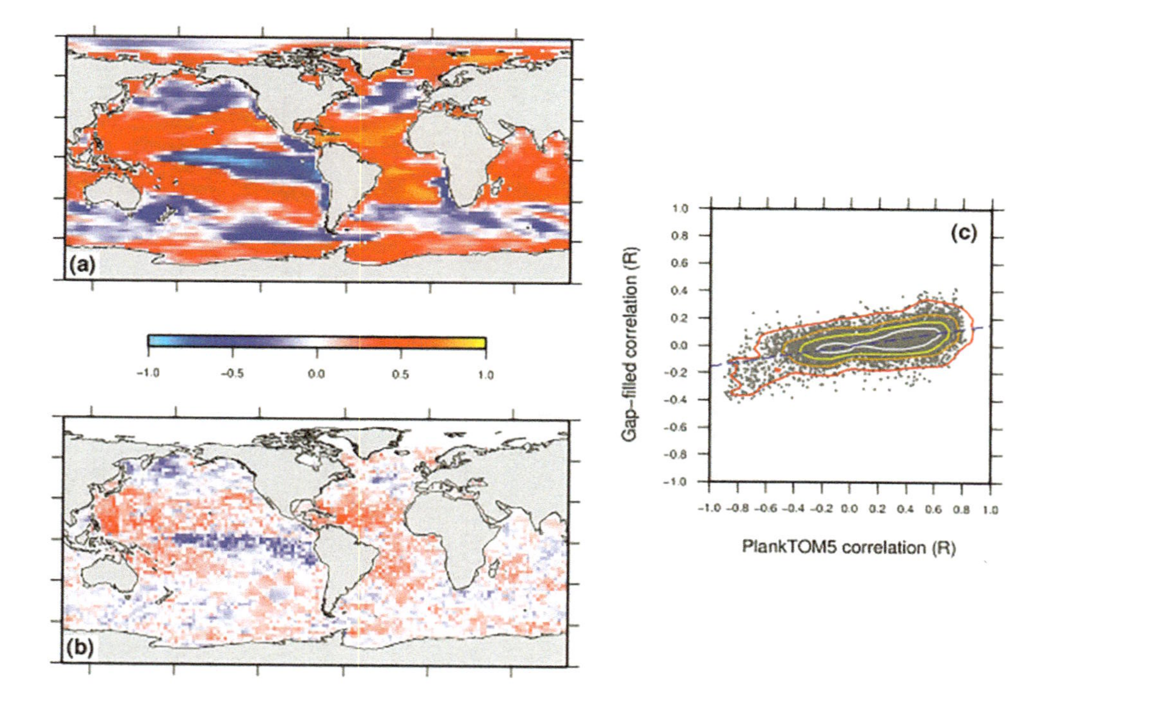

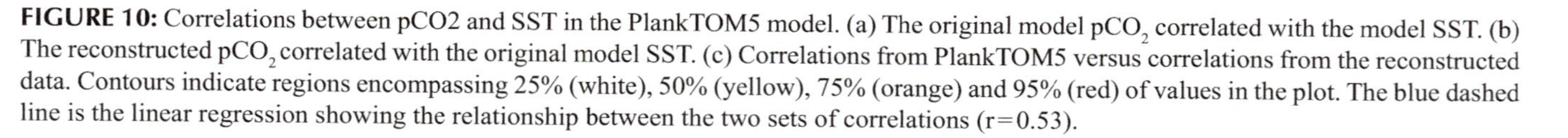

FIGURE 10: Correlations between pCO2 and SST in the PlankTOM5 model. (a) The original model pCO_2 correlated with the model SST. (b) The reconstructed pCO_2 correlated with the original model SST. (c) Correlations from PlankTOM5 versus correlations from the reconstructed data. Contours indicate regions encompassing 25% (white), 50% (yellow), 75% (orange) and 95% (red) of values in the plot. The blue dashed line is the linear regression showing the relationship between the two sets of correlations (r=0.53).

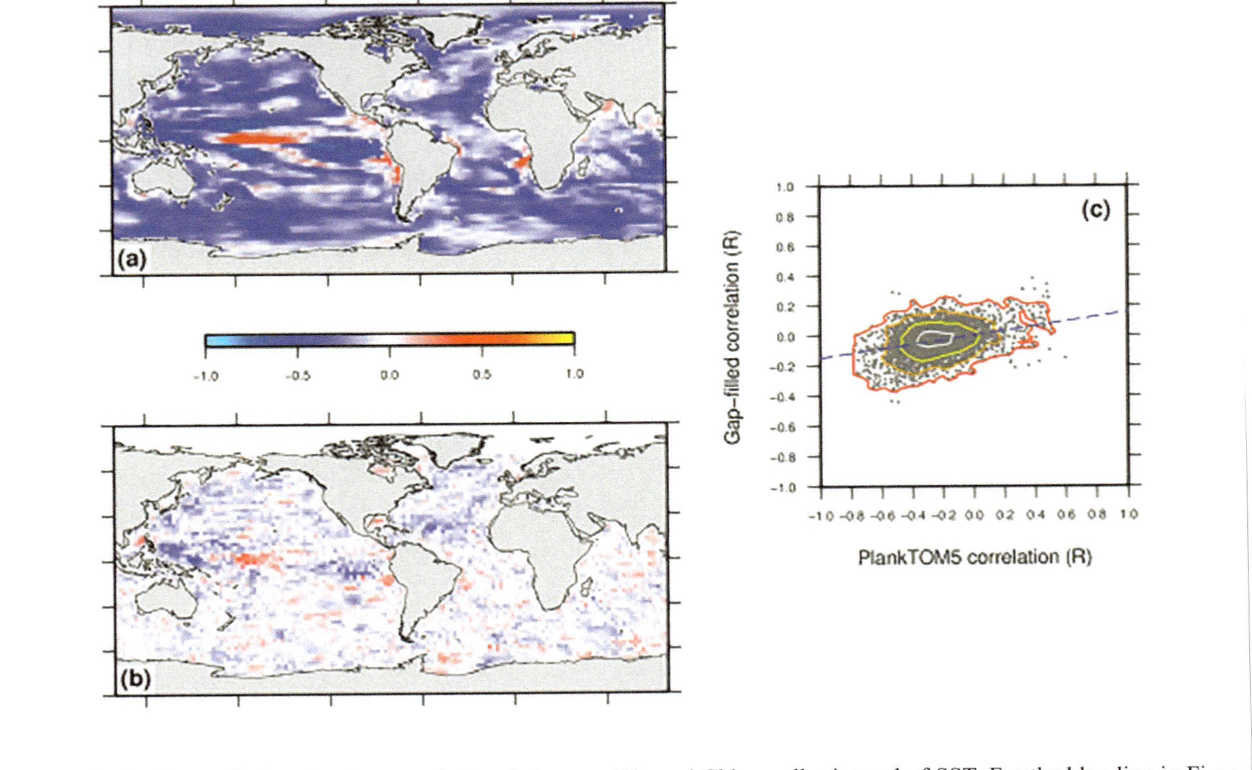

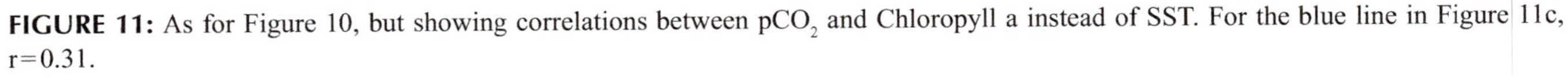
FIGURE 11: As for Figure 10, but showing correlations between pCO_2 and Chloropyll a instead of SST. For the blue line in Figure 11c, r=0.31.

TABLE 2: Comparison of the Mean Errors From Interpolating the Sampled Model Output (See Text) With Error Estimates From Regional Interpolation Studies

Study (Region)	Method	Period	Error (μatm)	This Study Error (μatm)	Uncertainty Estimate (μatm)
Wanninkhof et al. [1996] (Equatorial Pacific)	NO_3/SST regression	1985–1988a	1–32	4–13	6–18
Ono et al. [2004] (North Pacific)	SST/Chl regression	May 1997	21.0	9.0	9.0
Jamet et al. [2007] (North Atlantic)	SST/Chl/MLD regression	Winter 1994–1995	12.38–16.76b	6.0	7.9
		Spring 1994–1995	13.35–14.53b	8.0	6.8
		Summer 1994–1995	11.44–14.27b	8.0	8.9
		Autumn 1994–1995	8.98–17.33b	7.7	8.7
Gledhill et al. [2009] (Greater Caribbean Region)	Empirical CO2 gas solubility equations (SST/SSS)	1997–2006	9.0	6.0	9.0
Zhu et al. [2009] (Northern South China Sea)	SST regression	July 2000	25.1	20.3	26.7
	SST/Chl regression		4.6		
Watson et al. [2009] (North Atlantic)	SST/MLD regression	2000–2007c	1.8	1.0	7.9
	Neural network (SST/MLD)		0.77		
Telszewski et al. [2009] (North Atlantic)	Neural network (SST/MLD/Chl)	2004	8.1	7.1	8.2
		2005	12.6	7.1	8.4
		2006	12.5	6.2	8.2

TABLE 2: *Cont.*

Study (Region)	Method	Period	Error (µatm)	This Study Error (µatm)	Uncertainty Estimate (µatm)
Friedrich and Oschlies [2009a] (North Atlantic)	Neural network (SST/Chl)	2005	19.0	7.0	8.3
Friedrich and Oschlies [2009b] (North Atlantic)	Neural network (SST/SSS)	2005	14.4d 15.9d	6.7	8.3
Chierici et al. [2012] (Pacific Southern Ocean)	SST/Chl/MLD/PP regression	Dec 2006	14.0	15.0	15.0
Shadwick et al. [2010] (Scotia Shelf)	SST/Chl regression	May 2007 to Jun 2008	13.0	8.0	10.0

a Results from several individual months and locations are combined.
b Three different regressions were used, giving a range of errors.
c Annual mean values compared in both studies.
d Two source data sets: VOS ship lines and Argo floats respectively.

The South Atlantic and Eastern South Pacific showed similarly poor performance. In these regions the sampling of the original model was such that the correlation between pCO_2 and SST (Chla) was not representative of the complete model, so it is unsurprising that the relationship could not be reconstructed successfully. The other main region of difference was the Equatorial Pacific, which can be fully explained by a lack of observations. Here, the interannual signal was larger than the seasonal cycle in both pCO_2 and SST. While the curve fitting process can reconstruct a seasonal cycle, it cannot accurately reproduce interannual variability where there are missing observations. The corresponding difference in the correlation with Chla was neither as strong nor as widespread as that seen in SST, but the effect was still visible.

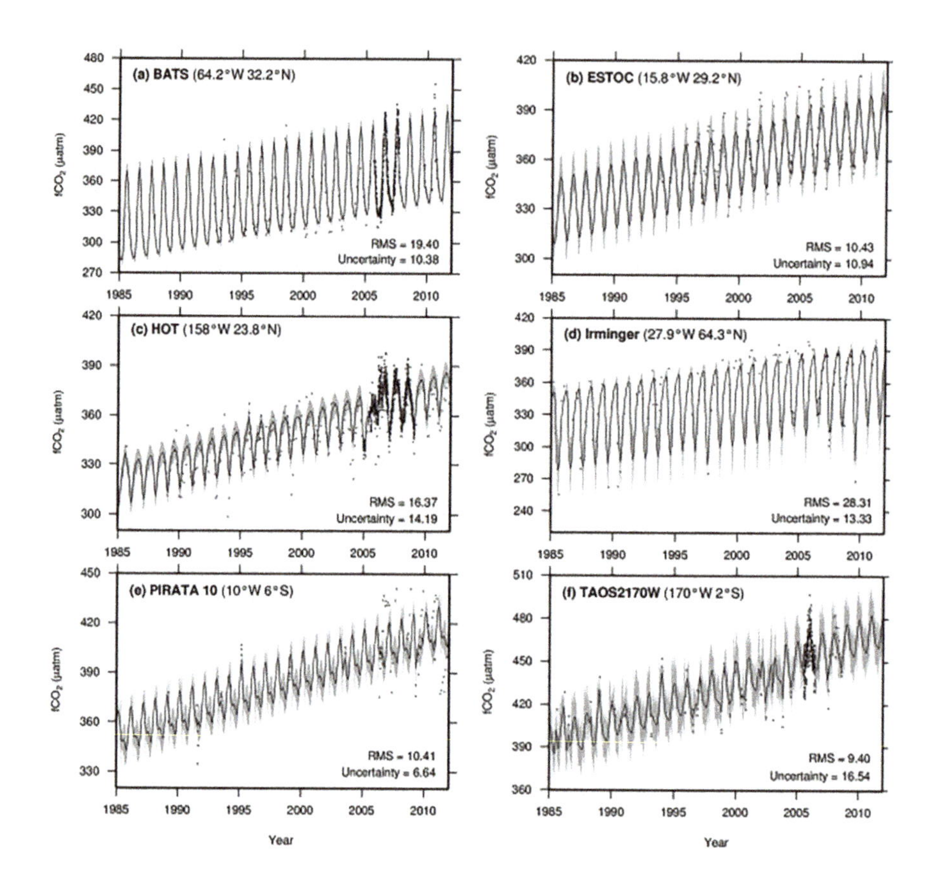

FIGURE 12: Comparison of CO_2 measurements from time series stations with the gap-filled values from the corresponding grid cell. Red dots are observations from time series data not included in the SOCAT database. Blue dots are the SOCAT measurements used in the gap-filling analysis presented here. The black line is the result of the gap-filling for the grid cell, with the calculated uncertainty in gray. "Uncertainty" text in each plot is the mean uncertainty calculated on the gap-filled values, and "RMS" is the root mean squared difference between the gap-filled values and the station measurements.

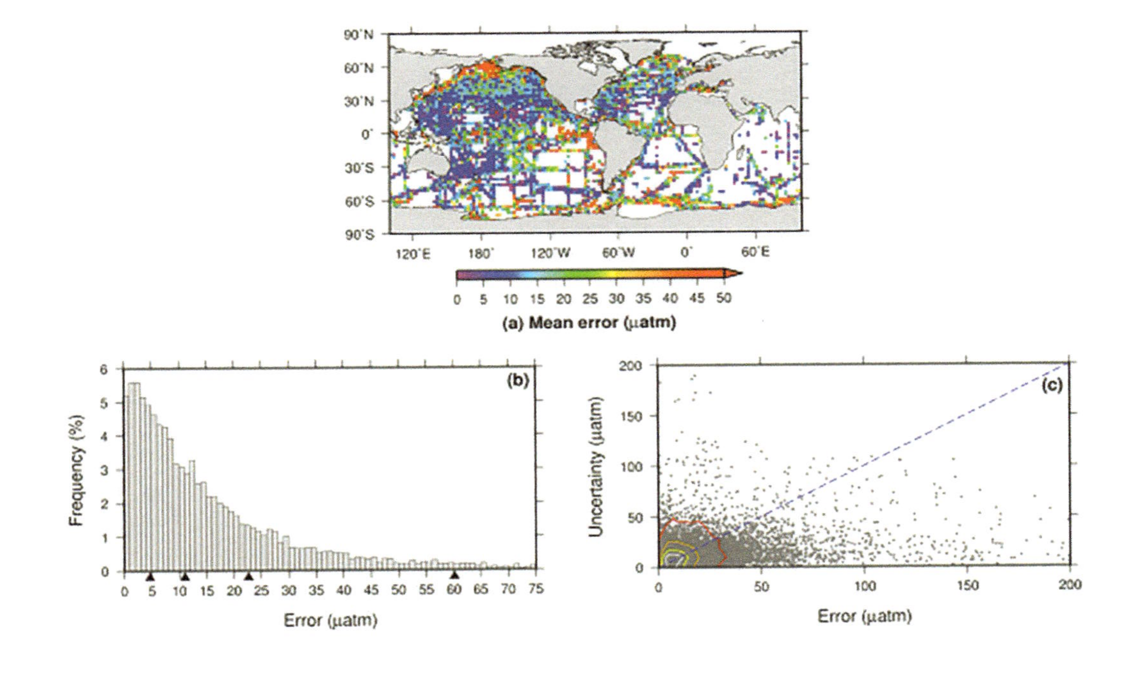

FIGURE 13: Errors between the interpolated fCO2 data calculated from SOCAT and measurements from LDEOv2012 that are not in the SOCAT database. (a) The mean error in each grid cell. (b) Histogram of all errors. 3.1% of errors are greater than 75 µatm. (c) Errors in the gap-filled reconstruction versus the uncertainties calculated for the corresponding data points. Contours indicate regions encompassing 25% (white), 50% (yellow), 75% (orange) and 95% (red) of values in the plot. The blue dashed line indicates the 1:1 relationship. Both axes are truncated to better show the most significant values.

We assessed the efficacy of the gap-filling method in relation to previous regional interpolation techniques by comparing the errors found in reconstructing the PlankTOM5 output with the errors reported for published studies (Table 2). The errors reported in those studies ranged from 0.77 to 32 μatm, while those found in this study were in the range 0.99–20 μatm for the corresponding times and regions, and were 4.2 μatm smaller on average. This shows that the global gap-filling method is comparable in performance to the regional studies. The errors in our interpolation (Table 2, column 5) were smaller than the uncertainties assigned to the interpolated values (Table 2, column 6) in the majority of cases.

2.3.2 COMPARISON TO OTHER CO_2 DATA

2.3.2.1 FIXED MOORINGS

There are a number of oceanographic CO_2 data sets from time series stations that are not included in the SOCAT database. We compared our interpolated data calculated from the SOCAT observations to the pCO_2 values from some of these stations. The SOCAT interpolated observations (Figure 12, red dots) in the corresponding time series boxes show significant differences to the time series observations themselves (in blue). These differences in the data were directly translated to the gap-filling method. Nevertheless the gap filling method achieved a fit close to the uncertainty at the European Station for Time Series in the Ocean (ESTOC) [González-Dávila and Santana-Casiano, 2009], at the TAO S2 170W in the Equatorial Pacific [Chavez, 2004], and to a lesser extent at the Hawaii Ocean Time-Series (HOT) [Dore et al., 2009]. At PIRATA 10 Lefèvre et al. [2007], however, the amplitude of the SOCAT interpolated observations was much smaller than that of the station observations, a bias which was translated in the gap-filling method to both the estimated values and an underestimated uncertainty. Finally, both the Bermuda Atlantic Time Series (BATS) [Bates, 2007] and Irminger station [Olafsson, 2007] have very large seasonal amplitudes of approximately 90 μatm (Figures 12a and 12d). This leads to large differences between the gap-filled time series and the station observations, because even small temporal offsets from the seasonal cycle fitted to the SOCAT observations can result in differences of tens of micro-atmospheres.

2.3.2.2 COMPARISON WITH LDEO DATA

The LDEOv2012 database [Takahashi and Sutherland, 2013] is another collection of global surface ocean CO_2 observations, constructed independently of SOCAT. This contains some observations that are not in SOCAT, so it was possible to compare these to our interpolated values.

We took the observations between 1985 and 2011 from the LDEOv2012 database and binned them on to a 2.5°×2.5° monthly grid, to match the resolution and coverage of our interpolated data set. We did the same for the complete SOCAT data set. We then removed from the gridded LDEO data any points where measurements were present in SOCAT. This left the gridded LDEO measurements that were not present in SOCAT. These were then compared to the corresponding points in the interpolated data set based on SOCAT, and the differences between them were assumed to be errors in the interpolated data (Figure 13a). 25% of the gap-filled values had an error of ≤4.7 μatm. 50% had errors ≤11.0 μatm, and 75% were within 22.6 μatm (Figure 13b). In relative terms, 25% of gap-filled values were within 1.3% of the true value, 50% were within 3.1%, and 75% were within 6.3%. As with the model evaluation above, we compared the differences between the interpolated values with the uncertainties calculated during the interpolation (Figure 13c). 25% of the uncertainties were within 7.1 μatm of the error, 50% were within 15.8 μatm, and 75% were within 30.2 μatm.

There were four main areas where the differences between the interpolated SOCAT data and the LDEO data were largest: the Equatorial Pacific, the Atlantic sector and coastal areas of the Southern Ocean, the Bering Sea, and the North Pacific between 30°N and 50°N. In the case of the Equatorial Pacific, the method struggles to capture the very high interannual variability in CO_2, as also seen in the model-based evaluation. The other regions also have very high variability [Naveira Garabato et al., 2004; Resplandy et al., 2014; Bates et al., 2011], and also poor sampling (Figure 1b).

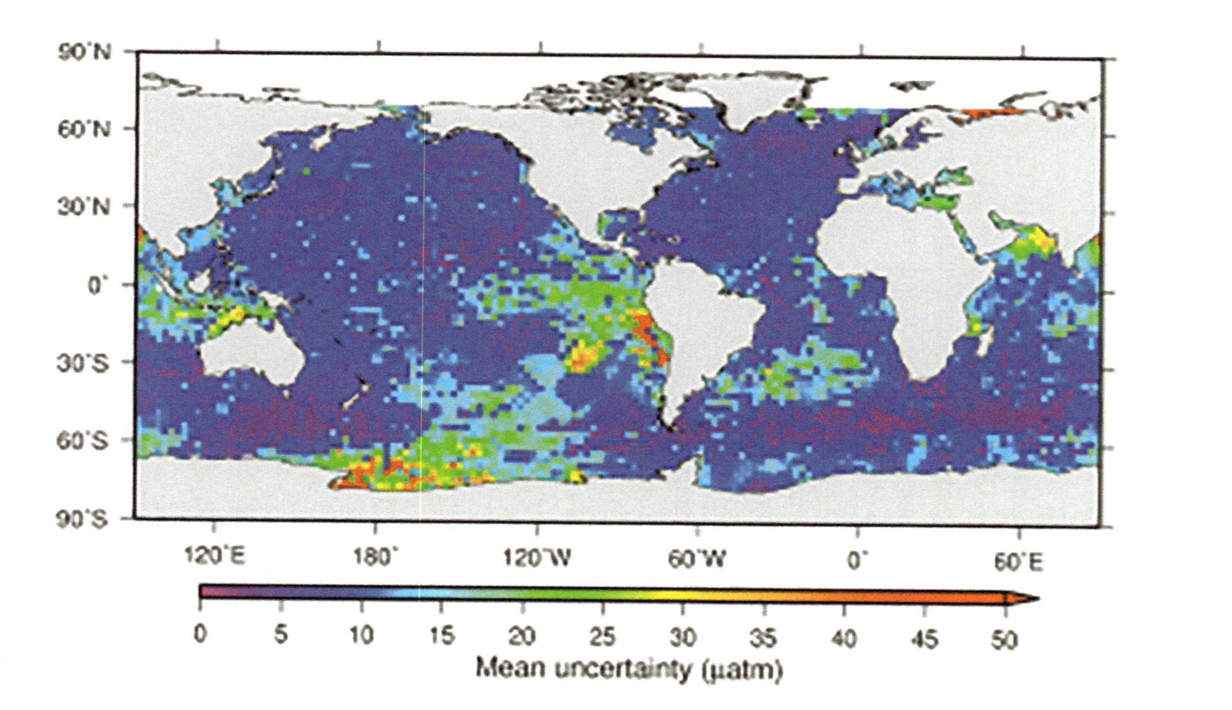

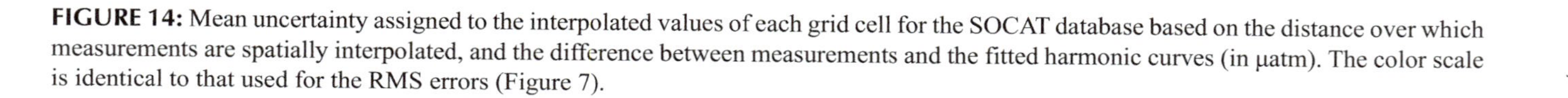

FIGURE 14: Mean uncertainty assigned to the interpolated values of each grid cell for the SOCAT database based on the distance over which measurements are spatially interpolated, and the difference between measurements and the fitted harmonic curves (in µatm). The color scale is identical to that used for the RMS errors (Figure 7).

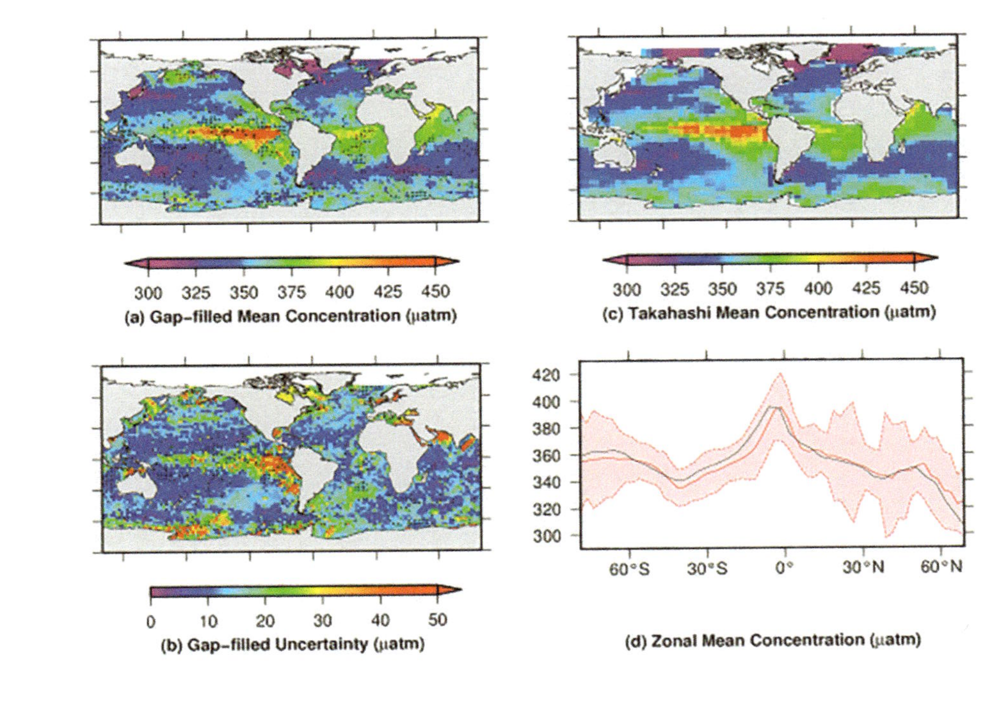

FIGURE 15: Comparison of the gap-filled annual mean fCO2 with the annual mean pCO2 from the Takahashi climatology [Takahashi et al., 2009]. (a) The mean concentration from the gap-filled SOCAT data. (b) The mean uncertainty of the values in Figure 15a. (c) The mean concentration from the Takahashi climatology. (d) The zonal mean concentration from Takahashi (black) and the gap-filled data (red). The red envelope indicates the uncertainty on the gap-filled values. In Figures 15a and 15b, black dots indicate cells where the gap-filled data are significantly different from Takahashi, i.e., the Takahashi value is not within the uncertainty range.

In the Southern Ocean there are very few observations to be used as input to the interpolation. In the Bering Sea there are a relatively high number of observations, but they are focused in a very small region. These values are interpolated over the entire Bering Sea, and the large spatial variability in this region means that the interpolated values do not reflect the true distribution of fCO_2. The range of errors between 20 and 25 μatm in the North Pacific was unexpected, because there is very good observation coverage here and the fCO_2 should be well constrained. Our investigations showed that the additional LDEO data in this region have much higher variability than the SOCAT data. The reason for this difference is unknown.

2.3.2.3 LIMITATIONS OF THE EVALUATION TECHNIQUES

Both of the evaluation techniques used above have limitations in their accuracy. Numerical models are of relatively coarse resolution, and are therefore unable to reproduce fine details on spatial and temporal scales. However, since the gap-filling method also produces coarse resolution results (monthly means on a 2.5°×2.5° grid), the problem has less impact that it would with a higher resolution interpolation.

In the interpretation of the comparison of interpolation results to observations from fixed stations and the LDEO database we need to consider that the comparison is between data from single geographical points and times, with values that are averages over much larger spatial and temporal scales. The latter eliminates the small scale variability captured by the former, and therefore cannot reproduce the observations accurately.

2.4 RESULTS

The gap-filling method allows us to produce monthly fCO_2 values for the years 1985–2011 on a 2.5°×2.5° grid south of 70°N based on the SOCAT v2 database. Uncertainties assigned to the gridded data (Figure 14) are predominantly in the range of 0 to 12 μatm. The uncertainties are typically a function of the number of available observations in each grid cell, with higher observation densities requiring less spatial interpolation (on which the uncertainty estimates are based). The smallest uncertainties are therefore in the North Pacific and North Atlantic. The uncertainties are largest in the Eastern Equatorial and South Pacific, South Atlantic and Southern Ocean. One exception to this is the Southern Ocean in the Antarctic Circumpolar Current, where strong zonal currents result in very similar fCO_2 in neighboring grid cells. The spatial interpolations therefore have very low uncertainty estimates there. The assigned uncertainties are typically larger than the predicted errors computed from model output (Figure 7), although there are exceptions, such as the Mediterranean and some coastal areas such as the Western Pacific coastline.

2.4.1 SEASONALITY

Figure 15 shows the annual mean fCO_2 for the year 2000, comparing our gap-filled results with the commonly used pCO_2 climatology for the year 2000 constructed from the LDEO database of pCO_2 observations using trend-based adjustments and lateral transport equations [Takahashi et al., 2002; Takahashi et al., 2009]. Again, we ignore differences between pCO_2 and fCO_2 as they are negligible (approximately 1 μatm). The gap-filled annual mean (Figure 15a) is very similar to the Takahashi climatology (Figure 15c): the two products match within the bounds of uncertainty over the great majority of the globe (Figure 15b). The RMS difference between the two maps is 9.7 μatm, with an overall pattern correlation of $r^2=0.78$. Examining the zonal mean concentration (Figure 15d) removes many of the effects of variability between individual cells, and provides a better picture of the coherence between the climatology and the gap-filled data. This shows that the overall structure of the two is very similar, with a zonal mean correlation of $r^2=0.93$.

We also analyzed the amplitude of the seasonal cycle in both data products (Figure 16). A traditional uncertainty cannot be given for the amplitude in the gap-filled data, since the possible range of amplitudes is defined by the uncertainties of the individual values. This is not evenly distributed, and so cannot be represented by a traditional ± value. Instead, we show the range between the largest and smallest possible amplitudes that could be calculated using the individual monthly values (Figure 16b). This is typically twice as large as one would expect a traditional ± uncertainty to be. The overall pattern is again similar between the two data sets, although there are more areas where the Takahashi amplitude is outside the uncertainty range of the gap-filled data (Figure 16a and 16b, black dots). The RMS error between the climatology and the gap-filled data are 23.9 μatm, with a zonal correlation of $r^2 = 0.56$ (Figure 16d). While not necessarily statistically significant, the regions of largest difference between the two data sets are the Equatorial Pacific, the Eastern South Pacific and regions of the Southern Ocean. The differences in seasonal amplitude in the Equatorial Pacific are large and account for some of the low correlation between the two methods. The Takahashi climatology excludes observations obtained during El Niño periods, but they are included in our method. Furthermore, the Equatorial Pacific has large uncertainty in our method because of the fine structure of ocean currents and relatively large interannual variability in this region [Cosca et al., 2003; Doney et al., 2009]. The differences in South Pacific and Southern Ocean are mostly due to low observation density (Figure 1), exacerbated in the Southern Ocean by its highly variable oceanic conditions [Naveira Garabato et al., 2004; Resplandy et al., 2014] that will lead to errors in both our method and the Takahashi et al. climatology. Amplitudes are more comparable elsewhere, with many consistent spatial structures in the North Pacific and North Atlantic.

2.4.2 LONG-TERM TRENDS

It is difficult to assess the accuracy of long-term trends in any observation-based fCO_2 database because they are known to be sensitive to the temporal and spatial scales over which they are calculated [e.g., McKinley et al., 2011, Fay and McKinley, 2013]. For this study, we assessed the trends over the 27 year period of the interpolation. Trends and their uncertainties (Figures 17a and 17b) were calculated for each grid cell using the same method as for the model validation (section 3.1). Trends were deemed insignificant when their uncertainty range crossed zero (Figure 17a, black dots). The fCO_2 trends were compared against the corresponding atmospheric trends (Figure 17c).

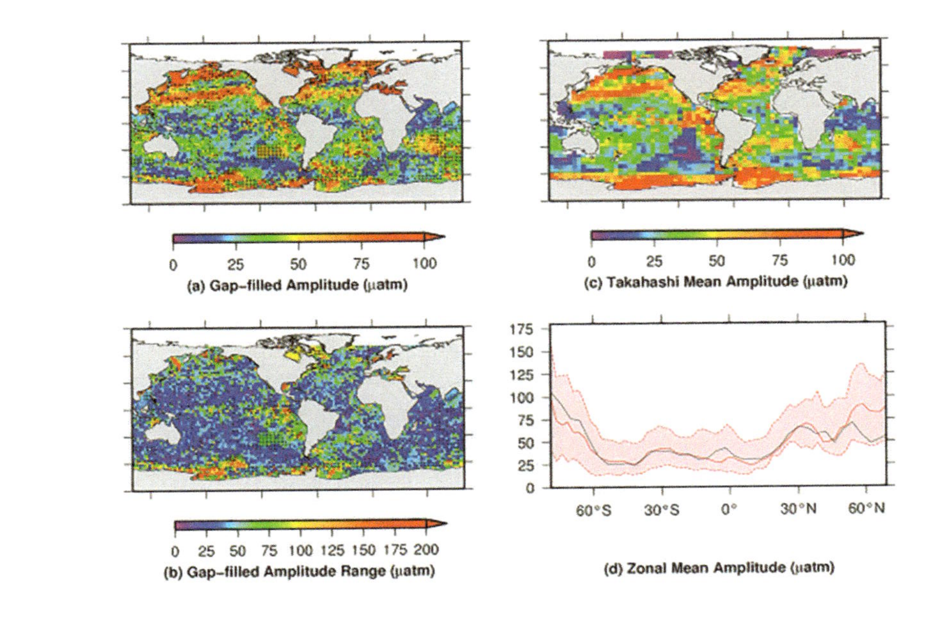

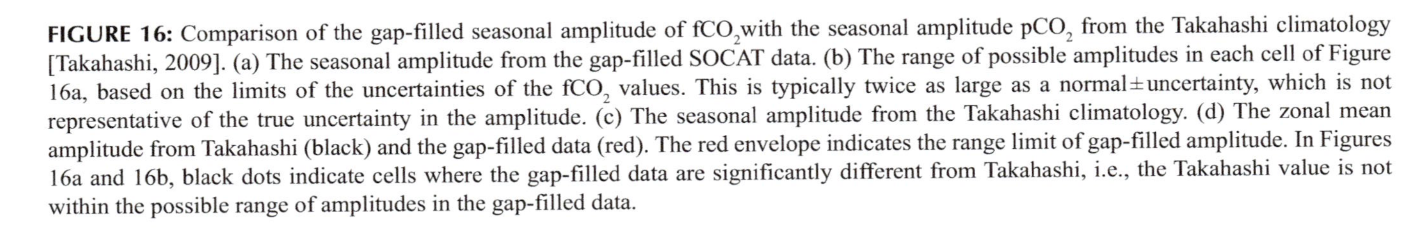

FIGURE 16: Comparison of the gap-filled seasonal amplitude of fCO_2 with the seasonal amplitude pCO_2 from the Takahashi climatology [Takahashi, 2009]. (a) The seasonal amplitude from the gap-filled SOCAT data. (b) The range of possible amplitudes in each cell of Figure 16a, based on the limits of the uncertainties of the fCO_2 values. This is typically twice as large as a normal$\pm$uncertainty, which is not representative of the true uncertainty in the amplitude. (c) The seasonal amplitude from the Takahashi climatology. (d) The zonal mean amplitude from Takahashi (black) and the gap-filled data (red). The red envelope indicates the range limit of gap-filled amplitude. In Figures 16a and 16b, black dots indicate cells where the gap-filled data are significantly different from Takahashi, i.e., the Takahashi value is not within the possible range of amplitudes in the gap-filled data.

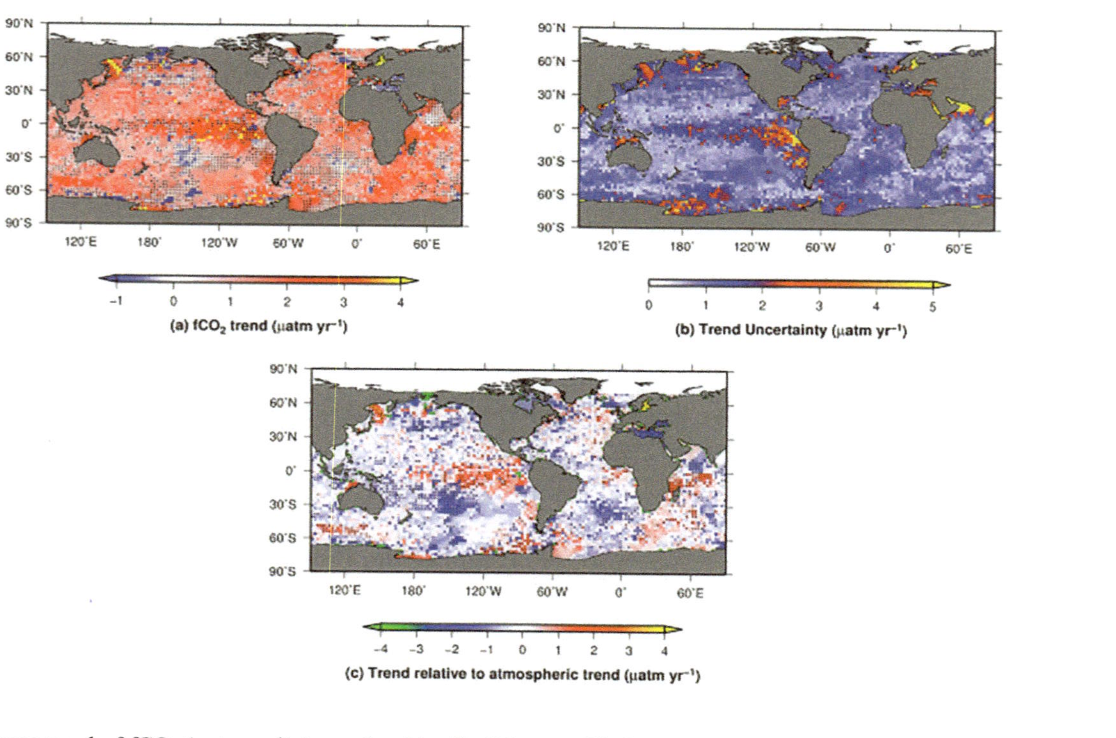

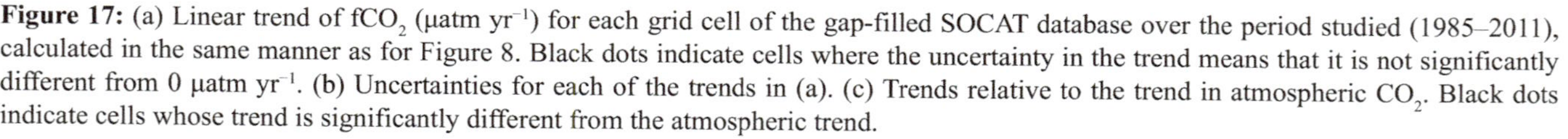

Figure 17: (a) Linear trend of fCO_2 (μatm yr^{-1}) for each grid cell of the gap-filled SOCAT database over the period studied (1985–2011), calculated in the same manner as for Figure 8. Black dots indicate cells where the uncertainty in the trend means that it is not significantly different from 0 μatm yr^{-1}. (b) Uncertainties for each of the trends in (a). (c) Trends relative to the trend in atmospheric CO_2. Black dots indicate cells whose trend is significantly different from the atmospheric trend.

Much of the ocean fCO_2 trend is slower than that of the trend in atmospheric CO_2, although there is large regional variability and uncertainty in the trends indicates that the difference is not significant in the majority of cases (Figure 17c). The global mean relative trend (ocean trend minus atmospheric trend) is -0.18 ± 0.76 μatm yr^{-1}. In the Atlantic between 35°N and 60°N there is some evidence of an east-west difference in the fCO_2 trend, with the eastern region (40°W to 0°; 0.04 ± 0.56 μatm yr^{-1}) increasing faster than the western (80°W to 40°W; -0.29 ± 0.85 μatm yr^{-1}). Similar but smaller effects have been observed in previous studies [Schuster et al., 2009; Takahashi et al., 2009; Watson et al., 2009]. This pattern, along with the widespread low fCO_2 trend in the South Atlantic, is also evident in Landschützer et al. [2014]. Trends in the North Pacific are almost exclusively slower than the atmospheric trend. In the midlatitudes (10°N to 35°N) the trends are only slightly lower (-0.14 ± 0.41 μatm yr^{-1}), similar to those found by Midorikawa et al. [2006]. The difference is much greater in the higher latitudes (35°N–60°N; -0.34 ± 1.06 μatm yr^{-1}), similar to Lenton et al. [2012] despite a region of very rapid increase in the Sea of Okhotsk to the north of Japan. The Equatorial Pacific (10°S to 10°N) is the region with least agreement with prior studies, reflecting its high uncertainty. We see significantly higher trends in the Eastern Equatorial Pacific (170°E to 120°W) than anywhere else on the globe (0.32 ± 0.75 μatm yr^{-1}), in disagreement with Feely et al. [2006]. Meanwhile, the Western Equatorial Pacific (100°E to 170°E) has slightly smaller trends (-0.37 ± 0.29 μatm yr^{-1}) than those seen in previous estimates [Feely et al., 2006; Ishii et al., 2009].

2.4.3 INTERANNUAL VARIABILITY

The interannual variability (IAV) in fCO_2 is only captured in our method where original measurements are used or interpolated, since the fitted curves will provide a climatological seasonal cycle and long-term trend only. Consequently, IAV cannot be fully resolved and is almost certainly smaller in magnitude than the true IAV. The spline fit applied in Figure 5d deviates from the harmonic curve fit where measurements are present. The difference between the harmonic fit and the spline represents the IAV that the method is able to capture. Averaging the IAV across ocean regions shows some key features of pCO_2 variation (Figure 18). In the Equatorial Pacific (Figure 18f), the progression of ENSO events is clearly visible. IAV of similar magnitude can be seen in the Southern Ocean (Figure 18h). In the North Atlantic, IAV is consistently discernible only after year 2000 (Figure 18a); There were no regular measurement programs prior to this, thus IAV could not be resolved consistently.

The cubic spline fitting performed in Step IV of the method (section 2.2.5) eliminates discontinuities in the data set, but also removes some of the IAV. The effect of this is negligible, however; the regional interannual variability shown in Figure 18 is between 0.5% and 2% smaller than it would be without the spline fitting.

2.4.4 COASTAL FCO_2

fCO_2 in coastal regions is much more variable (and therefore difficult to predict) in coastal regions than in the open ocean. However, at the 2.5°×2.5° used in this study individual grid cells are quite large (278 km across at the equator). Eliminating these grid cells would remove a relatively large portion of the ocean. While it would be possible to remove coastal measurements from the initial input to the interpolation by filtering them from the original SOCAT data set, tests have shown that these values still have considerable value in providing input to the seasonal curve fitting algorithms. The high variability of coastal fCO_2 is automatically incorporated into the uncertainty estimates for the method, since it is based partially on the difference between the observations and the fitted curve (section 2.2.4; Figure 5b). High variability therefore leads to large uncertainties in many coastal regions (Figure 14).

2.5 CONCLUSION

We have presented a gap-filling method adapted to the available global observations coverage of surface ocean fCO_2 values over the 1985–2011 period south of 70°N. The estimated accuracy of our results is comparable to or an improvement over estimates reported from other regional approaches (Table 2). The gap-filled fCO_2 data include gridded uncertainties based on the spatial and distance over which values have been interpolated and the closeness of temporal curve fits to the observations. These uncertainties can help guide data selection and interpretation in future studies using fCO_2 observations. The output of this method can be used to assess fCO_2 variability over multiple temporal and spatial scales, to help establish the optimal location and frequency for CO_2 observation programs, and to reduce the uncertainties in our knowledge of this key ocean variable. Our gap-filled data set can also provide prior estimates required by atmospheric inversion models, and data to evaluate process model simulations.

The technique developed here provides an alternative approach to those currently available in the literature as it relies neither on knowledge of other oceanic variables nor on the physical characteristics of the ocean. The independent statistical nature of the technique means that it can be readily applied to other environmental global data sets. It also means that the output directly depends on the quantity and quality of the data input, and that the absence of data in many regions and months can lead to an underestimation of the interannual variability and to excess spatial variability in the assessed trends. These biases would be reduced with the increased collection and inclusion of observations.

REFERENCES

1. Bacastow, R. B., C. D. Keeling, and T. P. Whorf (1985), Seasonal Amplitude Increase in Atmospheric CO2 Concentration at Mauna Loa, Hawaii, 1959-1982, J. Geophys. Res., 90(D6), 10,529–10,540, doi:10.1029/JD090iD06p10529.
2. Bakker, D. C. E., et al. (2014), An update to the Surface Ocean CO2 Atlas (SOCAT version 2), Earth Syst. Sci. Data, 6, 69–90, doi:10.5194/essd-6-69-2014.
3. Barnes, S. L. (1964), A technique for maximizing details in numerical weather map analysis, J. Appl. Meteorol., 3, 396–409, doi:10.1175/1520-0450(1964)003<0396:ATFMDI>2.0.CO;2.
4. Bates, N. R. (2007), Interannual variability of the oceanic CO2 sink in the subtropical gyre of the North Atlantic Ocean over the last 2 decades, J. Geophys. Res., 112, C09013, doi:10.1029/2006JC003759.
5. Bates, N. R., A. F. Michaels, and A. H. Knap (1996), Seasonal and interannual variability of oceanic carbon dioxide species at the U.S. JGOFS Bermuda Atlantic Time

series Study (BATS) site, Deep Sea Res., Part II, 43, 347–383, doi:10.1016/0967-0645(95)00093-3.
6. Bates, N. R., T. Takahashi, D. W. Chipman, and A. H. Knap, (1998), Variability of pCO2 on diel to seasonal timescales in the Sargasso Sea near Bermuda, J. Geophys. Res., 103(C8), 15,567–15,585, doi:10.1029/98JC00247.
7. Bates, N. R., J. T. Mathis, and M. A. Jeffries (2011), Air-sea CO2 fluxes on the Bering Sea shelf, Biogeosciences, 8(5), 1237–1253, doi:10.5194/bg-8-1237-2011.
8. Boutin, J., et al. (1999), Satellite sea surface temperature: A powerful tool for interpreting in situ pCO2 measurements in the equatorial Pacific Ocean, Tellus, Ser. B, 51, 490–508, doi:10.1034/j.1600-545 0889.1999.00025.x.
9. Buitenhuis, E. T., R. B. Rivkin, S. Sailley, and C. Le Quéré (2010), Biogeochemical fluxes through microzooplankton, Global Biogeochem. Cycles, 24, GB4015, doi:10.1029/2009GB003601.
10. Chambers, J. M., and T. J. Hastie (1991), Statistical Models in S, 608 pp., Chapman and Hall, Boca Raton, Fla.
11. Chavez, F. P. (2004), High-Resolution Ocean pCO2 Time-Series Measurements From Mooring TAO170W2S, Carbon Dioxide Inf. Anal. Cent., Oak Ridge Natl. Lab., U.S. Dep. of Energy, Oak Ridge, Tenn. [Available at http://cdiac.esd.ornl.gov/ftp/oceans/Moorings/TAO170W 04/.]
12. Chierici, M., S. R. Signorini, M. Mattsdotter-Björk, A. Fransson, and A. Olsen (2012), Surface water fCO2 algorithms for the high-latitude Pacific sector of the Southern Ocean, Remote Sens. Environ., 119, 184–196, doi:10.1016/j.rse.2011.12.020.
13. Cosca, C. E., R. A. Feely, J. Boutin, J. Etcheto, M. J. McPhaden, F. P. Chavez, and P. G. Strutton (2003), Seasonal and interannual CO2fluxes for the central and eastern equatorial Pacific Ocean as determined from fCO2-SST relationships, J. Geophys. Res., 108(C8), 3278, doi:10.1029/2000JC000677.
14. Cressman, G. P. (1959), An operational objective analysis system, Mon. Weather Rev., 87, 367–374, doi:10.1175/1520-0493(1959)087<0367:AOOAS>2.0.CO;2.
15. Dlugokencky, E. J., and P. P. Tans (2014), Trends in Atmospheric Carbon Dioxide, National Ocean & Atmospheric Administration Earth System Research Laboratory, Scripps Inst. of Oceanogr., San Diego, Calif. [http://www.esrl.noaa.gov/gmd/ccgg/trends/.]
16. Doney, S. C., B. D. Tilbrook, S. Roy, N. Metzl, C. Le Quéré, M. Hood, R. A. Feely, and D. C. E. Bakker (2009), Surface-ocean CO2 variability and vulnerability, Deep Sea Res., Part II, 56, 504–511, doi:10.1016/j.dsr2.2008.12.016.
17. Dore, J. E., R. Lukas, D. W. Sadler, M. J. Church, and D. M. Karl (2009), Physical and biogeochemical modulation of ocean acidification in the central North Pacific, Proc. Natl. Acad. Sci. U. S. A., 106,12,235–12,240, doi:10.1073/pnas.0906044106.
18. Fay, A. R., and G. A. McKinley (2013), Global trends in surface ocean pCO2 from in situ data, Global Biogeochem. Cycles, 27, 541–557, doi:10.1002/gbc.20051.
19. Feely, R. A., T. Takahashi, R. H. Wanninkhof, M. J. McPhaden, C. E. Cosca, S. C. Sutherland, and M.-E. Carr (2006), Decadal variability of the air-sea CO2 fluxes in the equatorial Pacific Ocean, J. Geophys. Res., 111, C08S90, doi:10.1029/2005JC003129.

20. Friedrich, T., and A. Oschlies (2009a), Neural network-based estimates of North Atlantic surface pCO2 from satellite data: A methodological study, J. Geophys. Res., 114, C03020, doi:10.1029/2007JC004646.
21. Friedrich, T., and A. Oschlies (2009b), Basin-scale pCO2 maps estimated from ARGO float data: A model study, J. Geophys. Res., 114, C10012, doi:10.1029/2009JC005322.
22. Gledhill, D. K., R. H. Wanninkhof, and C. M. Eakin (2009), Observing ocean acidification from space, Oceanography, 22, 48–59, doi:10.5670/oceanog.2009.96.
23. González-Dávila, M., and J. M. Santana-Casiano (2009), Sea Surface and Atmospheric fCO2 Data Measured During the ESTOC Time Series Cruises From 1995-2009, Oak Ridge Natl. Lab., U.S. Dep. of Energy, Oak Ridge, Tenn. [Available at http://cdiac.ornl.gov/ftp/oceans/ESTOC data/, Carbon Dioxide Information Analysis Center.]
24. Gruber, N., C. D. Keeling, and N. R. Bates (2002), Interannual variability in the North Atlantic Ocean carbon sink, Science, 298, 2374–2378, doi:10.1126/science.1077077.
25. Gurney, K. R., et al. (2002), Towards robust regional estimates of CO2 sources and sinks using atmospheric transport models, Nature, 415, 626–630, doi:10.1038/415626a.
26. Ishii, M., et al. (2009), Spatial variability and decadal trend of the oceanic CO2 in the western equatorial Pacific warm/fresh water, Deep Sea Res., Part II, 56, 591–606, doi:10.1016/j.dsr2.2009.01.002.
27. Jamet, C., C. Moulin, and N. Lefèvre (2007), Estimation of the oceanic pCO2 in the North Atlantic from VOS lines in-situ measurements: Parameters needed to generate seasonally mean maps, Ann. Geophys., 25, 2247–2257, doi:10.5194/angeo-25-2247-2007.
28. Jones, S. D., C. Le Quéré, and C. Rödenbeck (2012), Autocorrelation characteristics of surface ocean pCO2 and air-sea CO2 fluxes, Global Biogeochem. Cycles, 26, GB2042, doi:10.1029/2010GB004017.
29. Jones, S. D., C. Le Quéré, C. Rödenbeck, A. C. Manning, and A. Olsen (2015), Data and code archive for "A statistical gap-filling method to interpolate global monthly surface ocean carbon dioxide data," Pangaea, 849262, doi:10.1594/PANGAEA.849262.
30. Kalnay, E., et al. (1996), The NCEP/NCAR 40-year reanalysis project, Bull. Am. Meteorol. Soc., 77, 437–471, doi:10.1175/1520-0477(1996)077%3C0437:TNYRP%3E2.0.CO;2.
31. Keeling, C. D., R. B. Bacastow, A. F. Carter, S. C. Piper, T. P. Whorf, M. Heimann, W. G. Mook, and H. Roeloffzen (1989), A three-dimensional model of atmospheric CO2 transport based on observed winds: 1. Analysis of observational data, in The Pacific and the Western Americas, Geophys. Monogr.55, edited D. H. Peterson, 72 pp., AGU, Washington, D. C.
32. Körtzinger, A., U. Send, D. W. R. Wallace, J. Karstensen, and M. D. DeGrandpre (2008), Seasonal cycle of O2 and pCO2 in the central Labrador Sea: Atmospheric, biological, and physical implications, Global Biogeochem. Cycles, 22, GB1014, doi:10.1029/2007GB003029.

33. Landschützer, P., N. Gruber, D. C. E. Bakker, and U. Schuster (2014), Recent variability of the global ocean carbon sink, Global Biogeochem. Cycles, 28, 927–949, doi:10.1002/2014GB004853.
34. Lefèvre, N., and A. Taylor (2002), Estimating pCO2 from sea surface temperatures in the Atlantic gyres, Deep Sea Res., Part I, 49, 539–554, doi:10.1016/S0967-0637(01)00064-4.
35. Lefèvre, N., A. Guillot, L. Beaumont, and T. Danguy (2007), Variability of fCO2 in the Eastern Tropical Atlantic From Mooring PIRATA 6S 10W, Carbon Dioxide Inf. Anal. Cent., Oak Ridge Natl. Lab., U.S. Dep. of Energy, Oak Ridge, Tenn. [Available at http://cdiac.ornl.gov/ftp/oceans/Moorings/PIRATA.]
36. Le Quéré, C., et al. (2009), Trends in the sources and sinks of carbon dioxide, Nat. Geosci., 2, 831–836, doi:10.1038/ngeo689.
37. Le Quéré, C., et al. (2015), Global Carbon Budget 2014, Earth Syst. Sci. Data, 7, 47–85, doi:10.5194/essd-7-47-2015.
38. Lenton, A., N. Metzl, T. Takahashi, M. Kuchinke, R. J. Matear, T. Roy, S. C. Sutherland, C. Sweeney, and B. D. Tilbrook (2012), The observed evolution of oceanic pCO2 and its drivers over the last two decades, Global Biogeochem. Cycles, 26, GB2021, doi:10.1029/2011GB004095.
39. Levitus, S. (1982), Climatological atlas of the world ocean, NOAA Prof. Pap. 13, U. S. Gov. Print. Off., Washington, D. C.
40. Litt, E. J., N. J. Hardman-Mountford, J. C. Blackford, G. Mitchelson-Jacob, A. Goodman, G. E. Moore, D. G. Cummings, and M. Butenschon (2010), Biological control of pCO2 at station L4 in the Western English Channel over 3 years, J. Plankton Res., 32, 621–629, doi:10.1093/plankt/fbp133.
41. Lohrenz, S. E., and W.-J. Cai (2006), Satellite ocean color assessment of air-sea fluxes of CO2 in a river-dominated coastal margin, Geophys. Res. Lett., 33, L01601, doi:10.1029/2005GL023942.
42. Lüger, H., D. W. R. Wallace, A. Körtzinger, and Y. Nojiri (2004), The pCO2 variability in the midlatitude North Atlantic Ocean during a full annual cycle, Global Biogeochem. Cycles, 18, GB3023, doi:10.1029/2003GB002200.
43. Madec, G., and M. Imbard (1996), A global ocean mesh to overcome the North Pole singularity, Clim. Dyn., 12, 381–388, doi:10.1007/BF00211684.
44. Masarie, K. A., and P. P. Tans (1995), Extension and integration of atmospheric carbon dioxide data into a globally consistent measurement record, J. Geophys. Res., 100(D6), 11,593–11,610, doi:10.1029/95JD00859.
45. McKinley, G. A., A. R. Fay, T. Takahashi, and N. Metzl (2011), Convergence of atmospheric and North Atlantic carbon dioxide trends on multidecadal timescales, Nat. Geosci., 4, 606–610, doi:10.1038/ngeo1193.
46. Midorikawa, T. M. Ishii, K. Nemoto, H. Kamiya, A. Nakadate, S. Masuda, H. Matsueda, T. Nakano, and H. Y. Inoue (2006), Interannual variability of winter oceanic CO2 and air-sea CO2 flux in the western North Pacific for 2 decades, J. Geophys. Res., 111, C07S02, doi:10.1029/2005JC003095.
47. Mikaloff Fletcher, S. E., et al. (2006), Inverse estimates of anthropogenic CO2 uptake, transport, and storage by the ocean, Global Biogeochem. Cycles, 20, GB2002, doi:10.1029/2005GB002530.

48. Moran, P. A. P. (1950), Notes on continuous stochastic phenomena, Biometrika, 37, 17–23, doi:10.1093/biomet/37.1-2.17.
49. Naveira Garabato, A. C., K. L. Polzin, B. A. King, K. J. Heywood, and M. H. Visbeck (2004), Widespread intense turbulent mixing in the Southern Ocean, Science, 303, 210–213, doi:10.1126/science.1090929.
50. Olafsson, J. (2007), Irminger Sea Cruise Data From the 1991–2006 Cruises, CARINA Data Set, Carbon Dioxide Inf. Anal. Cent., Oak Ridge Natl. Lab., U.S. Dep. of Energy, Oak Ridge, Tenn. [Available at http://cdiac.ornl.gov/ftp/oceans/CARINA/IrmingerSea/.]
51. Olsen, A., J. A. Triñanes, and R. H. Wanninkhof (2004), Sea-air flux of CO2 in the Caribbean Sea estimated using in situ and remote sensing data, Remote Sens. Environ., 89, 309–325, doi:10.1016/j.rse.2003.10.011.
52. Olsen, A., K. R. Brown, M. Chierici, T. Johannessen, and C. Neill (2008), Sea-surface CO2 fugacity in the subpolar North Atlantic, Biogeosciences, 5, 535–547, doi:10.5194/bg-5-535-2008.
53. Ono, T., T. Saino, N. Kurita, and K. Sasaki (2004), Basin-scale extrapolation of shipboard pCO2 data by using satellite SST and Chla, Int. J. Remote Sens., 25, 3803–3815, doi:10.1080/01431160310001657515.
54. Park, G.-H., K. Lee, R. H. Wanninkhof, and R. A. Feely (2006), Empirical temperature-based estimates of variability in the oceanic uptake of CO2 over the past 2 decades, J. Geophys. Res., 111, C07S07, doi:10.1029/2005JC003090.
55. Park, G.-H., R. H. Wanninkhof, S. C. Doney, T. Takahashi, K. Lee, R. A. Feely, C. L. Sabine, J. A. Triñanes, and I. D. Lima (2010), Variability of global net sea-air CO2 fluxes over the last three decades using empirical relationships, Tellus, Ser. B, 62, 352–368, doi:10.1111/j.1600-0889.2010.00498.x.
56. Pfeil, B., et al. (2013), A uniform, quality controlled Surface Ocean CO2 Atlas (SOCAT), Earth Syst. Sci. Data, 5, 125–143, doi:10.5194/essd-5-125-2013.
57. Resplandy, L., J. Boutin, and L. Merlivat (2014), Observed small spatial scale and seasonal variability of the CO2 system in the Southern Ocean, Biogeosciences, 11(1), 75–90. doi:10.5194/bg-11-75-2014
58. Rödenbeck, C., R. F. Keeling, D. C. E. Bakker, N. Metzl, A. Olsen, C. L. Sabine, and M. Heimann (2013), Global surface-ocean pCO2 and sea air CO2 flux variability from an observation-driven ocean mixed-layer scheme, Ocean Sci., 9, 193–216, doi:10.5194/os-9-193-2013.
59. Sarma, V. V. S. S. (2003), Monthly variability in surface pCO2 and net air-sea CO2 flux in the Arabian Sea, J. Geophys. Res., 108(C8), 3255, doi:10.1029/2001JC001062.
60. Sasse, T. P., B. I. McNeil, and G. Abramowitz (2013), A new constraint on global air-sea CO2 fluxes using bottle carbon data, Geophys. Res. Lett., 40, 1594–1599, doi:10.1002/grl.50342.
61. Schuster, U., A. J. Watson, N. R. Bates, A. Corbière, M. González-Dávila, N. Metzl, D. Pierrot, and J. M. Santana-Casiano (2009), Trends in North Atlantic sea-surface fCO2 from 1990 to 2006, Deep Sea Res., Part II, 56, 620–639, doi:10.1016/j.dsr2.2008.12.011.
62. Shadwick, E. H., H. Thomas, A. Comeau, S. E. Craig, C. Hunt, and J. E. Salisbury (2010), Air-sea CO2 fluxes on the Scotian Shelf: Seasonal to multi-annual variability, Biogeosciences, 7, 3851–3867, doi:10.5194/bg-7-3851-2010.

63. Shim, J. H., D. Kim, Y. C. Kang, J. H. Lee, S.-T. Jang, and C.-H. Kim (2007), Seasonal variations in pCO2 and its controlling factors in surface seawater of the northern East China Sea, Cont. Shelf Res., 27, 2623–2636, doi:10.1016/j.csr.2007.07.005.
64. Takahashi, T., and S. C. Sutherland (2013), Global ocean surface water partial pressure of CO2 database: Measurements performed during 1957–2013 (Version 2012), ORNL/CDIAC-160, NDP-088, 20 pp., Carbon Dioxide Inf. Anal. Cent., Oak Ridge Natl. Lab., U.S. Dep. of Energy, Oak Ridge, Tenn.
65. Takahashi, T., et al. (2002), Global sea-air CO2 flux based on climatological surface ocean pCO2, and seasonal biological and temperature effects, Deep Sea Res., Part II, 49, 1601–1622, doi:10.1016/S0967-0645(02)00003-6.
66. Takahashi, T., S. C. Sutherland, R. A. Feely, and C. E. Cosca, (2003a), Decadal variation of the surface water pCO2 in the western and central equatorial Pacific, Science, 302, 852–856, doi:10.1126/science.1088570.
67. Takahashi, T., et al. (2003b), Climatological mean and decadal change in surface ocean pCO2, and net sea-air CO2 flux over the global oceans, Deep Sea Res., Part II, 56, 554–577, doi:10.1016/j.dsr2.2008.12.009.
68. Takahashi, T., et al. (2009), Climatological mean and decadal change in surface ocean pCO2, and net seaair CO2 flux over the global oceans, Deep Sea Res., Part II, 56, 554–577, doi:10.1016/j.dsr2.2008.12.009.
69. Telszewski, M., et al. (2009), Estimating the monthly pCO2 distribution in the North Atlantic using a self-organizing neural network, Biogeosciences, 6, 1405–1421, doi:10.5194/bg-6-1405-2009.
70. Valsala, V., and S. Maksyutov (2010), Simulation and assimilation of global ocean pCO2 and air-sea CO2 fluxes using ship observations of surface ocean pCO2 in a simplified biogeochemical offline model, Tellus, Ser. B, 62, 821–840, doi:10.1111/j.1600-0889.2010.00495.x.
71. Wanninkhof, R. H., R. A. Feely, H. Chen, C. E. Cosca, and P. P. Murphy (1996), Surface water fCO2 in the eastern equatorial Pacific during the 1992-1993 El Niño, J. Geophys. Res., 101(C7), 16,333–16,343, doi:10.1029/96JC01348.
72. Watson, A. J., et al. (2009), Tracking the variable North Atlantic sink for atmospheric CO2, Science, 326, 1391–1393, doi:10.1126/science.117739.
73. Wong, C. S., J. R. Christian, S.-K. Emmy Wong, J. Page, L. Xie, and S. Johannessen (2010), Carbon dioxide in surface seawater of the eastern North Pacific Ocean (Line P), 1973-2005, Deep Sea Res., Part I, 57, 687–695, doi:10.1016/j.dsr.2010.02.003.
74. Zeng, J., Y. Nojiri, P. Landschützer, M. Telszewski, and S.-I. Nakaoka (2014), A global surface ocean fCO2 climatology based on a feed-forward neural network, J. Atmos. Oceanic Technol., 31, 1838–1849, doi:10.1175/JTECH-D-13-00137.1.
75. Zhu, Y., S. Shang, W. Zhai, and M. Dai (2009), Satellite-derived surface water pCO2 and air-sea CO2 fluxes in the northern South China Sea in summer, Prog. Nat. Sci., 19, 775–779, doi:10.1016/j.pnsc.2008.09.004.

CHAPTER 3

The Seasonal Sea-Ice Zone in the Glacial Southern Ocean as a Carbon Sink

ANDREA ABELMANN, RAINER GERSONDE, GREGOR KNORR, XU ZHANG, BERNHARD CHAPLIGIN, EDITH MAIER, OLIVER ESPER, HANS FRIEDRICHSEN, GERRIT LOHMANN, HANNO MEYER, AND RALF TIEDEMANN

3.1 INTRODUCTION

The past four climate cycles are characterized by a repetitive pattern of gradually declining and rapidly increasing atmospheric CO_2 concentrations, ranging between ~180 p.p.m. during glacials and ~280 p.p.m. during interglacials [1]. Although multiple processes on land and in the ocean are involved in the modulation of the observed CO_2 variability [2], physical and biological processes in the Southern Ocean (SO) have been identified to be the key in these changes [3]. This view is supported by the tight relationship between CO_2 and Antarctic temperature development [4]. Most important are changes in ocean ventilation/stratification, sea-ice extent, wind patterns, atmospheric transport of micronutrients (for example, iron) and biological productivity and export, according to proxy and model-based studies [3, 5, 6, 7, 8, 9, 10, 11]. Despite the scientific progress, the different hypotheses on the SO's sensitivity to modulate the carbon cycle and the identification of involved processes remain under debate. In the SO, the availability of silicon nutrients (Si), the consumption by primary producers (diatoms) and cycling pathways are key for effective carbon sequestration [8, 12]. The widespread deposition of biogenic opal, which consists primarily in diatoms but also in radiolarians and, to a minor extent, in sponge spicules, allows for the application of specific opal-based proxies to trace these processes and related environmental conditions. However, controversial views exist for the interpretation of the proxies used to trace past productivity and their impact on the carbon cycle [3, 5, 13]. Similarly, glacial–interglacial changes in surface ocean stratification, which control ocean atmosphere exchange and the availability of nutrients, have been discussed contentiously. This has resulted in different notions of the impact of physical and biological processes in ice-free and ice-covered areas on the glacial–interglacial climate evolution [3, 5, 13]. Isotope records of diatom-bound nitrogen ($\delta^{15}N$) are interpreted to indicate a low-productivity glacial seasonal sea-ice zone (SIZ) resulting from constricted nutrient supply to the surface ocean, owing to permanent and enhanced near-surface stratification [3, 13]. Further information on surface water (euphotic zone) conditions comes from oxygen isotopes ($\delta^{18}O$) of diatoms, used to identify meltwater supply from the Antarctic continent [14, 15, 16, 17]. Silicon isotope ($\delta^{30}Si$) measurements on diatoms and sponge spicules provide insights into the development of silicon utilization in surface waters [18, 19, 20] and the silicon inventory of the deep ocean [19, 21, 22]. A yet unexploited window into subsurface and deeper water conditions presents the isotope signal from radiolarians (protozooplankton). In combination with the diatom isotope data, these signals provide an enhanced framework to detect changes of upper and lower water column conditions, and thus the pattern and glacial–interglacial variability of stratification and nutrient exchange.

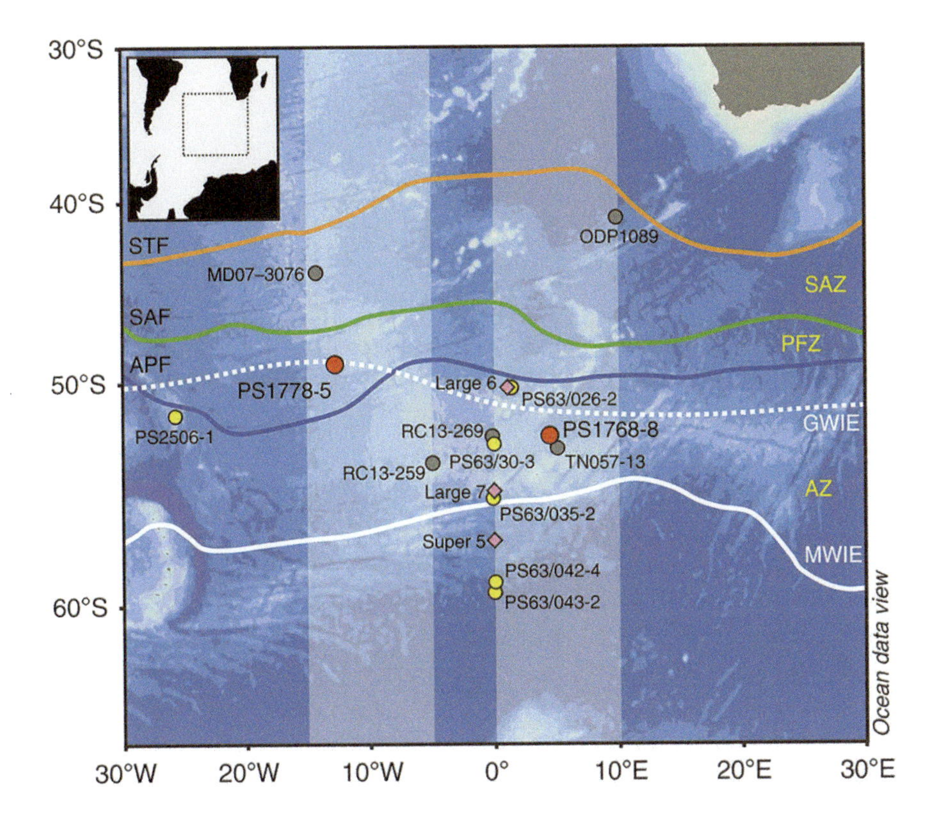

FIGURE 1: Map of the study area. The map shows the location of sediment core PS1768-8 in the sea-ice free Antarctic Zone (AZ) of the Atlantic sector of the Southern Ocean and core PS1778-5 from the Polar Front Zone (PFZ). Also indicated are sites of water column sampling [29] (pink diamonds), surface sediment sampling (yellow dots), and locations of cores discussed in the text (grey dots). Locations of the modern winter ice edge (MWIE) and the GWIE were derived from data in refs 34, 67, respectively. Oceanic fronts from ref. 68: Antarctic Polar Front (APF), Subantarctic Front (SAF) and Subtropical Front (STF); the latter two delimit the Subantarctic Zone (SAZ). Light-blue-shaded areas represent the zones of modelled transects (Fig. 5b–e).

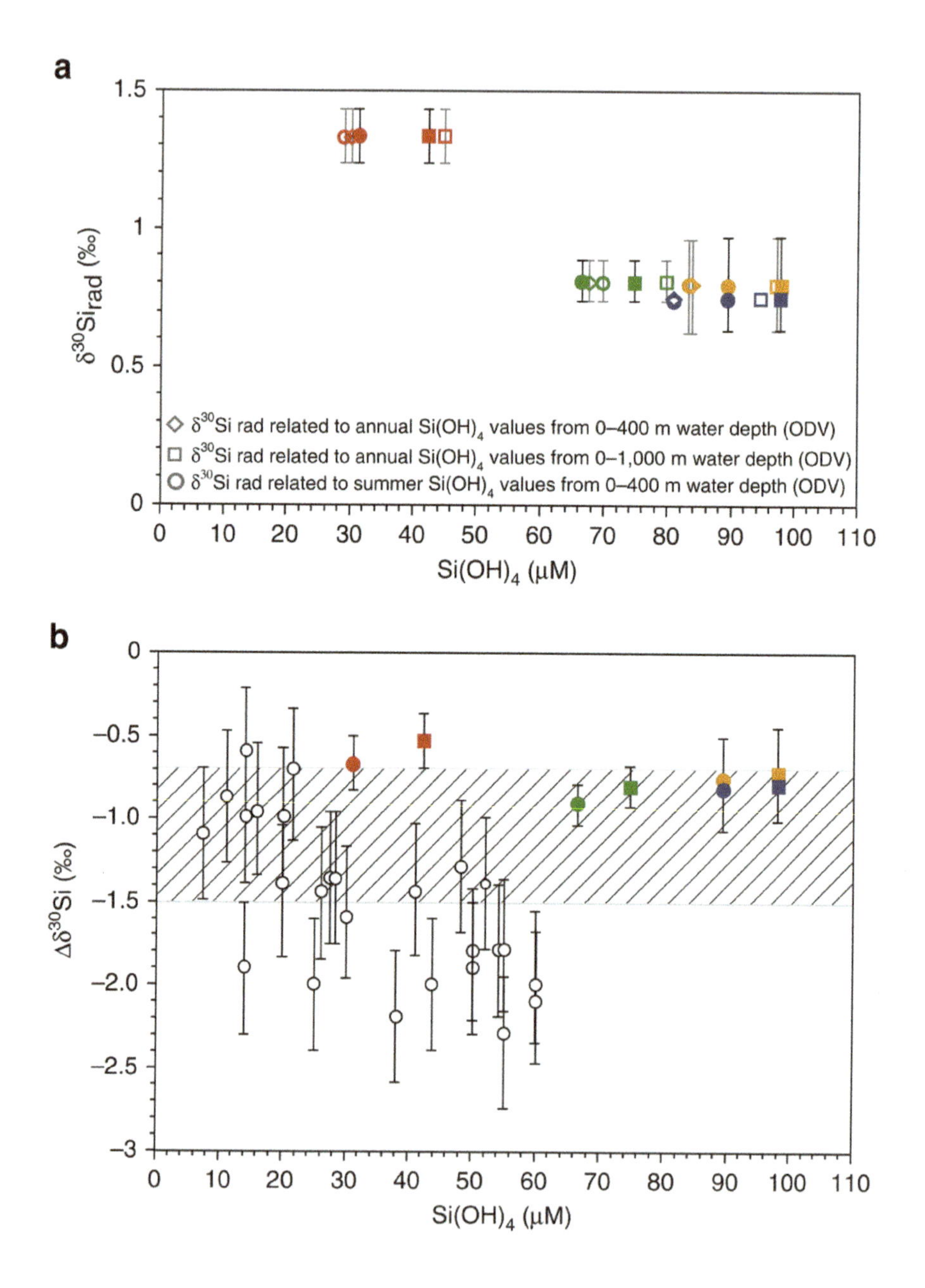
a
1.5
1
0.5
0
δ30Sirad (‰)
◇ δ30Si rad related to annual Si(OH)4 values from 0–400 m water depth (ODV)
□ δ30Si rad related to annual Si(OH)4 values from 0–1,000 m water depth (ODV)
○ δ30Si rad related to summer Si(OH)4 values from 0–400 m water depth (ODV)
0 10 20 30 40 50 60 70 80 90 100 110
Si(OH)4 (μM)
b
0
−0.5
−1.0
−1.5
−2.0
−2.5
−3
Δδ30Si (‰)
0 10 20 30 40 50 60 70 80 90 100 110
Si(OH)4 (μM)

FIGURE 2: $\delta^{30}Si$ data of radiolarians ($\delta^{30}Si_{rad}$) and $\Delta\delta^{30}Si$ offsets in surface sediments compared with $Si(OH)_4$ concentrations and to $\Delta\delta^{30}Si$ offsets of diatoms. (a) $\delta^{30}Si$ data of radiolarians ($\delta^{30}Si_{rad}$) in four surface sediments from the study area compared with Si(OH)4 concentrations in sea water. Colours refer to different surface sediments (red: PS63/026-2, green: PS63/035-2, blue: PS63/042-2, orange: PS63/043-2) (for site location, see Fig. 1 and Supplementary Table 3). Filled circles and squares present the $\delta^{30}Si_{rad}$ values (Supplementary Table 3) plotted versus $Si(OH)_4$ concentrations measured at different water depth intervals (filled circles: upper ~300–400 m and filled squares: upper 1,000 m) at nearby seawater sampling stations [29] (Fig. 1 and Supplementary Table 3), whereas the open symbols display the $\delta^{30}Si_{rad}$ values plotted versus annual and summer $Si(OH)_4$ concentrations at different water depth intervals in the study area according to the World Ocean Atlas [33] (Supplementary Table 3). (b) $\Delta\delta^{30}Si_{rad}$ offsets between $\delta^{30}Si$ values of seawater samples [27, 28, 29, 69' and $\delta^{30}Si$ values of radiolarians ($\delta^{30}Si_{rad}$) and diatoms ($\delta^{30}Si_{diat}$), respectively, versus seawater $Si(OH)_4$ concentrations at the seawater sampling stations. Coloured symbols display different surface sediments (see Fig. 2a). Different symbols (circles and squares) display the $\delta^{30}Si$ offsets between $\delta^{30}Si_{rad}$ values of the four surface sediment samples and seawater $\delta^{30}Si$ values at different water depth intervals (filled circles: upper ~300–400 m and filled squares: upper 1,000 m, see Fig. 2a) at the nearby oceanographic stations [29]. The dashed area illustrates the diatom $\delta^{30}Si$ fractionation offset of -1.1 ± 0.4‰ obtained from diatom culture studies [24]\. The white dots with error bars display the $\Delta\delta^{30}Si$ offsets between $\delta^{30}Si_{diat}$ and $\delta^{30}Si$ of seawater obtained from field data [27, 28, 29, 69]. Error bars show $\pm2\sigma$ s.d.

Here we apply the new approach to combine $\delta^{18}O$ and $\delta^{30}Si$ measurements of diatom and radiolarian opal to two late Quaternary sediment cores (PS1768-8 and PS1778-5) from the sea ice-free Antarctic Zone and from the Polar Front Zone of the Atlantic sector of the SO, respectively (Fig. 1). We are aware that the calibration of the new proxies requires further investigations, especially with respect to the isotope fractionation of radiolarians. Here we attenuate the lack of data on radiolarian fractionation by combining $\delta^{30}Si$ measurements from surface sediments and water-column samples available from the study area. As least information is available on oxygen isotope fractionation of radiolarians, we primarily base our $\delta^{18}O$-related interpretations on the signals from diatom opal. Altogether, the combination of our opal isotope results with other proxy data and climate simulations using a fully coupled global atmosphere–ocean general circulation model [23] (AOGCM) enables the establishment of coherent paleoceanographic scenarios. This combined data/modelling interpretation implies that the glacial near-surface stratification in the SIZ was variable. Relatively deep mixing during fall and winter allowed for surface-ocean refueling with nutrients from a potentially enriched deep reservoir, which generated a carbon sink in the glacial SIZ.

3.2 RESULTS

3.2.1 OPAL-BASED ISOTOPE PROXIES

A critical requirement for appropriate analyses and interpretation of $\delta^{18}O$ and $\delta^{30}Si$ in diatom ($\delta^{18}O_{diat}$ and $\delta^{30}Si_{diat}$) and radiolarian ($\delta^{18}O_{rad}$ and $\delta^{30}Si_{rad}$) opal is the extraction and separation of both microfossil groups (Methods, Supplementary Figs 1–4 and Supplementary Table 1). Our diatom fraction (10–40 μm) used for isotope measurements to reconstruct surface water conditions is dominated by two species: *Eucampia antarctica* in the lower part of the cores and *Thalassiosira lentiginosa* in the upper core portions. The shift in species composition is abrupt and its timing is unrelated to the glacial–interglacial change of the opal isotope signals (Methods and Supplementary Fig. 5), which suggests that the diatom isotope signals are not biased by species-related effects. Except for two studies from the North Pacific, this is in line with other investigations, indicating vital effects to be either non-existent or within the analytical reproducibility [14, 17, 24]. In contrast to the diatoms, the species composition of the individual radiolarian fractions does not significantly vary throughout the investigated core sections, so that species-related isotope effects in the different fractions remain unlikely. The radiolarian fraction >250 μm (PS1768-8) consists of two large-sized species (*Actinomma antarctica* and *Spongotrochus glacialis* adult), mainly dwelling in the upper 100–400 m of the water column, thus representing surface–subsurface conditions (Supplementary Table 2). Radiolarians assembled in the 125- to 250- μm fraction (PS1768-8) and the >125-μm fraction (PS1778-5) display a more diverse species composition, also including species with a deeper habitat (>400 m) [25] (Methods). We also rule out seasonal effects, because sediment trap studies show that the diatom and radiolarian export in the SIZ of the SO occur synchronously and are restricted to spring–summer [26].

The $\delta^{18}O$ signal in diatoms generally depends on both temperature and $\delta^{18}O$ in seawater [17]. A robust relationship between diatom $\delta^{30}Si$ and silicic acid utilization is derived from culture experiments and field data [19, 20, 24, 27, 28, 29, 30]. Diatom culture studies point to a mean fractionation factor of −1.1‰ (ref. 24) and show limited variation with species and growth rate19. Larger variability in diatom $\delta^{30}Si$ fractionation was derived from field data ranging between approximately −0.6‰ and −2.3‰, which reflects the natural variation but also the methodological challenge of calculating fractionation offsets [19] (Fig. 2b). Although the oxygen and silicon isotope fractionation of diatoms is rather well investigated, the knowledge concerning the isotope fractionation of radiolarians is less developed. The generation of such data is complicated by the lack of successful radiolarian culturing experiments [31] and isotope measurements of radiolarians collected in the water column. A first approach to obtain information on this important issue relies on the modelling of a fractionation offset ranging between −1.1‰ and −2.1‰ derived from deglacial $\delta^{30}Si_{rad}$ values [32]. To assist the interpretation of our results, we moved a step forward and estimated the radiolarian fractionation offset ($\Delta\delta^{30}S_{irad}$) by using $\delta^{30}S_{irad}$ values from four surface sediment samples from the Atlantic sector in combination with $\delta^{30}Si_{Si(OH)4}$ values from surface and deeper waters close to the surface sediment sample sites [29] (Figs 1, 2a,b, and Supplementary Table 3). Considering that the fractionation of diatoms and sponges is suggested to occur in equilibrium with the surrounding water [19, 24], it is reasonable to assume that this is also true for the fractionation of radiolarians. For the calculation of $\Delta\delta^{30}Si_{rad}$, we used the following equation adapted from ref. 21

$$\varepsilon \sim \Delta^{30}Si_{rad} = \delta^{30}Si_{rad} - \delta^{30}Si_{Si(OH)4} \quad (1)$$

where ε is the fractionation factor by opal-producing organisms, $\delta^{30}Si_{rad}$ is the silicon isotope composition of radiolarian opal and $\delta^{30}Si_{Si(OH)4}$ is the silicon isotope composition of sea water. The $\Delta\delta^{30}Si_{rad}$ values were calculated with $\delta^{30}Si_{Si(OH)4}$ values averaged from two different water depth intervals (0 to ~300–400 m and 0 to ~1,000 m), to cover all possible depth ranges of the included species (Fig. 2b and Supplementary Table 3). The obtained $\Delta\delta^{30}Si_{rad}$ values range between −0.5‰ and −0.9‰, and show a linear relationship with the $Si(OH)_4$ concentrations. The $\delta^{30}Si_{rad}$ fractionation offset calculated in our study is more positive than the fractionation applied in ref. 32 (−1.1‰ to −2.1‰). However, both fractionation estimates are in the range of the observed diatom fractionation (Fig. 2b). The modern $\delta^{30}Si_{rad}$ values display an inverse trend to the $Si(OH)_4$ concentrations [29, 33] in the upper 400 and upper 1,000 m of the water column (Fig. 2a and Supplementary Table 3). Higher $\delta^{30}Si_{rad}$ values of approximately +1.4‰ correspond to lower $Si(OH)_4$ concentrations of 30–45 μM l^{-1}, whereas lower $\delta^{30}Si_{rad}$ values of +0.7‰ and +0.8‰ correlate with higher $Si(OH)_4$ concentrations of 67–98 μM l^{-1}, which is comparable to the relationship between $\delta^{30}Si$ values and $Si(OH)_4$ concentrations documented for diatoms and sponges.

To test the reliability of the available information on diatom and radiolarian fractionation we calculated $\delta^{30}Si_{Si(OH)4}$ from $\delta^{30}Si$ values of radiolarians and diatoms averaged over the Holocene in both cores (Methods and Supplementary Table 4) and related the obtained $\delta^{30}Si_{Si(OH)4}$ data to $\delta^{30}Si_{Si(OH)4}$ and $Si(OH)_4$ concentrations reported from modern water column studies29 (Fig. 3a). In our $\delta^{30}Si_{Si(OH)4}$ calculation we considered a fractionation of −1.1‰ for diatom $\delta^{30}Si$ data [24]. Considering the remaining uncertainty in the definition of radiolarian fractionation offsets, we tested the applicability of three offset values. This includes $\Delta\delta^{30}Si_{rad}$ of −0.8‰ (average estimated offset from this study, Supplementary Tables 3 and 4), −1.5‰ (average estimated offset from ref. 32) and −1.2‰ representing an average over both. We note that the Holocene $\delta^{30}SiS_{i(OH)4}$ values reconstructed from δ30Si of diatoms and surface–subsurface dwelling radiolarians (>250 μm fraction) are in the range of $\delta^{30}Si_{Si(OH)4}$ values reported from the modern mixed layer (ML) in the Atlantic sector of the SO [29] (Fig. 3a). The $\delta^{30}Si_{Si(OH)4}$ values reconstructed from the $\delta^{30}Si$ data in the radiolarian fractions 125–250 μm and >125 μm, which also include species with a deeper habitat, are shifted towards lower values. These $\delta^{30}Si_{Si(OH)4}$ values are in the range of $\delta^{30}Si_{Si(OH)4}$ data and $Si(OH)_4$ concentrations from the Circumpolar Deep Water (CDW) [28, 29] (Fig. 3a). While the application of $\Delta\delta^{30}Si_{rad}$ values of −1.2‰ and −1.5‰ leads to realistic seawater $Si(OH)_4$ concentrations, estimates calculated with a fractionation offset of -0.8‰

tend to result in overestimated $Si(OH)_4$ concentration. This is most apparent for the result from the 125- to 250-μm radiolarian fraction (PS1768-8) reaching values comparable to those in modern Northwest Pacific Deep Water [20], which exceed CDW concentrations (Fig. 3a). Although information on isotope fractionation in radiolarians remains incomplete and requires additional efforts (for example, in water column studies and new approaches for radiolarian culturing), our $\Delta\delta^{30}Si_{rad}$ calculations and their relation to modern $Si(OH)_4$ concentrations point to a similar fractionation in diatoms and radiolarians. This assumption represents a step towards quantification of past $Si(OH)_4$ concentration and its variability in surface and subsurface to intermediate-deeper water.

3.2.2 DOWN-CORE DATA INTERPRETATION

During the last glacial, core PS1768-8 (52°35.61′S, 4°28.5′E, water depth 3,270 m) was positioned in the northern glacial SIZ and core site PS1778-5 (49°00.7′S, 12°41.8′W, water depth 3,380 m) was in the area of the glacial winter sea-ice edge (GWIE) [34] (Fig. 1 and Supplementary Fig. 6). This is in agreement with glacial-time winter sea-ice concentrations (WSICs) based on a new transfer function [35], which display glacial sea-ice concentrations ~60% in the PS1768-8 record (Fig. 4e) and ~40% at PS1778-5 (Fig. 4i). We assigned sea-ice concentrations of 40%–50% to be indicative of the average paleo-sea-ice edge, because these values are in the middle of the abrupt decline of Antarctic sea-ice concentration, which marks the modern sea-ice edge [35, 36]. A similar definition of the average sea-ice edge was proposed based on microwave remote-sensing observations [37]. Our sediments document the last glacial, the glacial–interglacial transition and the early part of the Holocene (Fig. 4). In the absence of biogenic carbonate, which hampers the development of continuous foraminiferal oxygen isotope records and carbonate-based AMS14C data series in the studied cores, the generation of age models for both cores considers the dating strategy and stratigraphic data from a compilation of last glacial sea-surface temperature and sea-ice records from the Atlantic sector of the SO38 (Methods and Supplementary Figs 7,8).

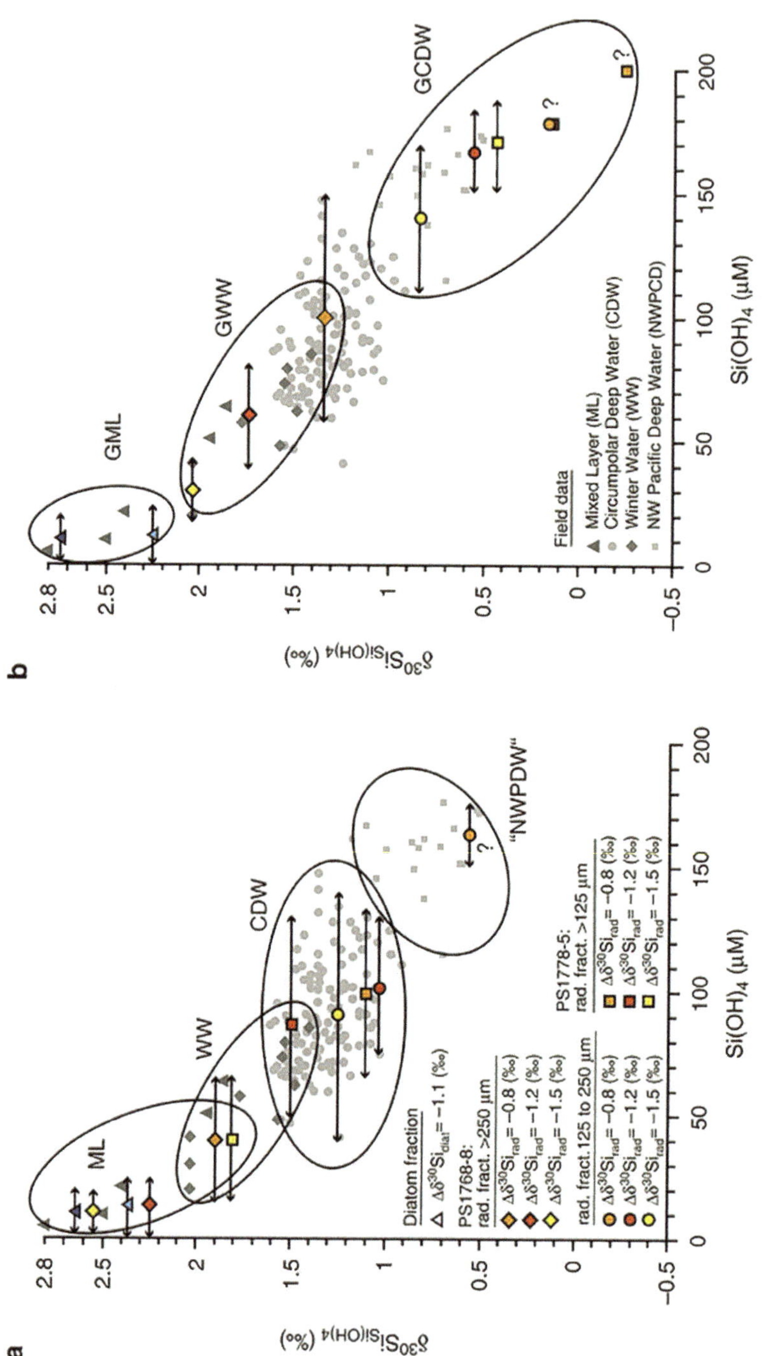
a
b
ML
WW
CDW
"NWPDW"
GML
GWW
GCDW
Diatom fraction
Δδ30Sidiat = −1.1 (‰)
PS1768-8:
rad. fract. >250 µm
Δδ30Sirad = −0.8 (‰)
Δδ30Sirad = −1.2 (‰)
Δδ30Sirad = −1.5 (‰)
rad. fract. 125 to 250 µm
PS1778-5:
rad. fract. >125 µm
Field data
Mixed Layer (ML)
Circumpolar Deep Water (CDW)
Winter Water (WW)
NW Pacific Deep Water (NWPCD)
δ30SiSi(OH)4 (‰)
Si(OH)4 (µM)

FIGURE 3: Reconstructed $\delta^{30}Si_{Si(OH4)}$ values and $Si(OH)_4$ seawater concentrations compared with values in modern water masses. (a) Average Holocene and (b) glacial-time $\delta^{30}Si_{Si(OH4)}$ values and $Si(OH)_4$ concentrations in seawater reconstructed from $\delta^{30}Si_{rad}$ data and $\delta^{30}Si_{diat}$ values in cores PS1768-8 and PS1778-5 compared with modern seawater $\delta^{30}Si_{Si(OH4)}$ values and $Si(OH_{)4}$ concentrations in different water masses [29] (Supplementary Table 4). The blue triangles indicate the $\delta^{30}Si_{Si(OH4)}$) values reconstructed from the $\delta^{30}Si_{diat}$ data of the two cores by using a $\Delta\delta^{30}Si$ offset of −1.1‰ (light blue: core PS1768-8 and dark blue: core PS1778-5), the coloured diamonds indicate the $\delta^{30}Si_{Si(OH4)}$ values reconstructed from the $\delta^{30}Si_{rad}$ data of the size fraction >250 μm in core PS1768-8, the coloured dots indicate the $\delta^{30}Si_{Si(OH4)}$ values reconstructed from the $\delta^{30}Si_{rad}$ data of the size fraction 125–250 μm in core PS1768-8 and the coloured squares indicate the $\delta^{30}Si_{Si(OH4)}$ values reconstructed from the $\delta^{30}Si_{rad}$ data of the size fraction >125 μm in core PS1778-5. The $\Delta\delta^{30}Si$ offsets applied to the $\delta^{30}Si_{rad}$ data were −0.8‰ (orange), −1.2‰ (red) and −1.5‰ (yellow) (see Methods). The obtained $\delta^{30}Si_{Si(OH4)}$ values were related to $\delta^{30}Si_{Si(OH4)}$ values from water stations and to their ranges in $Si(OH)_4$ concentrations [20, 28, 29]. The modern $\delta^{30}Si$ values and $Si(OH)_4$ concentrations were measured on seawater samples from the mixed layer (ML; grey triangles) [29], Winter Water (WW; grey diamonds) [29], CDW (grey dots) [28, 29] and Northwest Pacific Deep Water ('NWPDW'; grey squares) [20]. The reconstructed glacial-time $\delta^{30}Si_{Si(OH4)}$ values and $Si(OH)_4$ concentrations (b) may allow to discriminate Glacial Mixed Layer (GML), Glacial WW (GWW) and Glacial CDW (GCDW). Double-headed arrows display the range in Si(OH)4 concentrations that can be attributed to the reconstructed $\delta^{30}Si_{Si(OH4)}$ values. It is worth noting that $\delta^{30}Si_{Si(OH4)}$ estimations based on $\Delta\delta^{30}Si_{rad}$ of −0.8‰ may tend to overestimated $Si(OH)_4$ concentrations.

In both cores, PS1768-8 and PS1778-5, the $\delta^{18}O_{diat}$ and $\delta^{18}O_{rad}$ signals display a similar pattern with decreasing values from the last glacial period to the early Holocene (Fig. 4c,g). The $\delta^{18}O_{diat}$ and $\delta^{18}O_{rad}$ values range between +45.1‰ and +41.7‰, thus being close to the $\delta^{18}O_{diat}$ values obtained from other SO sediment cores [14, 15, 16]. The major shifts from glacial to Holocene $\delta^{18}O$ values occur in close relation to sea-surface water temperature (SST) increase overprinting the ice volume signal (Supplementary Figs 7 and 8). The early–middle Holocene diatom and radiolarian $\delta^{18}O$ records from PS1768-8 display a trend similar to a planktic foraminifer $\delta^{18}O$ record from the nearby core TN057-13 (ref. 15, Fig. 1 and Supplementary Fig. 7). At both sites, our $\delta^{18}O_{diat}$ records are inconsistent with distinct glacial freshwater supply resulting from iceberg melting. This differs from the $\delta^{18}O_{diat}$ records reported from cores TN057-13 (ref. 15) and RC13-259 (ref. 14) recovered in our study area (Fig. 1). These records are characterized by generally lower glacial values and higher $\delta^{18}O_{diat}$ values during the last deglaciation and the Holocene. In TN057-13, the $\delta^{18}O_{diat}$ record is punctuated by $\delta^{18}O_{diat}$ decreases that are in close correlation with increased values of ice rafted debris (IRD) [15]. According to a more recent geochemical study, IRD in TN057-13 primarily consists of volcanic tephra from the South Sandwich Islands transported by sea ice to the studied site [39]. Measurements of $\delta^{18}O$ on samples that contain, besides biogenic silica, also non-biogenic components such as terrigenous minerals and volcanic tephra, may result in lower $\delta^{18}O$ values and thus bias the isotopic signal towards freshwater-related values [17, 40]. Therefore, the apparent anti-correlation between the content of IRD (mostly tephra) and $\delta^{18}O_{diat}$ values in core TN057-13 may not only be explained by salinity decrease due to iceberg melting but also by a contribution from $\delta^{18}O$-depleted tephra to the $\delta^{18}O_{diat}$ signal. In contrast, IRD deposition in core PS1768-8 decreases between 18 and 16 cal. ka BP. Thus, the $\delta^{18}O_{diat}$ values show no correlation with IRD (Supplementary Fig.7), which suggests that the $\delta^{18}O_{diat}$ data presented here may be more reliable than those of ref. 15. This is confirmed by the purity of the cleaned samples in this study, which is exceptionally high, ranging between 97.9% and 99.8% SiO_2 (Methods and Supplementary Table 1), ruling out bias of our $\delta^{18}O_{diat}$ values.

The $\delta^{30}Si$ signal of diatoms in core PS1768-8 ranges generally around +1‰ in the glacial and increases to about +1.2‰ in the Holocene (Fig. 4d). The values correspond well to those reported from the nearby core RC13-269 (ref. 41 and Fig. 1). In the northern core PS1778-5, we observe distinctly higher glacial $\delta^{30}Si_{diat}$ values around +1.6‰ and a slight decrease by ~0.05‰ towards the Holocene, which is however within the analytical error (Fig. 4h).

In comparison with the diatom records, the contrasts between glacial and interglacial $\delta^{30}Si_{rad}$ values are generally more pronounced. During glacial conditions, the silicon isotope signals of diatoms and radiolarians in core PS1768-8 display offsets, with ~0.6‰ lower δ30Sirad values in the >250 μm fraction that represents surface–subsurface conditions and ~1.7‰ lower $\delta^{30}Si_{rad}$ values in the 125- to 250-μm fraction, which also includes signals from intermediate-deeper waters. Even more pronounced is the offset between the glacial diatom and radiolarian $\delta^{30}Si$ signals from the northern core PS1778-5 where the radiolarian fraction >125 µm combines species dwelling at surface–subsurface and intermediate-deeper water depth (Methods). The glacial $\delta^{30}Si_{rad}$ values range around −1.1‰ and thus are ~2.7‰ lower than the diatom values. During the deglacial transition, surface–subsurface $\delta^{30}Si_{rad}$ values in the southern core PS1768-8 increase to $\delta^{30}Si_{diat}$ values and remain close to the diatom values during the Holocene (Fig. 4d). Such deglacial convergence of the diatom and radiolarian records is also observed in core PS1778-5, but an offset of ~0.9‰ between $\delta^{30}Si_{diat}$ and $\delta^{30}Si_{rad}$ persists into the early Holocene (Fig. 4h).

We interpret the glacial offsets between the $\delta^{30}Si_{diat}$ and $\delta^{30}Si_{rad}$ records as resulting from the presence of a glacial surface-water stratification separating the diatom and radiolarian habitats at the time of their production (spring–summer). Considering that the $\delta^{18}O_{diat}$ records are not indicative of significant freshwater supply, we suggest that the stratification is primarily induced by sea-ice melt during spring, as melting sea ice has no significant effect on the oxygen isotopic composition of surface waters [42] but largely affects surface ocean salinity and thus the surface water structure even beyond the winter sea-ice edge. The effect of sea-ice melting during spring is assisted by seasonal warming and weakening winds as observed in the modern SO. Assuming that the offsets between the $\delta^{30}Si_{diat}$ and $\delta^{30}Si_{rad}$ records reflect stratification in the upper water column, they would point to glacial surface waters with increased silicic acid consumption and subsurface and intermediate deepwaters with higher silicic acid availability. The deglacial convergence between the $\delta^{30}Si_{diat}$ and $\delta^{30}Si_{rad}$ records may point to a deepening of the spring–summer ML depth (MLD), leading to the same Si pool for surface–subsurface-dwelling radiolarians (>250 µm fraction) and surface-dwelling diatoms at site PS1768-8. Considering that this interpretation is strongly based on a radiolarian isotope signal obtained from a nearly monospecific >250 µm fraction, we are confident that the $\delta^{30}Si_{rad}$ signal represents environmental change rather than fractionation effects.

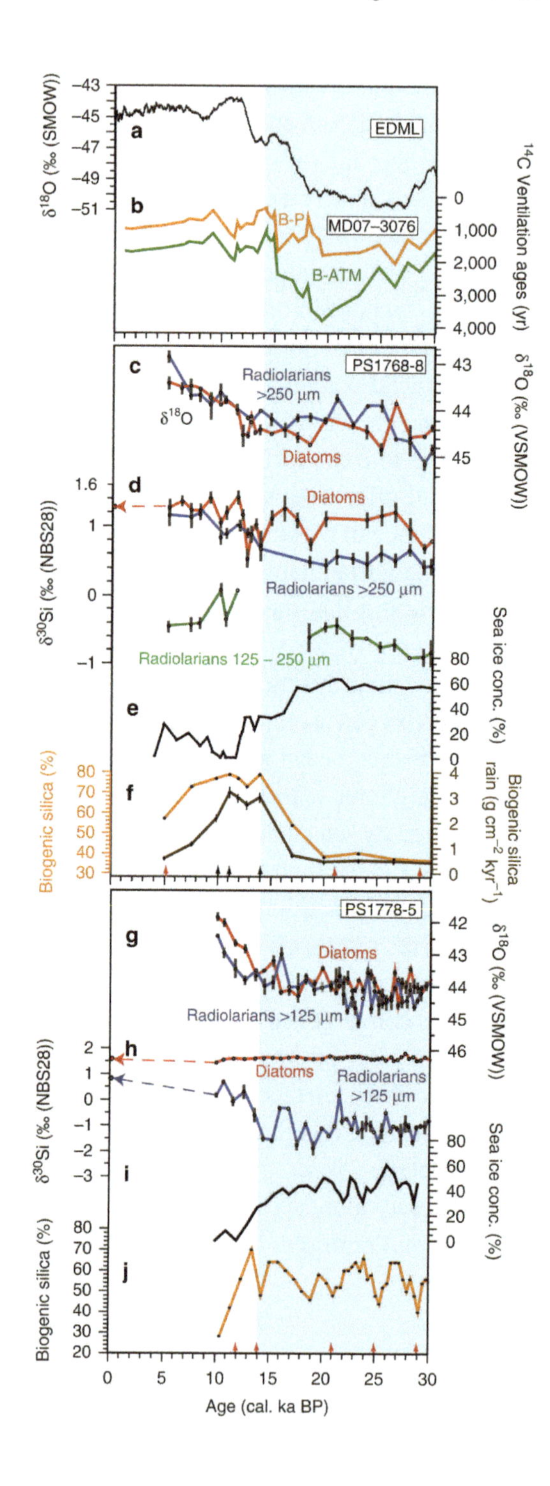
EDML
MD07–3076
B-P
B-ATM
PS1768-8
Radiolarians >250 μm
δ18O
Diatoms
Radiolarians 125 – 250 μm
PS1778-5
Radiolarians >125 μm
δ18O (‰ (SMOW))
14C Ventilation ages (yr)
δ18O (‰ (VSMOW))
δ30Si (‰ (NBS28))
Sea ice conc. (%)
Biogenic silica (%)
Biogenic silica rain (g cm−2 kyr−1)
Age (cal. ka BP)

FIGURE 4: Oxygen and silicon isotope records of diatoms and radiolarians from cores PS1768-8 and PS1778-5 compared with other paleoclimatic records over the last 30 kyr. (a) $\delta^{18}O$ record of Antarctic ice core drilled at EPICA Dronning Maud Land (EDML) site [70] against AICC2012 (Antarctic Ice Core Chronology 2012)62. (b) ^{14}C ventilation ages plotted as benthic-planktic foraminifera age difference (B-P) and benthic-atmospheric difference (B-ATM) from core MD07-3076 (Fig. 1) [57]. (c–f) Proxies from core PS1768-8 located in the glacial SIZ. (c) $\delta^{18}O$ and (d) $\delta^{30}Si$ records from one diatom and two radiolarian fractions (>250 μm and 125–250 μm) measured at the same aliquot of biogenic opal. The red dashed arrow points to the $\delta^{30}Si_{diat}$ value obtained from a nearby seafloor surface-sediment sample assumed to be of modern age (Supplementary Table 7). Error bars indicate range of replicate and triplicate measurements (Supplementary Tables 8–12). (e) Diatom transfer function-based estimates of WSIC [35]. (f) Biogenic silica percentages and biogenic silica rain rates [50]. (g–j) Proxies from core PS1778-5 located close to the GWIE. (g) $\delta^{18}O$ and (h) $\delta^{30}Si$ records of diatoms and radiolarians (>125 μm fraction) measured at the same aliquot of biogenic opal. The red and blue dashed arrows point to the $\delta^{30}Si_{diat}$ and $\delta^{30}Si_{rad}$ values obtained from a nearby seafloor surface sediment sample assumed to be of modern age (Supplementary Table 7) (i) Diatom transfer function-based estimates of WSIC. (j) Biogenic silica percentages. Arrows in f and j indicate age pointers (black arrows mark AMS14C dates and red arrows mark ages obtained by diatom and radiolarian biofluctuation stratigraphy, see Supplementary Tables 5 and 6). Blue-shaded area delineates Marine Isotope Stage (MIS) 2 and the late part of MIS 3.

3.2.3 MODELLING GLACIAL SEA ICE AND MLD VARIABILITY

To further evaluate the physical changes associated with sea-ice variations, we re-analysed two model simulations with conditions during the Last Glacial Maximum (LGM) [23] and interglacial periods [43], respectively, and performed two additional sensitivity experiments to test the impact of a deglacial CO_2 rise and a poleward wind field shift in the SO. All model results are based on simulations using the same fully coupled AOGCM [23]. The model configuration includes the atmosphere component ECHAM5 (ref. 44) at T31 resolution (~3.75°) with 19 vertical layers, complemented by a land-surface scheme including dynamical vegetation (JSBACH) [45]. The ocean component MPI-OM [46], including the dynamics of sea ice formulated using viscous-plastic rheology [47], has an average horizontal resolution of 3° × 1.8° with 40 uneven vertical layers. The performance of this climate model was evaluated for SO Holocene

[43] and glacial [23] conditions, showing that the glacial and interglacial (pre-industrial) sea-ice field (Supplementary Fig. 9) and general ocean circulation can be simulated reasonably well, providing a suitable reference to explore the underlying physical mechanism accounting for the proxy records developed in this study. The model has also been applied to analyse glacial millennial-scale variability [48] and warm climates in the Miocene [49]. For further details of the model and experimental configuration, see Methods and Supplementary Figs 9–11.

By re-analysing model results for the LGM [23] from two latitudinal transects (centred at the respective core locations) across the SO in 10° longitudinal windows (Fig. 1), we identify strong seasonal changes in sea-ice cover in the area of both core locations (Fig. 5b,c) consistent with the interpretation of a link between vertical stratification changes and seasonal sea-ice variations. These variations result in annual sea-surface salinity and accompanying MLD variations that favour an increase of the MLD during the season of sea ice growth, followed by an MLD decrease when sea ice declines. Hence, the glacial simulation shows a relatively shallow glacial ML during austral spring–summer, reaching minimum values of 40–60 m in the study area (Fig. 5), which can be attributed to the melting of sea ice during this time. Such a pattern would separate the main habitat depth of diatoms from the deeper living radiolarians as suggested by the glacial diatom-radiolarian $\delta^{30}Si$ offset (Fig. 4d,h). An MLD deepening is simulated for fall–winter seasons, which is promoted by enhanced vertical mixing during sea-ice formation (Fig. 5).

To test the robustness of this interpretation, zonal heterogeneities in sea-ice distribution need to be taken into account. Therefore, we evaluated the seasonal MLD changes for both the exact latitudes of the core locations (Fig. 5, dashed lines) and the latitudes where the physical conditions coincide with the proxy-based LGM WSIC (Fig. 5, solid lines). The respective WSIC amounts on average to 60% at PS1768-8 and to 41.6% at PS1778-5 (Supplementary Fig. 6). This approach ties the proxy-based information from the sediment cores to physical conditions simulated by the model. Based on this approach, we can estimate that during glacial spring–summer the MLD reached minimum values between 40 and 60 m at both studied sites and increased during sea-ice formation in fall–winter, reaching a maximum depth of up to 350 m at the site located in the area of the GWIE (Fig. 5a).

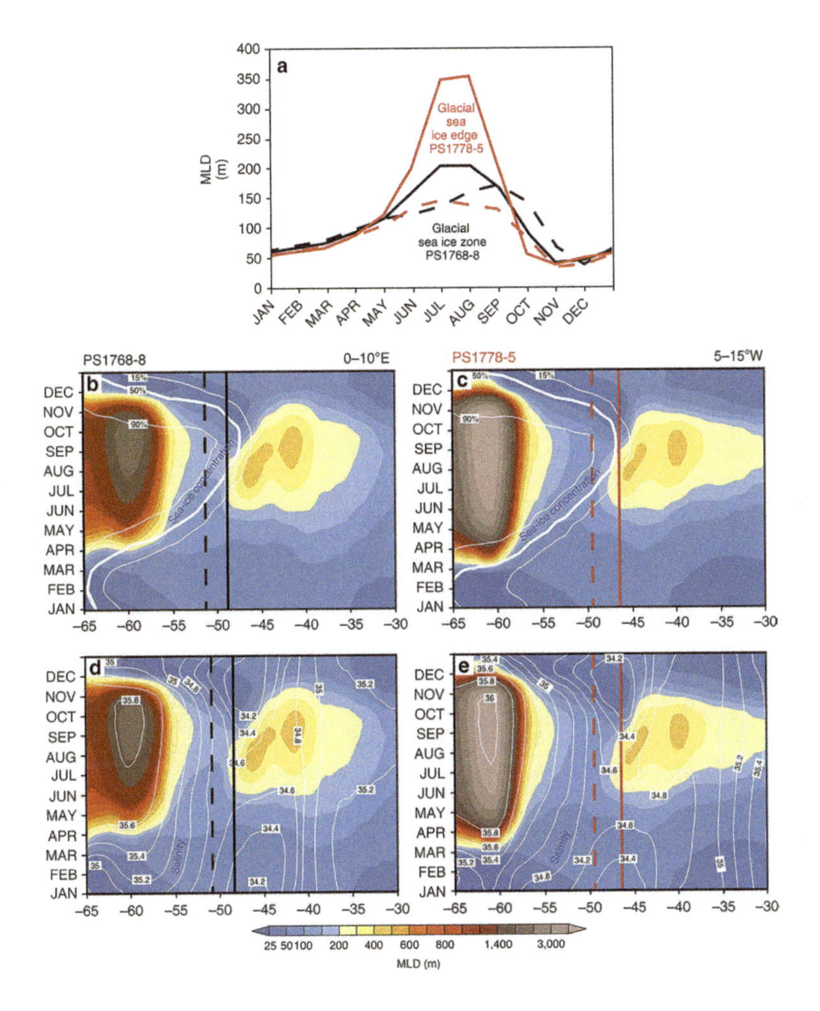

FIGURE 5: Modelled monthly averaged MLD, sea-ice concentration and salinity in the Atlantic sector of the Southern Ocean during the LGM. (a–e) Monthly averaged MLDs for PS1768-8 and PS1778-5 overlaid in b and c with sea-ice concentration (%), and in d and e with sea-surface salinity (psu). Shown are latitudinal transects zonally averaged between 0 and 10°E (for PS1768-8), and 5 and 15°W (for PS1778-5). Dashed lines indicate the respective changes at the exact latitudinal positions of the cores, whereas the solid lines show the changes at the latitudinal position, where the physical conditions coincide with the reconstructed proxy-based LGM WSIC. The transect zone and core locations are indicated in Fig. 1. The relatively deep glacial MLD at ca. 60–65°S in the Weddell Sea (b–e) are mainly attributed to strengthened brine rejection associated with enhanced in-situ sea-ice formation and northward sea-ice export, which are key processes for SO deep water formation during the LGM (for example, see ref. 23).

3.3 DISCUSSION

The combination of proxy data and AOGCM modelling implies that the sea surface of the glacial SIZ and the GWIE, at least in the Atlantic sector, was distinctly stratified during austral spring–summer with a relatively shallow MLD (40–60 m; Fig. 5). Increased glacial deposition of iron [50, 51] released from the melting winter sea ice transformed the glacial SIZ into a seasonally high productivity region governed by primary producers with low Si:N demand [5], leading to enhanced utilization of nitrate [3, 13], slightly reduced consumption of silicic acid and low opal export [50] (Fig. 4f). A similar productivity regime was triggered by iron-fertilization experiments in the modern SO, showing that involved diatoms (for example, Chaetoceros) follow a 'boom-and-bust' life cycle strategy characterized by rapid biomass build-up during favourable growth conditions, succeeded by mass mortality and rapid population decline. Such a productivity regime results in enhanced organic carbon but low biogenic opal export to the deep ocean and thus leads to the decoupling of biogenic carbon and opal export [12]. In contrast to modern conditions, this would convert the glacial SIZ, which was enlarged during the glacial [34, 52], into an efficient carbon sink during spring–summer. Authigenic uranium concentrations, a potential proxy for organic carbon deposition, support this view, because they display increased values in sediments deposited in the glacial SIZ [50].

For the area of the GWIE (PS1778-5), modelling suggests even longer-lasting spring–summer stratification compared with the glacial SIZ (PS1768-8) (Fig. 5). However, in this environment a different productivity-export regime developed. This regime is characterized by the production of thick-shelled diatoms (Si:N ratios >4, ref. 53) leading to enhanced silicic acid utilization ($\delta^{30}Si_{diat}$=+1.62‰ to +1.86‰) at the sea surface and high percentages of biogenic opal in the sediments [5, 50] (Fig. 4h,j). Species involved (for example, *Fragilariopsis kerguelensis*) [5] follow a 'persistence' strategy and are hallmarked by enhanced ability to withstand grazing pressure. They are most prominent producers in the modern iron-limited open SO and present efficient silica sinkers [12] with a major contribution to the modern Antarctic opal belt [5]. Sediment records suggest that this regime extended from the glacial SIZ into the area of the modern Subantarctic Zone as mirrored by increased opal flux and dominant deposition of *F. kerguelensis* [5, 50].

The northward displacement of the zone of enhanced opal burial (opal belt) together with winter sea-ice expansion [50] during glacial periods was identified to represent a phenomenon most difficult to explain [3]. Queries concern the northward transfer of required nutrients from the glacial Antarctic Zone, assuming that surface water stratification resulted in reduced nutrient supply and almost complete consumption of available nutrients in this zone during the glacial [3, 13]. Suggested explanations address increased leakage of silica into the Subantarctic region [3, 54] and northward-shift in wind-driven upwelling [13]. We emphasize a process (nutrient supply by winter mixing) that has been previously rejected [13]. Our modelling suggests that the shallow spring–summer stratification was disrupted by a significant MLD increase during glacial austral winter. The simulations indicate that the winter MLD is more than double the summer MLD in the area of PS1768-8 (~200 m) and up to 350 m close to the GWIE (Fig. 5a). Around the GWIE and north of it, thus in the zone of enhanced glacial opal export [50], modelling indicates even deeper ML conditions during glacial winter (Fig. 5). Such a deep MLD would allow for efficient nutrient refueling of glacial winter surface waters. A deep winter mixing would provide high nutrient (for example, $Si(OH)_4$) availability at the onset of spring–summer production when surface water stratification is suggested to develop. A similar winter mixing process between ML, Winter Water and CDW is also suggested to take place in the modern SO (refs 29, 30). During spring–summer stratification, nutrient supply to the ML may be accomplished by diapycnal diffusion as described from the modern SO (ref. 55).

Amplification of the nutrient injection into the glacial surface ocean may stem from the presence of a deep reservoir that was more enriched in $Si(OH)_4$ compared with modern conditions. The $\delta^{30}Si_{diat}$ values at both studied sites point to similar isotopic compositions of the glacial ML and the modern ML (Fig. 3). However, the glacial radiolarian-derived $\delta^{30}Si_{Si(OH)4}$ signal that can be related to surface–subsurface conditions (fraction >250 μm) is close to modern $\delta^{30}Si_{Si(OH)4}$ values reported from the nutrient replete Winter Water and CDW in the Atlantic sector south of the

Antarctic Polar Front. In contrast to the Holocene, the glacial radiolarian-derived $\delta^{30}Si_{Si(OH)4}$ values from the fractions 125–250 μm and >125 μm, which reflect intermediate and deeper water conditions as well, are distinctly lower. These lower values fall in the range of $\delta^{30}Si_{Si(OH)4}$ values reported from the modern deep Northwest Pacific (approximately +0.5‰), which are related to the highest $Si(OH)_4$ concentrations observed in the World Ocean (~160–180 $\mu mol\,l^{-1}$) [20] (Fig. 3b). Although our estimates still bear uncertainties, our reconstructions using different radiolarian fractionation offsets consistently result in lower glacial $\delta^{30}Si_{Si(OH)4}$ values and thus higher glacial $Si(OH)_4$ concentrations in SO intermediate deepwaters compared with the Holocene.

The suggested presence of higher $Si(OH)_4$ concentrations in glacial circumpolar deep waters may be challenged by sponge spicule-based $\delta^{30}Si$ records from the Scotia Sea, interpreted to indicate that the glacial deep $Si(OH)_4$ concentrations were not different from modern conditions [21]. However, the records exhibit very negative $\delta^{30}Si$ values (−3‰ to −3.5‰) and thus are in a range where the application of this proxy to sponge spicules is prone to larger uncertainties [56]. Another sponge spicule-derived $\delta^{30}Si$ record from the area straddled by the Subtropical Front (ODP1089, Fig. 1), indicating no significant glacial–interglacial contrast in bottom water $Si(OH)_4$ concentration in the Atlantic sector [22], is not in conflict with our results. Indeed, this observation allows for approximation of the northern extent of higher $Si(OH)_4$ concentrations trapped in the glacial SO. Presuming a similar relationship between modern and glacial biogenic opal deposition and decline of high silicic acid concentration throughout the water column (Supplementary Fig. 12), the northward displacement of the biogenic opal belt by ~5° in latitude [50] would place the glacial silicic acid front in the area between 50 and 45°S, but not as far north as the area of site ODP1089 (Fig. 1). This is in line with a northward migration of the Subantarctic Front as postulated from the mapping of ^{14}C reservoir ages [57]. Further support comes from the pattern of LGM surface water temperature [34] and model simulations, indicating a frontal northward displacement by ~5°–7° in latitude from interglacial to glacial conditions (Supplementary Fig. 10).

A glacial SO trapping nutrients more efficiently than at present is consistent with the scenario of an Antarctic deep water body, whose age relative to the atmosphere was more than two times older than during the Holocene, and which was presumably CO_2 enriched [57] and characterized by increased salinity as mirrored by our model [23] and suggested by proxy data [58]. Possible mechanisms that have an impact on the nutrient trapping may include wind-generated changes of upwelling and downwelling in the SO, sea-ice extent variability and the availability of iron [54]. A glacial northward export of the nutrient Si (Silicic Acid Leakage Hypothesis) would be in some conflict with a glacial SO nutrient enrichment, but data suggest that enhanced Si leakage was confined to the deglacial period [19]. Indeed, the sponge spicule-derived $\delta^{30}Si$ record from the Subantarctic Atlantic (ODP1089, Fig. 1) indicates a bottom water spillout of the SO reservoir during this time, marked by a distinct $\delta^{30}Si$ excursion towards more negative values (increased $Si(OH)_4$ concentrations) [22]. Such $Si(OH)_4$ export would support the hypothesis that an expansion of Si(OH)4-enriched Antarctic Bottom Water was the source for maximum opal fluxes (diatom blooms) in the coastal upwelling area off northwest Africa during the last deglaciation [59].

The glacial–interglacial transition is characterized by successive changes, starting with the retreat of the sea ice that is accompanied by an increase in opal sedimentation between 18,000 and 16,000 years ago (Fig. 4e,f). This is followed by a steep increase in ocean ventilation at about 15,000 years ago [58] (Fig. 4b), which marks the rapid intensification in Atlantic thermohaline circulation at the onset of the Bølling [60]. Assuming that the deglacial convergence between our $\delta^{30}Si_{diat}$ and $\delta^{30}Si_{rad}$ records reflects a deepening of the spring–summer MLD as proposed above, the MLD deepening would coincide with increased biogenic opal rain (Fig. 4d,f). Owing to a data gap in our >250 μm radiolarian record in PS1768-8, we cannot document exactly the onset of the change in surface water structure in the glacial SIZ. However, considering that the available record is closely tied to the retreat of sea ice and increasing deposition of biogenic opal (Fig. 4d–f), we speculate that the process of MLD deepening was initiated around 18,000 years ago, and that the MLD possibly reached its maximum thickness at around 14,000 years. In core PS1768-8, this is documented by the lowest $\delta^{30}Si_{diat}$ values in the studied sediment interval (Fig. 4d). These $\delta^{30}Si_{diat}$ minima that are only recorded by surface dwellers (diatoms) during a time of enhanced ventilation and maximum biogenic opal export suggest that the supply of silicic acid exceeded the consumption by diatoms during this time interval. This points to the injection of nutrient-rich deep waters into the euphotic zone [5], interpreted to result from enhanced wind-driven upwelling governing biogenic opal production and export [6].

The deglacial MLD deepening leads to conditions that persist in the Holocene (PS1768-8) as recorded by diatom and surface–subsurface radiolarian (>250 μm) $\delta^{30}Si$, which can be related to modern ML conditions in the Atlantic sector of the SO (Fig. 3a). The Holocene down-core data are comparable to available $\delta^{30}Si_{diat}$ and $\delta^{30}Si_{rad}$ data from surface sediments (assumed to reflect modern conditions) in the study area (Fig. 4d,h).

It has been postulated that the destratification and enhanced upwelling during the last deglaciation allowed for a CO_2 release from the deep SO, providing a direct link to the coinciding increase in atmospheric CO_2 (ref. 6). The primary mechanisms proposed to drive the destratification is a southward shift in the Southern Westerlies winds in response to a displacement of the Earth's thermal equator, the Intertropical Convergence Zone [6]. Another mode of operation can be derived from our sensitivity experiments applying a prescribed atmospheric CO_2 increase from 180 to 240 p.p.m.v. (representing a surrogate for deglacial warming) and a poleward shift of the Southern Westerlies wind belt by 3° (Supplementary Figs 10 and 11). The sensitivity experiments show that the poleward shift of the Westerlies has a negligible effect on the position of the Antarctic Polar Front and MLD in comparison with the changes induced by the increase in atmospheric CO_2 (Supplementary Figs 10 and 11). This suggests that the destratification during the last deglaciation can be primarily attributed to sea-ice margin retreat induced by atmospheric warming and an associated southward shift of the seasonal sea-ice melting zone.

The view that sea ice presents the major player in governing SO glacial surface-water structure and related ocean–atmosphere exchange, nutrient cycling and biological productivity and export regimes entrains major implications to be considered for the estimation of SO effects on the climate system. Quantification of the impact of physical and biological processes in the SO on the glacial carbon cycle requires consideration of seasonal variability in sea-ice extent and related seasonal and spatial variability in surface ocean mixing rates. Other factors to be taken into account are the development of specific productivity regimes making the SIZ an efficient carbon sink and the area north of the sea-ice edge a region that primarily affects the Si cycle, and the potential establishment of a nutrient-enriched deep SO reservoir.

3.4 METHODS

3.4.1 STRATIGRAPHY

To generate reliable age models for the studied cores, which lack continuous foraminiferal isotope records and carbonate-based AMS^{14}C data series, we considered the strategy and stratigraphic data presented in a compilation of last glacial sea-surface temperature and sea-ice records from the Atlantic sector of the SO [38]. We have revised the correlation of 12 cores including PS1768-8 and PS1778-5 using 53 AMS^{14}C dates from 8 cores (with 3 AMS^{14}C dates available from PS1768-8) in addition to intercore correlation based on different parameters. The AMS^{14}C dates were converted to calendar years and presented as cal. ka BP (103 years before present). Parameters used for intercore correlation include the abundance pattern of the radiolarian *Cycladophora davisiana* and the diatom *E. antarctica*, together with foraminiferal oxygen isotope records, if available. Correlations were performed with AnalySeries 2.0 (ref. 61). The age assignment of the radiolarian and diatom abundance pattern was inferred from AMS^{14}C dates obtained from 12 cores from the study area [38]. Stratigraphic pointers for both cores (including the AMS^{14}C dates from PS1768-8) and their definition are presented in Supplementary Tables 5 and 6. Although this approach to construct age models for SO records bears uncertainties due to interpolations, especially in (mostly glacial) intervals that could not be dated by continuous ^{14}C dates or δ^{18}O data, we are confident that our age model is robust enough to allow for appropriate documentation of the environmental development from the last glacial into the present interglacial. According to our age models, the summer SST and WSIC began to shift towards Holocene values between 18 and 16 cal. ka BP (Fig. 4 and Supplementary Figs 7 and 8). This timing fits well with the onset of Southern Hemisphere warming and the start of

CO_2 release documented in Antarctic ice cores, dated independently of our approach [62, 63]. Increase and decline of biogenic opal sedimentation is similar to the pattern recorded from the nearby core TN057-13 (ref. 6). However, in contrast to TN057-13, the biogenic opal, summer SST and WSIC records in our core PS1768-8 display no distinct variations, which can clearly be attributed to short-term climate variability (for example, the Antarctic Cold Reversal) during the last deglacial period. This may be attributed to three- to fourfold higher sedimentation rates at site TN057-13 (ref. 6) compared with site PS1768-8.

3.4.2 ISOTOPES IN BIOGENIC OPAL

A prerequisite for measuring $\delta^{18}O$ and $\delta^{30}Si$ in diatom and radiolarian opal is the careful extraction and separation of sufficient purified material from both microfossil groups (Supplementary Figs 1–4). This is because the different life strategies and depth habitats of the two microfossil groups and different species within these groups, as well as contamination by sponge spicules and non-biogenic components (for example, rock fragments and clay minerals), may affect the isotope signal. For our study we have applied a new method, which allows for the separation of pure diatom and radiolarian fractions from the same sample aliquot. With our technical setup, an average of 2 mg purified diatom or radiolarian opal per sample is needed to obtain combined $\delta^{18}O$ and $\delta^{30}Si$ measurements from the same sample aliquot. This represents the lowest amount yet used for such combined measurements in comparison with other measuring techniques [64]. Considering that replicates and triplicates should be measured, as far as sample availability allows, to test the reproducibility of the measurements, the amount of opal to be separated and to be enriched increases accordingly. This precludes in general the separation of radiolarians by picking single radiolarian skeletons, because such a procedure takes an extraordinary expenditure of time. To generate well-established and large sample sets, a routine preparation method has been developed allowing for the removal of non-biogenic components and biogenic carbonate followed by the separation and enrichment of radiolarians and diatoms in specific size fractions.

Sample preparation includes wet-chemical cleaning and extraction of radiolarian and diatoms through different sieving and settling techniques (Supplementary Fig. 1). The samples are first washed with HCl and H_2O_2 to remove carbonates and organic material. Mineral grains are removed through density separations (on average ten density treatments for silica-rich sediments from the SO) using specific sodium polytungstate solutions. The separation of radiolarians from diatoms and sponge spicules is accomplished through several sieving and settling steps combined with ultrasonic treatment. Our tests show that after ultrasonic treatment radiolarians are still intact, whereas diatoms break up and can be removed through sieving. Repetitions of ultrasonic treatment and sieving steps lead to the separation and enrichment of radiolarians and diatoms in different size fractions. For isotope determinations, we use the pure 10–40 μm diatom fraction (both cores), the >250-μm and the 125- to 250-μm pure radiolarian fraction for core PS1768-8. As there was not enough material, we could only use one radiolarian fraction (>125 μm) for core PS1778-5 and one radiolarian fraction (>250 μm or >125 μm) for each surface sediment sample (Supplementary Tables 3 and 7). The smaller size fraction 40–125 μm contains radiolarians and diatoms of similar size, which are difficult to separate from each other and thus were not used for isotope measurements. Microscopic slides for determining the species composition of diatoms and radiolarians were prepared after completion of the preparation and separation of the preparation line (Supplementary Fig. 1).

For the combined oxygen and silicon isotope measurements, the cleaned radiolarian and diatom samples were dehydrated at 1,100 °C by inert gas flow dehydration under a He flow and further reacted to SiF_4 and O_2 by laser fluorination under BrF_5 atmosphere. The liberated oxygen was cleaned of any byproducts and analysed with a PDZ Europa 2020 mass spectrometer according to the method described in refs 65, 66. The $^{18}O/^{16}O$ reference ratio of known isotopic composition is measured in analogy to the $^{18}O/^{16}O$ sample and the final $\delta^{18}O$ value was calculated relative to Vienna standard mean ocean water. For silicon isotope measurements, the separated and cleaned SiF_4 gas was directed into glass vials and measured separately with a Finnigan MAT 252 mass spectrometer and measured against the $^{30}Si/^{28}Si$ reference ratio of a SiF_4 gas of known isotopic composition [18]. The final $\delta^{30}Si$ value was calculated relative to NBS-28. To test the reproducibility of the measurements at least, replicates and triplicates were measured on all samples with sufficient amount of material (Supplementary Tables 3 and 7–12). The analytical precision of silicon isotope measurements was better than ±0.12‰ (δ^{30}Si; 1σ) for all used working standard materials [18]. The overall precision for all working standards used for oxygen isotope measurements lies between±0.2‰ and 0.3‰ (1σ) (refs 40, 66).

3.4.3 PURITY OF SAMPLES PREPARED FOR ISOTOPE MEASUREMENTS

In general, contamination of biogenic opal samples by minerals (for example, quartz, feldspar, micas, clay minerals, rock fragments and volcanic tephra) can bias, especially the δ^{18}O signal towards lower values [17, 40], a pattern that may lead to misinterpretation of the isotope results. Considering that the contaminants may be silicates that are difficult to remove with the generally applied cleaning methods, the purity of the samples cleaned for isotope measurements needs to be tested. We can document that the cleaning procedure applied in our study leads to very well-purified diatom and radiolarian samples. Our testing of samples from core PS1768-8 using inductively coupled plasma optical emission spectrometry and energy-dispersive X-ray spectrometry indicates (1) a high degree of purification with the near absence of elemental compositions, indicating non-biogenic components, and (2) SiO_2 contents ranging between 98.5% and 99.5% (inductively coupled plasma optical emission spectrometry), and 97.9% and 99.8% (energy-dispersive X-ray spectrometry) (Supplementary Table 1).

3.4.4 EFFECT OF DISSOLUTION ON THE ISOTOPE SIGNALS

Only few studies concern the potential impact of diagenesis on the opal isotope signal, which come to opposing results concerning the effect of dissolution on the δ^{30}Si signal [19]. The diatoms extracted in this study are mainly composed of heavily silicified diatoms such as *E. antarctica* and *T. lentiginosa*, and are not affected by dissolution (Supplementary Fig.

3). The (rare) occurrence of very well-preserved thinly silicified diatoms (for example, *Rhizosolenia* sp.) confirms the excellent preservation of the diatom assemblage (Supplementary Fig. 3). This also concerns the radiolarian fractions, which are mainly composed of large and heavily silicified specimens, which are well preserved, although they were treated in the ultrasonic bath for several hours (Supplementary Fig. 4).

3.4.5 SPECIES COMPOSITION OF DIATOM AND RADIOLARIAN FRACTIONS

For the isotopic measurements, we used a pure diatom fraction (10–40 μm) for the representation of surface-water conditions. This fraction mainly consists of the species *E. antarctica* (glacial indicator) and *T. lentiginosa* making up between 70% and 98%, and 78% and 97% of the species composition in the extracted diatom fractions of PS1768-8 and PS1778-5, respectively (Supplementary Fig. 5). The amount of these species in the original diatom assemblages is on average only between 20% and 29%. The difference in diatom species composition between the original sample and the extracted fraction results from our techniques for purification and extraction of a specific size class allowing for the generation of samples containing the least possible number of species. The shift from *E. antarctica*-dominated fractions to fractions with increased *T. lentiginosa* abundances occurs abruptly between three sample depths and is unrelated to the more gradual change in isotope signal across the glacial–interglacial shift (Supplementary Fig. 5). Small-sized sea-ice-related diatoms (for example, Fragilariopsis curta) that are not or only rarely included in the 10–40 μm diatom fraction do not affect the isotope signal.

For the isotopic measurements on radiolarians, we used three radiolarian fractions to document surface–subsurface and intermediate deepwater conditions. In core PS1768-8, we measured the >250 μm fraction, which is composed of two radiolarian species: *A. antarctica*, which accounts for >90% of radiolarians in this fraction, and *S. glacialis* (adult forms), which ranges between 1% and 10% in this fraction. As different radiolarian species live in different water depths, we used data from a plankton study in the Atlantic sector of the SO (ref. 25), to get information about the depth habitat of radiolarians in the upper 1,000 m of the water column (Supplementary Fig. 13 and Supplementary Table 2). Both species occur in the upper 200–400 m with *S. glacialis* predominantly occurring in the upper 100 m and *A. antarctica* in the upper 300–400 m. As there was not enough material to separate the >250-μm fraction from core PS1778-5, we used here the >125-μm fraction for radiolarian isotope measurements. This fraction is also dominantly composed of the two species *A. antarctica* and *S. glacialis*, but in contrast to the >250-μm fraction from core PS1768-8 the >125-μm fraction also contains deeper living species (for example, *Spongopyle osculosa, Spongogurus pylomaticus* and *Cromyecheinus antarctica*) that occur in water depths >400 m (ref. 25). As the $\delta^{30}Si_{rad}$ values of the >125-μm fraction exhibit distinctly lower values (glacial average −1.06‰) than the $\delta^{30}Si_{rad}$ values from the >250-μm fraction from core PS1768-8 (average glacial values +0.54‰), we segregated the 125- to 250-μm fraction from core PS1768-8 to get more information about this latitudinal difference in the $\delta^{30}Si$ signature. The $\delta^{30}Si_{rad}$ measurements from this fraction show distinctly lower values (glacial average −0.67‰) than the $\delta^{30}Si_{rad}$ values from the >250-μm fraction. The fraction 125–250 μm also includes deeper dwelling radiolarians similar as in the >125-μm fraction from core PS1778-5. This suggests that the presence of deeper dwelling radiolarians shifts the $\delta^{30}Si$ signal to lower values. *C. davisiana*, a species typical for glacial assemblages, is not included to the fractions >125 μm because of its small size, which impedes a potential impact of this species on glacial results.

3.4.6 ESTIMATION OF $\Delta^{30}SI_{SI(OH)4}$ AND $SI(OH)_4$ CONCENTRATIONS

The estimation of Holocene and glacial silicic acid changes were carried out as follows:

1. The average $\delta^{30}Si$ values of the diatom (10–40 μm) and radiolarian fractions (>250, 125–250 and >125 μm) for the Holocene (until 12 cal. ka BP) and the last glacial period (~19–29 cal. ka BP) were calculated (Supplementary Table 4).

2. Based on the average diatom and radiolarian $\delta^{30}Si$ values, the $\delta^{30}Si_{Si(OH)4}$ values for the Holocene and last glacial period were calculated using the formula:

$$\Delta\delta^{30}Si = \delta^{30}Si_{Si(OH)4} \tag{2}$$

$$\delta^{30}Si_{Si(OH)4} = \delta^{30}Si_{BSi} - \Delta\delta^{30}Si \tag{3}$$

where $\delta^{30}Si_{BSi}$ is the silicon isotope composition of biogenic silica (diatoms or radiolarians), $\delta^{30}Si_{Si(OH)4}$ is the silicon isotope composition of the input sea water (Supplementary Table 4) and $\Delta\delta^{30}Si$ is the fractionation offset. The calculations were performed using different $\Delta\delta^{30}Si$ values: $\Delta\delta^{30}Si_{diat}$=−1.1‰ (ref. 24), $\Delta\delta^{30}Si_{rad}$=−0.8‰ (average $\Delta\delta^{30}Si_{rad}$ this study, Supplementary Table 4), $\Delta\delta^{30}Si_{rad}$=−1.2‰ (average $\Delta\delta^{30}Si_{rad}$ of this study and ref. 32), $\Delta\delta^{30}Si_{rad}$=−1.5‰ (average $\Delta\delta^{30}Si_{rad}$ of ref. 32).

3. The reconstructed $\delta_{30}Si_{Si(OH)4}$ values were related to $\delta_{30}Si_{Si(OH)4}$ values and $Si(OH)_4$ concentrations from different water masses reported from modern water-column studies [20, 28, 29]. We note that our reconstructed $\delta_{30}Si_{Si(OH)4}$ values reflect a rather broad range in $Si(OH)_4$ concentrations (Fig. 3 and Supplementary Table 4). In spite of this large range in variability, our calculated $\delta_{30}Si_{Si(OH)4}$ values for the diatoms and different radiolarian fractions are in the range of $\delta_{30}Si_{Si(OH)4}$ values of specific water masses and reflect their range in $Si(OH)_4$ concentrations [20, 28, 29] (Fig. 3 and Supplementary Table 4).

3.4.7 RECONSTRUCTION OF WSIC

WSIC (%) was estimated from the diatom assemblage composition preserved in the sediment records using the transfer function technique. For our study we selected the estimations obtained with the Imbrie and Kipp transfer function method using a setup with 172 reference sites, 28 diatom taxa/taxa groups, logarithmic-transformed diatom data, quadratic regression and a three-factor model with a root mean square error of prediction of 7.3% (ref. 35). Similar patterns of sea-ice concentration were also obtained with three other transfer function techniques, all showing the onset of sea-ice retreat after the last glacial period at around 18,000 years ago [35]. Although the diatom signal used for the estimate of WSIC is based on a signal produced during the diatom spring–summer bloom, experimental (sediment trap) data and statistical test show that this signal reflects the occurrence probability of winter sea ice, which is well correlated with the WSIC at a given site (ref. 35 and references included therein).

3.4.8 MODEL SETUP AND EXPERIMENTAL DESIGN

For our simulations we use the experimental settings of the LGMW simulations in ref. 23, which is integrated for 4,000 years from a cold ocean state, to evaluate MLD characteristics under LGM (21 ka) and present-day conditions, as well as for the sensitivity experiments. The thermodynamics of sea ice relate changes in sea-ice thickness to a balance of radiant, turbulent and oceanic heat fluxes. The effect of snow accumulation is taken into account, along with snow–ice transformation when the snow–ice interface sinks below the sea level because of snow loading. The impact of ice growth and ice melting is included in the model, assuming a sea-ice salinity of 5 PSU46. In experiment WIND, the implementation of the poleward wind field shift (3° southwards) of the Westerlies in the SO (experiment WIND, Supplementary Figs 10 and 11) has been performed in analogy to ref. 43. In experiment CO_2, a deglacial CO_2 increase from 180 to 240 p.p.m.v. has been applied. Both experiments have been integrated for 600 years. All figures show climatological mean characteristics averaged over a period of 100 years at the end of each simulation.

REFERENCES

1. Petit, J. R. et al. Climate and atmospheric history of the past 420,000 years from the Vostok ice core, Antarctica. Nature 399, 429–436 (1999).

2. Archer, D. E., Winguth, A., Lea, D. & Mahowald, N. What caused the glacial/interglacial atmospheric pCO2 cycles? Rev. Geophys. 38, 159–186 (2000).
3. Sigman, D. M, Hain, M. P & Haug, G. H. The polar ocean and glacial cycles in atmospheric CO2 concentration. Nature 466, 47–55 (2010).
4. Parrenin, F. et al. Synchronous change of atmospheric CO2 and Antarctic temperature during the last deglacial warming. Science 339, 1060–1063 (2013).
5. Abelmann, A., Gersonde, R., Cortese, G., Kuhn, G. & Smetacek, V. Extensive phytoplankton blooms in the Atlantic Sector of the glacial Southern Ocean. Paleoceanography 21, PA1013 doi:10.1029/2005PA001199 (2006).
6. Anderson, R. F. et al. Wind-driven upwelling in the Southern Ocean and the deglacial rise in atmospheric CO2. Science 323, 1443–1448 (2009).
7. Bouttes, N. et al. Impact of oceanic processes on the carbon cycle during the last termination. Clim. Past 8, 149–170 (2012).
8. Ragueneau, O. et al. A review of the Si cycle in the modern ocean: recent progress and missing gaps in the application of biogenic opal as a paleoproductivity proxy. Global Planet. Change 26, 317–365 (2000).
9. ISIStephens, B. B. & Keeling, R. F. The influence of Antarctic sea ice on glacial-inter- glacial CO2 variations. Nature 404, 171–174 (2000).
10. Kurahashi-Nakamura, T., Abe-Ouchi, A., Yamanaka, Y. & Misumi, K Compound effects of Antarctic sea ice on atmospheric pCO2 change during glacial–interglacial cycle. Geophys. Res. Lett. 34, L20708 doi:10.1029/2007GL030898 (2007).
11. CASSun, X. & Matsumoto, K. Effects of sea ice on atmospheric pCO2: A revised view and implications for glacial and future climates. J. Geophys. Res. 115, G02015 (2010).
12. Assmy, P. et al. Thick-shelled, grazer-protected diatoms decouple ocean carbon and silicon cycles in the iron-limited Antarctic Circum Polar Current. PNAS 110, 20633–20638 (2013).
13. CASPubMedRobinson, R. S. & Sigman, D. M. Nitrogen isotopic evidence for a poleward decrease in surface nitrate within the ice age Antarctic. Quat. Sci. Rev. 27, 1076–1090 (2008).
14. Shemesh, A., Burckle, L. H. & Hays, J. D. Late Pleistocene oxygen isotope records of biogenic silica from the Atlantic sector of the Southern Ocean. Paleoceanography 10, 179–196 (1995).
15. ISIShemesh, A. et al. Sequence of events during the last deglaciation in Southern Ocean sediments and Antarctic ice cores. Paleoceanography 17, 599–605 (2002).
16. Pike, J., Swann, G. E. A., Leng, M. J. & Snelling, A. M. Glacial discharge along the West Antarctic Peninsula during the Holocene. Nat. Geosci. 6, 199–202 (2013).
17. CASSwann, G. E. A. & Leng, M. J. A review of diatom δ18O in palaeoceanography. Quat. Sci. Rev. 28, 384–398 (2009).
18. Maier, E. et al. Combined oxygen and silicon isotope analysis of diatom silica from a deglacial subarctic Pacific record. J. Quat. Sci. 28, 571–581 (2013).
19. Hendry, K. R. & Brzezinski, M. A. Using silicon isotopes to understand the role oft the Southern Ocean in modern and ancient biogeochemistry and climate. Quat. Sci. Rev. 89, 13–26 (2014).
20. Reynolds, B. C., Frank, M. & Halliday, A. N. Silicon isotope fractionation during nutrient utilization in the North Pacific. Earth Planet. Sci. Lett. 244, 431–443 (2006).

21. CASHendry, K. R., Georg, R. B., Rickaby, R. E. M., Robinson, L. F. & Halliday, A. N. Deep ocean nutrients during the Last Glacial Maximum deduced from sponge silicon isotopic compositions. Earth Planet. Sci. Lett. 292, 290–300 (2010).
22. CASEllwood, M. J., Wille, M. & Maher, W. Glacial silicic acid concentrations in the Southern Ocean. Science 330, 1088–1091 (2010).
23. CASPubMedZhang, X., Lohmann, G., Knorr, G. & Xu, X. Different ocean states and transient characteristics in Last Glacial Maximum simulations and implications for deglaciation. Clim. Past 9, 2319–2333 (2013).
24. ISIDe la Rocha, C. L., Brzezinski, M. A. & DeNiro, M. J. Fractionation of silicon isotopes by marine diatoms during biogenic silica formation. Geochim. Cosmochim. Acta 61, 5051–5056 (1997).
25. Abelmann, A. & Gowing, M. M. Spatial distribution pattern of living polycystine radiolarian taxa - baseline study for paleoenvironmental reconstructions in the Southern Ocean (Atlantic sector). Marine Micropaleontol. 30, 3–28 (1997).
26. Abelmann, A. & Gersonde, R. Biosiliceous particle flux in the Southern Ocean. Marine Chem. 35, 503–536 (1991).
27. CASVarela, D. E., Pride, C. J. & Brzezinski, M. A. Biological fractionation of silicon isotopes in Southern Ocean surface waters. Glob. Biogeochem. Cycles 18, GB1047 (2004).
28. CASCardinal, D et al. Relevance of silicon isotopes to Si-nutrient utilization and Si-source assessment in Antarctic waters. Glob. Biogeochem. Cycles 19, doi:10.1029/2004GB002364 (2005).
29. Fripiat, F. et al. Silicon pool dynamics and biogenic silica export in the Southern Ocean inferred from Si-isotopes. Ocean Sci. 7, 533–547 (2011).
30. CASEgan, K. et al. Diatom silicon isotopes as a proxy for silicic acid utilisation: a Southern Ocean core top calibration. Geochim. Cosmochim. Acta 96, 174–192 (2012).
31. CASKrabberød, A. K. et al. Radiolaria divided into Polycystina and Spasmaria in combined 18S and 28S rDNA Phylogeny. PLoS ONE 6, 23526 (2011).
32. CASHendry, K. R. et al. Silicon isotopes indicate enhanced carbon export efficiency in the North Atlantic during deglaciation. Nat. Commun. 5, 3107 (2014).
33. CASPubMedGarcia, H. E. et al. World Ocean Atlas 2009. NOAA Atlas NESDIS 4, 398 (2010).
34. Gersonde, R., Crosta, X., Abelmann, A. & Armand, L. Sea-surface temperature and sea ice distribution of the Southern Ocean at the EPILOG last Glacial Maximum – a circum-Antarctic review based on siliceous microfossil records. Quat. Sci. Rev. 24, 869–896 (2005).
35. ISIEsper, O. & Gersonde, R. New tools for the reconstruction of Pleistocene Antarctic sea ice. Palaeogeogr. Climatol. Ecol. 399, 260–283 (2014).
36. Reynolds, R. W. et al. An improved in situ and satellite SST analysis for climate. J. Clim. 15, 1609–1625 (2002).
37. ISICampbell, W. J. et al. Variations of mesoscale and large-scale sea ice morphology in the 1984 Marginal Ice Zone Experiment as observed by microwave remote sensing. J. Geophys. Res. 92, 6805–6824 (1987).

38. Gersonde, R. et al. Last Glacial sea-surface.temperatures and sea-ice extent in the Southern Ocean (Atlantic-Indian sector) - A multiproxy approach. Paleoceanography 18, doi:10.1029/2002PA000809 (2003).
39. Nielsen, S. H. et al. Origin and significance of ice-rafted detritus in the Atlantic sector of the Southern Ocean. Geochem. Geophys. Geosyst. 8, Q12005 (2007).
40. CASChapligin, B. et al. Assessment of purification and contamination correction methods for analysing the oxygen isotope composition from biogenic silica. Chem. Geol. 300–301, 185–199 (2012).
41. De La Rocha, C. L., Brzezinski, M. A., DeNiro, M. J. & Shemesh, A. Silicon-isotope composition of diatoms as indicator of past oceanic change. Nature 395, 680–683 (1998).
42. Craig, H. & Gordon, L. I. in Stable Isotopes in Oceanographic Studies and Paleotemperatures ed. Tongiorgi E. 9–130Consiglio Nazionale della Ricerche Laboratorio di Geologia Nucleare (1965).
43. Wei, W., Lohmann, G. & Dima, M. Distinct modes of internal variability in the Global Meridional Overturning Circulation associated to the Southern Hemisphere westerly winds. J. Phys. Oceanogr. 42, 785–801 (2012).
44. ISIRoeckner, E. et al. Sensitivity of simulated climate to horizontal and vertical resolution in the ECHAM5 Atmosphere Model. J. Climate 19, 3771–3791 (2006).
45. Raddatz, T. et al. Will the tropical land biosphere dominate the climate-carbon cycle feedback during the twenty first century? Clim. Dyn. 29, 565–574 (2007).
46. ISIMarsland, S. J., Haak, H., Jungclaus, J. H., Latif, M. & Röske, F. The Max-Planck-Institute global ocean/sea ice model with orthogonal curvilinear coordinates. Ocean Modell. 5, 91–127 (2003).
47. Hibler, W. D. A dynamic thermodynamic sea ice model. J. Phys. Oceanogr. 9, 815–846 (1979).
48. ISIZhang, X., Lohmann, G., Knorr, G. & Purcell, C. Abrupt glacial climate shifts controlled by ice sheet changes. Nature 512, 290–294 (2014).
49. Knorr, G. & Lohmann, G. Climate warming during Antarctic ice sheet expansion at the Middle Miocene transition. Nat. Geosci. 7, 376–381 (2014).
50. Frank, M. et al. Similar glacial and interglacial export bioproductivity in the Atlantic sector of the Southern Ocean: multiproxy evidence and implications for glacial atmospheric CO2. Paleoceanography 15, 642–658 (2000).
51. Kohfeld, K. E. et al. Southern Hemisphere westerly wind changes during the last glacial maximum: Paleo-data synthesis. Quat. Sci. Rev. 68, 76–95 (2013).
52. Collins, L. G., Pike, J., Allen, C. & Hodgson, D. A. High-resolution reconstruction of southwest Atlantic sea-ice and its role in the carbon cycle during marine isotope stages 3 and 2. Paleoceanography 27, PA3217 (2012).
53. Hoffmann, L. J., Peeken, I. & Lochte, K. Effects of iron on the elemental stoichiometry during EIFEX and in the diatoms Fragilariopsis kerguelensis and Chaetoceros dichaeta. Biogeosciences 4, 569–579 (2007).
54. CASMatsumoto, K., Chase, Z. & Kohfeld, K. Different mechansims of silicic acid leakage and their biogeochemical consequences. Paleoceanography 29, 238–254 (2014).

55. Law, C. S., Abraham, E. R., Watson, A. J. & Liddicoat, M. I. Vertical eddy diffusion and nutrient supply to the surface mixed layer of the Antarctic Circumpolar Current. J. Geophys. Res. 108, 3272 (2003).
56. Wille, M. et al. Silicon isotopic fractionation in marine sponges: a new model for understanding silicon isotopic fractionation in sponges. Earth Planet. Sci. Lett. 292, 281–289 (2010).
57. Skinner, L. C., Fallon, S., Waelbroeck, C., Michel, E. & Barker, S. Ventilation of the deep Southern Ocean and deglacial CO2 rise. Science 328, 1147–1151 (2010).
58. Adkins, J. K., McIntyre, K. & Schrag, D. P. The salinity, temperature, and δ18O of the glacial deep ocean. Science 298, 1769–1773 (2002).
59. Meckler, A. N. et al. Deglacial pulses of deep-ocean silicate into the subtropical North Atlantic ocean. Nature 495, 495–499 (2013).
60. Knorr, G. & Lohmann, G. Rapid transition in the Atlantic thermohaline circulation triggered by global warming and meltwater during the last deglaciation. Geochem. Geophys. Geosyst. 8, doi: 10.1029/2007GC001604 (2007).
61. Paillard, D., Labeyrie, L. & Yiou, P. Macintosh program performs time-series analysis. Eos Trans. AGU 77, 379 (1996).
62. Veres, D. et al. The Antarctic ice core chronology (AICC2012): an optimized multi-parameter and multi-site dating approach for the last 120 thousand years. Clim. Past 9, 1733–1748 (2013).
63. ISIMarcott, S. A. et al. Centennial-scale changes in the global carbon cycle during the last glaciation. Nature 514, 616–619 (2014).
64. Leng, M. J. & Sloane, H. J. Combined oxygen and silicon isotope analysis of biogenic silica. J. Quat. Sci. 23, 313–319 (2008).
65. Chapligin, B. et al. A high-performance, safer and semiautomated approach for the δ18O analysis of diatom silica and new methods for removing exchangeable oxygen. Rapid Comm. Mass Spectrom. 24, 2655–2664 (2010).
66. CASChapligin, B. et al. Inter-laboratory comparison of oxygen isotope compositions from biogenic silica. Geochim. Cosmochim. Acta 75, 7242–7256 (2011).
67. CASComiso, J. C. in Sea Ice, an Introduction to its Physics, Chemistry, Biology and Geology eds Thomas D. N., Diekmann G. S. 112–142Blackwell (2003).
68. Orsi, A. H., Whitworth, T. III & Nowlin, W. D. jr On the meridional extend and fronts of the Antarctic Circumpolar Current. Deep Sea Res. I 42, 641–673 (1995).
69. Hendry, K. R. & Robinson, L. F. The relationship between silicon isotope fractionation in sponges and silicic acid concentration: Modern and core-top studies of biogenic opal. Geochim. Cosmochim. Acta 81, 1–12 (2012).
70. CASEPICA Community Members. One-to-one coupling of glacial climate variability in Greenland and Antarctica. Nature 444, 195–198 (2006).

There are several supplemental files that are not available in this version of the article. To view this additional information, please use the citation on the first page of this chapter.

CHAPTER 4

On the Influence of Interseasonal Sea Surface Temperature on Surface Water pCO_2 at 49.0°N/16.5°W and 56.5°N/52.6°W in the North Atlantic Ocean

NSIKAK U. BENSON, OLADELE O. OSIBANJO, FRANCIS E. ASUQUO, AND WINIFRED U. ANAKE

4.1 INTRODUCTION

Carbon dioxide (CO_2) dominance in the atmosphere (mainly from anthropogenic sources) over other greenhouse gases has resulted in increasing pCO_2 in the surface ocean leading to measurably decreased pH (ocean acidification) (Canadell et al., 2007; Hopkins et al., 2010; Keeling and Whorf, 2005; Levine et al., 2008; Sabine and Feely, 2007). The world oceans are major natural sinks of atmospheric CO_2. However, the North

On the Influence of Interseasonal Sea Surface Temperature on Surface Water pCO2 at 49.0°N/16.5°W and 56.5°N/52.6°W in the North Atlantic Ocean. Journal of Oceanography and Marine Science *5,7 (2014), DOI: 10.5897/JOMS2014.0114.*

Atlantic Ocean is generally regarded as a primary gate for CO_2 entering the global ocean due to its subpolar climate.

In the open ocean, it has been established that significant correlation exists between surface water pCO_2 and sea surface temperature (SST). However, the sea surface pCO_2–SST relationships are primarily governed by a combination of processes, such as biological activity, physical transport-upwelling of nutrients, and thermodynamics (e.g. temperature effects on CO_2 dissociation and solubility) (Körtzinger et al., 2008a, b; Chen et al., 2007). Takahashi et al. (2002, 2009) have elucidated the mechanism and the role of thermodynamic effects on the uptake of CO_2 by the global oceans. Recently, several studies have reported the low uptake of pCO_2 in the North Atlantic Ocean suggesting a gradual weakening of an active carbon storehouse (Corbière et al., 2007; Omar and Olsen, 2006; Schuster and Watson, 2007; Schuster et al., 2009; Ullman et al., 2009).

The phenomenal CO_2 low uptake is due to several factors including rising sea surface temperatures (Corbière et al., 2007), deep convection and re-stratification periods (Körtzinger et al., 2008a, b; Straneo, 2006) and variations in biological productivity (Behrenfeld et al., 2006; Lefèvre et al., 2004).

The role of temperature-controlled and biological processes in regulating ocean pCO_2 have been intensively investigated and reported (Feely et al., 2002; Friederich et al., 2008; Körtzinger et al., 2008a; Takahashi et al., 1993; 2002; Watson et al., 1991). According to Shim et al. (2007), the temporal change in surface water temperature could be a major factor that drives a seasonal variation in surface pCO_2. However, thermodynamic effect is caused by the dependence of CO_2 solubility and dissociation constants on temperature (Rangama et al., 2005). Many global biogeochemical cycles (notably CO_2 and CH_4 cycles), mediated by biological processes are highly dependent on temperature. The solubility of CO_2 and the dissociation of carbonic acid in seawater are moderated by temperature. It has been established that as temperature decreases the solubility of gases increases; this infers greater gas solubility for seawater in high latitudes. In this paper, observed data from two North Atlantic time series sites are employed in an attempt to assess the interseasonal sea surface temperature variations and anomalies at the eastern and western basins of the North Atlantic Ocean, and examine its effect on seasonal pCO_2 distribution at these sites.

4.1.1 DESCRIPTION OF MOORING STATIONS

The Porcupine Abyssal Plain (PAP) observatory (Figure 1), located in the Northeast Atlantic oceanographic region is a major long-term ocean observatory operated since 1989 for international and interdisciplinary scientific research and monitoring, which are focused on physical-biogeochemical observations. It is approximately 4800 m deep and is geographically positioned between the North Atlantic Current (NAC) and Azores Current (AC), and lies south of the main stream of the NAC, where it is subject to return flows from the West and Northwest. It is also characterized by significant presence of mesoscale eddies and deep winter mixing with strong interannual variability between 300 and 800 m (Longhurst, 2007). On the other hand, the K1 Central Labrador Sea (K1 CELAS) mooring site (Figure 1) is a deep-water formation oceanographic site of research importance that examines complex oceanic processes from surface waters to the seafloor by recording biological, chemical and physical parameters, as well as investigations of trends and variability in deep convection activity (Avsic et al., 2006).

4.2 METHODS AND DATA ANALYSIS

Detailed sampling and analytical procedures for SST and pCO_2 data generation at the KI CELAS and PAP observatories have been reported previously (Körtzinger et al., 2008a, b). These involved measurements of pCO_2 with an autonomous sensor (SAMI-CO_2, Sunburst Sensors LLC, Missoula, Montana, United States), while temperature measurements were carried out with an SBE-37 MicroCAT recorder (Sea-Bird Electronics Inc., Bellevue, Washington, United States). The data used in this study were obtained during three consecutive deployments between July 2003 and July 2005 for the PAP and K1 CELAS mooring stations. Details of mooring coordinates, nominal depth of deployment of sensors used, sampling interval, deployment and recovery dates, parameters measured and the total number of observational data successfully recovered are presented in Tables 1 and 2.

4.3 RESULTS AND DISCUSSION

4.3.1 DISTRIBUTION OF SURFACE WATER PCO_2 AND SST

The distributions of sea surface temperature and pCO_2 for the PAP-2 to PAP-4 deployments at the PAP and K1 CELAS time series observatories together with the atmospheric CO_2 are shown in Figure 2. The pCO_2 distribution across the spatial gradients at both the eastern (PAP) and western (K1 CELAS) sites of the subpolar North Atlantic Ocean showed distinct and consistent seasonal variability.

The surface water pCO_2 cycle is characteristically marked by a minimum and maximum pCO_2 distribution pattern for the summertime and wintertime, respectively. It is also well depicted in Figure 2 that the pCO_2 distribution patterns at both oceanographic sites are in antiphase to the temperature signal.

4.3.2 SST ANOMALIES

The PAP and K1 CELAS monthly anomalies of SST derived as a difference between monthly averages and the long-term mean temperatures of each oceanographic station are presented here. Negative SST anomalies (>2.5°C) were generally observed during late fall through the wintertime into springtime, whereas positive SST anomalies (<3.2°C) characterized the summertime at all the PAP sites (Figure 3a). A similar trend was observed for the K1 Central Labrador Sea data although with a relatively smaller but significant negative SST anomaly of approximately 1.36°C and high SST positive anomaly of 3.6°C (Figure 3a).

A comparison of the monthly anomalies of observed pCO_2 with respect to SST anomalies at both sites are shown in Figure 3b and c. For pCO_2 anomalies calculated based on average monthly observations at the PAP and K1 CELAS sites indicate positive anomalies of approximately 38 and 25 μatm respectively. These positive deviations coincided with the highest SST positive anomaly obtained for the PAP observations during spring/

summer period (black triangles) Figure 3b, while it corresponded with the lowest SST negative anomaly for the K1 CELAS observed data during fall/ winter period (transparent red triangles) Figure 3c. This implies that a positive pCO_2–SST relationship exists for the observed data obtained from the PAP time series site, while an inverse correlation may be established for K1 CELAS site. However, it should be noted that the pCO_2–SST anomalies comparison did not suggest a clear and consistent relationship especially for the PAP location. For instance, positive pCO_2 deviations derived for summer and early fall of 2003 (July – October) coincided with positive SST anomalies, whereas the SST anomalies indicated an opposite behavior with marked negative pCO_2 anomalies during the same period in 2004 (corresponding to 3rd PAP deployment) (Figure 3b). This variation might be attributed to thermodynamic effect or other physical processes such as mixing and stratification that might have resulted in negative pCO_2 anomalies with corresponding positive SST anomalies. Moreover, for the winter / springtime, positive pCO_2 anomalies at the PAP site are generally associated with significant negative SST anomalies which suggest biologically driven pCO_2 variability. Negative SST anomalies are usually associated with enhanced nutrient and dissolved inorganic carbon (DIC) inputs, which could invariably lead to increase in primary productivity (Borges et al., 2007; Boyd et al., 2001).

4.3.3 DEPENDENCE OF SURFACE WATER PCO_2 ON TEMPERATURE

4.3.3.1 CORRELATION BETWEEN PCO_2 AND SST

The Pearson correlation analysis were carried out to establish the interannual / inter-seasonal relationships between observed pCO_2 and SST data obtained from the PAP and K1 CELAS sites. More so, given the large number of data, the interseasonal test of linear fits between pCO_2 and SST were evaluated based on average monthly data collated from hourly measurements. The derived linear fits generally indicated strong but negative correlations between pCO_2 and SST.

4.3.3.2 CORRELATION BETWEEN PCO$_2$ AND SST AT PAP SITE

Figures 4a, b and c show the results of fitting linear model to describe the relationship of observed PAP site pCO_2 as a function of sea surface temperature for the second (PAP2), third (PAP3), and forth (PAP4) deployments on an annual timescale.

A better mechanistic understanding of how changes in SST and other processes may have influenced sea surface pCO_2 was evaluated using seasonal observed data during each deployment. Sea surface pCO_2–SST correlations generally showed negative correlations between these two variables ($r^2 = 25.85$, 74.13, 5.75, $p < 0.0001$) (Figures 4a to c) at the PAP site suggesting that SST had a non-dominance influence on pCO_2 variability. This also suggests that a large part of the observed variation may be attributed to non thermal processes such as the enhanced biological activity associated with physical transport – upwelling of nutrient enriched water into the euphotic zone, mixing or stratification. This argument is supported by the derived pCO_2–SST correlations for summer – fall 2003 observed data that characteristically indicated a strong influence of SST on pCO_2 variability (Table 3). It should be noted however, that changes in sea surface temperature principally influenced the surface water pCO_2 cycle at the PAP site during deployments in 2004 to 2005, with insignificant biological effect except during wintertime. In general, the correlations between temperature and pCO_2 based on observed data suggest that pCO_2 seawater patterns in the Northeast subpolar Atlantic Ocean is due to the counteracting effects of temperature, mixing and strong to moderate biological production. This observation is consistent with earlier reports by Körtzinger et al. (2008a) and Takahashi et al. (2002, 2009).

4.3.3.3 INTER-RELATIONSHIP BETWEEN PCO$_2$ AND SST AT LABRADOR SEA SITE

Figure 5a and b illustrate the relationships of pCO_2 as a function of SST during the period. A closer inspection of the derived linear fits reveal that the sea surface pCO_2–SST correlations were characteristically more variable and generally depicted the irrefutable effect of temperature and biology on pCO_2, although it is clear that the pCO_2 cycle is strongly governed by thermodynamic forcing than biological effect. The pCO_2–SST correlation obtained for observed data during the deployment in 2004 indicated that there is a good linear relationship between sea surface pCO_2 and in-situ SST (Table 3).

A similar relationship was found for sea surface pCO_2–SST correlations obtained for 2005 K1 deployment, but with a moderately strong relationship ($r^2 = 0.53$, $p < 0.0000$) (Figure 5b). On an annual to seasonal timescale, the distribution pattern in surface seawater pCO_2 might not be controlled by a seasonal change in temperature only but also by biology as well as mixing within the subsurface and stratification of the epipelagic zone of the K1 CELAS site (Körtzinger et al., 2008b).

However, it is obvious that the thermodynamic effects and other physical processes are the dominating variability driver compared to weak biology signature at this subpolar NW Atlantic site. A comparison of the seasonal sea surface pCO_2–SST correlations at the K1 CELAS site also reveal significant but negative correlations between these two parameters during autumn 2004 and spring 2005. This implies that other physical processes such as turbulence, mixing and stratification primarily govern the variability in pCO_2 distribution. On the other hand, the pCO_2–SST correlation obtained for summer 2005 indicates a positively significant correlation implying that mainly thermodynamic effects induce pCO_2 variability. A summary of pCO_2–SST relationships at both the Northeast and Northwest sites of the Atlantic Ocean is presented in Table 3.

4.4 CONCLUDING REMARKS

This study has demonstrated that the surface water pCO_2 distribution across the spatial gradients over a seasonal timescale at both the eastern (PAP) and western (K1 CELAS) basins of the subpolar North Atlantic Ocean is relatively consistent, however with distinct seasonal large variability. At the PAP site, consistent undersaturation of oceanic surface water was observed relative to the atmospheric CO_2, while a similar trend occurred over the western area at the K1 CELAS location but with some degree of supersaturation between February and March 2005. On a seasonal timescale, the surface water pCO_2 cycle is characteristically marked by minimum and maximum pCO_2 distribution pattern for the summertime and wintertime respectively. It is obvious that the pCO_2 distribution pattern in the NE PAP and NW CELAS sites of the subpolar North Atlantic were in antiphase to the temperature signal. Investigation of sea surface pCO_2–SST correlations generally indicated moderate to strong but negative correlations between pCO_2 and SST. Thus we conclude that the variation in surface ocean pCO_2 may not be controlled by change in sea surface temperature only, but by biological activities and other physical processes. In the Northeastern basin, the variability in pCO_2 distribution is primarily governed by other physical processes such as mixing and stratification during the autumn and springtime, while the pCO_2–SST relationship obtained for summertime indicates that pCO_2 variability is induced mainly by thermodynamic effects.

REFERENCES

1. Avsic T, Karstensen J, Send U, Fischer J (2006). Interannual variability of newly formed Labrador Sea Water from 1994 to 2005. Geophys. Res. Lett. 33:L21S02.
2. Behrenfeld MJ, O'Malley RT, Siegel DA, McClain CR, Sarmiento JL, Feldman GC, Milligan AJ, Falkowski PG, Letelier RM, Boss ES (2006). Climate-driven trends in contemporary ocean productivity. Nature 444:752-755.
3. Borges AV, Tilbrook B, Metzl N, Lenton A, Delille B (2007). Inter-annual variability of the carbon dioxide oceanic sink south of Tasmania. Biogeosciences Discuss. 4:3639-3671.
4. Boyd PW, Crossley AC, DiTullio GR, Griffiths FB, Hutchins DA, Queguiner B, Sedwick PN, Trull TW (2001). Control of phytoplankton growth by iron supply and irradiance in the subantarctic Southern Ocean: Experimental results from the SAZ Project. J. Geophys. Res. 106(C12)31:573-583.
5. Canadell JG, Le Que´re´ C, Raupach MR, Field CB, Buitenhuis ET, Ciais P, Conway TJ, Gillett NP, Houghton RA, Marland G (2007). Contributions to accelerating atmospheric CO2 growth from economic activity, carbon intensity, and efficiency of natural sinks. Proc. Natl. Acad. Sci. 104(47):18866-18870.
6. Corbière A, Metzl N, Reverdin G, Brunet C, Takahashi T (2007). Interannual and decadal variability of the oceanic carbon sink in the North Atlantic subpolar gyre. Tellus 59B(2):168-178.
7. Chen F, Cai W-J, Benitez-Nelson C, Wang Y (2007). Sea surface pCO2-SST relationships across a cold-core cyclonic eddy: Implications for understanding regional variability and air-sea gas exchange. Geophys. Res. Lett. 34:L10603.

8. Feely RA, Boutin J, Cosca CE, Dandonneau Y, Etcheto J, Inoue HY, Ishii M, Le Quéré C, Mackey DJ, McPhaden M, Metzl N, Poisson A, Wanninkhof R (2002). Seasonal and interannual variability of CO2 in the equatorial Pacific. Deep Sea Res. Part II: Top. Stud. Oceanogr. 49(13-14):2443-2469.
9. Friederich GE, Ledesma J, Ulloa O, Chavez FP (2008). Air-sea carbon dioxide fluxes in the coastal southeastern tropical Pacific. Prog. Oceanogr. 79:156-166.
10. Hopkins FE, Turner SM, Nightingale PD, Steinke M, Bakker D, Liss PS (2010). Ocean acidification and marine trace gas emissions. Proc. Natl. Acad. Sci. 107(2):760-765.
11. Keeling CD, Whorf TP (2005). Atmospheric CO2 records from sites in the SIO air sampling network. In Trends: A Compendium of Data on Global Change. Carbon Dioxide Information Analysis Center, Oak Ridge National Laboratory, U.S. Department of Energy, Oak Ridge, Tenn., U.S.A.
12. Körtzinger A, Send U, Lampitt RS, Hartman S, Wallace DWR, Karstensen J, Villagarcia MG, Llinas O, DeGrandpre MD (2008a). Seasonal pCO2 cycle at 49°N / 16.5°W in the northeast Atlantic Ocean and what it tells us about biological productivity. J. Geophys. Res.113, C04020.
13. Körtzinger A, Send U, Wallace DWR, Kartensen J, DeGrandpre M (2008b). Seasonal cycle of O2 and pCO2 in the central Labrador Sea: Atmospheric, biological, and physical implications. Global Biogeochem. Cycles 22, GB1014, doi:10.1029/2007 GB003029.
14. Lefèvre N, Watson AJ, Olsen A, Ríos AF, Pérez FF, Johannessen T (2004). A decrease in the sink for atmospheric CO2 in the North Atlantic. Geophys. Res. Lett. 31:L07306, doi:10.1029/2003GL018957.
15. Levine NM, Doney SC, Wanninkhof R, Lindsay K, Fung IY (2008). Impact of ocean carbon system variability on the detection of temporal increases in anthropogenic CO2. J. Geophys. Res. 113:C03019.
16. Longhurst AR (2007). Ecological Geography of the Sea, 2nd ed., 542 pp. Academic, Boston, Mass.
17. Omar AM, Olsen A (2006). Reconstructing the time history of the air-sea CO2 disequilibrium and its rate of change in the eastern subpolar North Atlantic, 1972-1989. Geophys. Res. Lett. 33:L04602
18. Rangama Y, Boutin J, Etcheto J, Merlivat L, Takahashi T, Delille B, Frankignoulle M, Bakker DCE (2005). Variability of the net air-sea CO2 flux inferred from shipboard and satellite measurements in the Southern Ocean of Tasmania and New Zealand. J. Geophys. Res. 110, C09005.
19. Sabine CL, Feely RA (2007). The Oceanic Sink for Carbon Dioxide, in: Reay, D.S., Hewitt, C.N., Smith, K.A., Grace, J. (Eds), Greenhouse Gas Sinks. Athenaeum Press Ltd, Gateshead, UK.
20. Schuster U, Watson AJ (2007). A variable and decreasing sink for atmospheric CO2 in the North Atlantic. J. Geophys. Res. 112:C11006.
21. Schuster U, Watson AJ, Bates NR, Corbiere A, Gonzalez-Davila M, Metzl N, Pierrot D, Santana-Casiano M (2009). Trends in North Atlantic sea-surface fCO2 from 1990 to 2006. Deep-Sea Res.

22. Shim J, Kim D, Kang YC, Lee JH, Jang S-T, Kim C-H (2007). Seasonal variations in pCO2 and its controlling factors in surface seawater of the northern East China Sea. Cont. Shelf Res. 27:2623-2636.
23. Straneo F (2006). Heat and Freshwater Transport through the Central Labrador Sea. J. Phys. Oceanogr. 36(4):606-628.
24. Takahashi T, Olafsson J, Goddard J, Chipman DW, Sutherland SC (1993). Seasonal variation of CO2 and nutrients in the high-latitude surface oceans: A comparative study. Global Biogeochem. Cycles 7: 843-878.
25. Takahashi T, Sutherland SC, Sweeney C, Poisson A, Metzl N, Tilbrook B, Bates N, Wanninkhof R, Feely RA, Sabine C, Olafsson J, Nojiri Y (2002). Global sea to air CO2 flux based on climatological surface ocean pCO2, and seasonal biological and temperature effects. Deep Sea Res. Part II: Top. Stud. Oceanogr. 49:1601-1622.
26. Takahashi T, Sutherland SC, Wanninkhof R, Sweeney C, Feely RA, Chipman DW, Hales B, Friederich G, Chavez F, Sabine C, Watson A, Bakker DCE, Schuster U, Metzl N, Inoue HY, Ishii M, Midorikawa T, Nojiri Y, Körtzinger A, Steinhoff T, Hoppema M, Olafsson J, Arnarson TS, Tilbrook B, Johannessen T, Olsen A, Bellerby R, Wong CS, Delille B, Bates NR, de Baar HJW (2009). Climatological mean and decadal change in surface ocean pCO2, and net sea–air CO2 flux over the global oceans. Deep-Sea Research II.
27. Ullman D, McKinley GA, Bennington V, Dutkiewicz S (2009). Trends in the North Atlantic carbon sink: 1992-2006. Global Biogeochem. Cycles 23, GB4011.
28. Watson AJ, Robinson C, Robinson JE, Williams PJL, Fasham MJR (1991). Spatial variability in the sink for atmospheric carbon dioxide in the North Atlantic. Nature 350:50-53.

There are several figures and tables that are not available in this version of the article. To view this additional information, please use the citation on the first page of this chapter.

CHAPTER 5

Carbon Export by Small Particles in the Norwegian Sea

GIORGIO DALL'OLMO AND KJELL ARNE MORK

5.1 INTRODUCTION

The export of particles from the surface to the deep ocean affects nutrient distributions, sustains mesopelagic and bathypelagic organisms, and, ultimately, contributes to controlling the Earth's climate [Trull et al., 2008]. Despite the importance of this process, however, modern estimates of global carbon export vary between 5 and 12 Pg C yr^{-1} [Laws et al., 2000; Henson et al., 2011]. This large range of variation reflects an ignorance arising from lack of observations and from substantial disagreement among the different methods and assumptions used to estimate the magnitude of export. To predict how oceanic carbon uptake will evolve in the future and to interpret past records of CO_2, we need to understand the factors that control the uncertainty of our current estimates of global carbon export.

In situ optical scattering instruments mounted on autonomous platforms such as Bio-Argo floats can provide observations that are complementary to existing methods for estimating carbon export. These autonomous platforms are advantageous because they allow measurements over sustained periods of time in remote and/or hazardous ocean regions [Bishop and Wood, 2009].

Particulate optical backscattering (bbp) is sensitive to marine particles in the approximate size range 0.2–20 μm [Stramski et al., 2004] (see also supporting information) and is correlated to the concentration of particulate organic carbon (POC) [e.g., Cetinić et al., 2012]. When bbp is measured from autonomous platforms, one can estimate the vertically-resolved and time-varying fields needed to study the dynamics of POC in the water column [Briggs et al., 2011].

Here we focus on bbp profiles measured by Bio-Argo floats in the Norwegian Sea. We show that these signals can track the integrated stocks of small marine particles in the euphotic and mesopelagic zones. We further estimate vertically resolved POC export fluxes and transfer efficiencies.

5.2 METHODS

The Lofoten Basin is a deep (~3000 m) region of the open Norwegian Sea (Figure 1). In this basin the bathymetry-driven middepth flow typically forces Argo floats to remain in the basin of deployment, even after multiple years of operation [Voet et al., 2010]. Thus, this deep basin is ideal to conduct pilot studies using biogeochemical floats.

Two Teledyne Webb Autonomous Profiling Explorer floats equipped with conductivity-temperature-depth and WETLabs sensors for measuring chlorophyll fluorescence (chl) and optical backscattering at 700 nm collected data from 2010 to 2012 in the Lofoten Basin (Figure 1). The floats were programmed to surface every 10 days during the winter and every 5 days in the summer (see also supporting information).

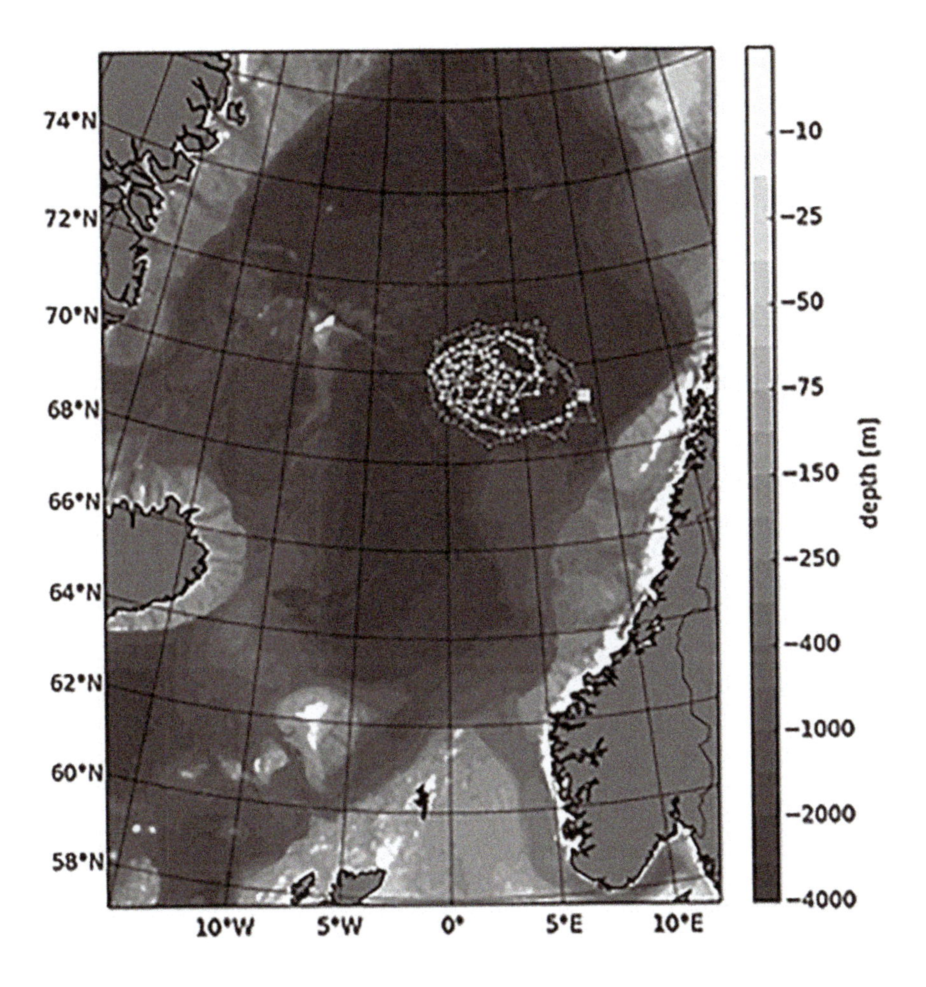

FIGURE 1: Bathymetric map of the Nordic Seas with the locations of profiles from Argo floats 6900798 and 6900799 (yellow and magenta symbols and lines, respectively) in the Lofoten Basin. Larger symbols indicate the locations of the last measurements. Profile times ranged from every 5 days in the spring-summer to every 10 days during late autumn and winter.

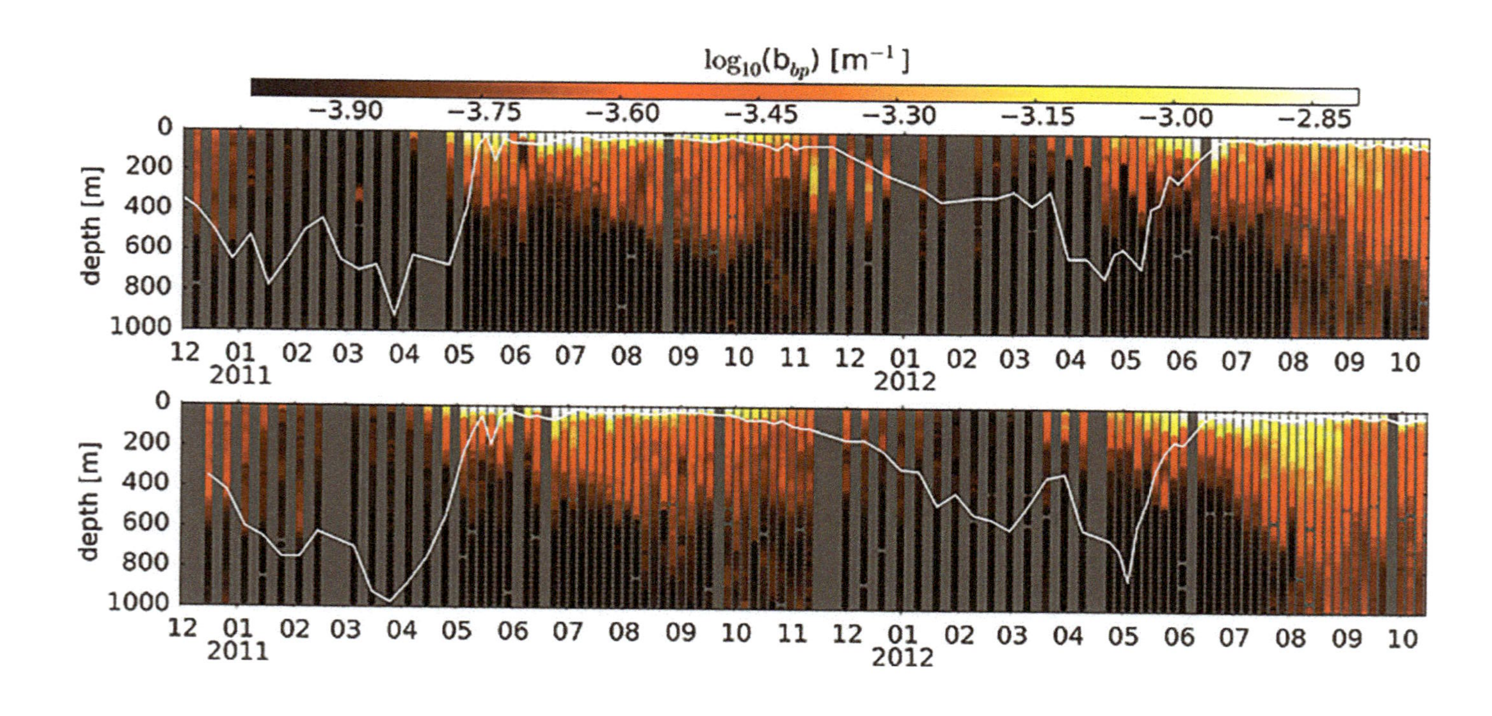

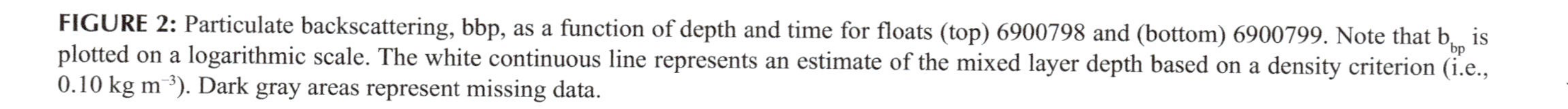

FIGURE 2: Particulate backscattering, bbp, as a function of depth and time for floats (top) 6900798 and (bottom) 6900799. Note that b_{bp} is plotted on a logarithmic scale. The white continuous line represents an estimate of the mixed layer depth based on a density criterion (i.e., 0.10 kg m^{-3}). Dark gray areas represent missing data.

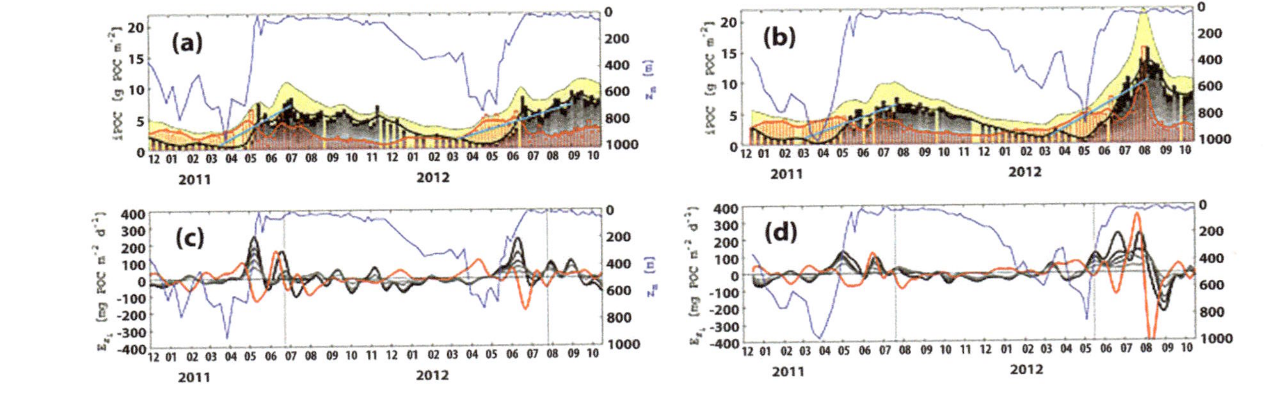

FIGURE 3: (a and b) Vertically integrated POC stocks (iPOC) in various layers of the water column, for floats 6900798 and 6900799, respectively. Bars: data; solid lines: smoothed data. Red: iPOC in the upper layer. Black: $iPOC_{zp}^{1000}$. Gray: $iPOC_{zp+zi}^{1000}$, with z_i=[50,100,200,300,400] m and darker grays correspond to smaller z_i. Yellow shaded area: smoothed $iPOC_0^{1000}$. Light blue lines: linear regressions used to estimate seasonal E_0. (c and d) Corresponding instantaneous POC fluxes (E_{zi}). Red: particle accumulation rates in the productive layer. Black and gray: E_{zi} just below and at different depths below the productive layer (colors represent depth as for iPOC). Vertical dashed lines: times when the examples instantaneous T_{zi} presented in Figure 4 were computed. Blue dashed lines are mixed layer depths.

Manufacturer-supplied offsets and scaling factors were applied to raw fluorescence and backscattering data. Particulate backscattering (bbp) was then computed following established protocols [e.g., Dall'Olmo et al., 2012]. To account for biases in the dark offsets, the minimum bbp value measured by each float (~$5\times10{-}4\,m^{-1}$) was removed from the corresponding time series (note that this subtraction may cause an underestimation of POC, see below). High-frequency spikes were further removed from the b_{bp} data using a median filter. Finally, b_{bp} was used to estimate the concentration of particulate organic carbon (POC) using empirical conversion factors published for the North Atlantic. Different values were used within and below the mixed layer: 37,530 and 31,620 mg POC m^{-2}, respectively [Cetinić et al., 2012].

Estimates of mixed layer depth (z_m) were computed as the depth at which potential density decreased by 0.10 kg m^{-3} from the surface layer [e.g., Bagoien et al., 2012]. The depth of the bottom of the euphotic zone (z_{eu}) was computed based on an empirical relationship that was established between float-based surface chl and Moderate Resolution Imaging Spectroradiometer (MODIS) AQUA remote sensing estimates of z_{eu} (supporting information). Finally, the thickness of the layer where particles can potentially be produced by photosynthesis (z_p) was computed as the maximum of z_{eu} and z_m.

Particulate organic carbon values were integrated over various depths including the sampled water column (0–1000 m, $iPOC_0^{1000}$), the "productive layer" (0–z_p m, $iPOC_0^{zp}$), the "mesopelagic layer" (z_p–1000 m, $iPOC_{zp}^{1000}$), as well as over progressively deeper parts of the mesopelagic layer ($iPOC_{zp+zi}^{1000}$, with z_i = [50, 100, 200, 300, 400] m).

Assuming spatial homogeneity and a negligible POC flux at 1000 m, the time rate of change of iPOC integrated from z_i meters below z_p to 1000 m, $E_{zi} = \delta iPOC_{zp+zi}^{1000}/\delta t$, corresponds to the net rate at which POC accumulates in the water column below depths z_p+z_i. E_{zi} can thus be interpreted as the net POC flux at a depth z_p+z_i, and E_0 indicates the net POC flux just below the productive layer (i.e., export production). Instantaneous values of E_{zi} were computed from iPOC values, after application of a pseudo-Gaussian smoothing filter (solid lines in Figure 3). To provide a first-order estimate of seasonal small-particle accumulation in the mesopelagic region, we regressed $iPOC_{zp}^{1000}$ between the time when $iPOC_0^{1000}$

was minimal (approximately before the maximum in z_m) and the time when $iPOC_{zp}^{1000}$ reached its seasonal maximum (summer–autumn). A linear least squared model forced through the starting point was used to compute seasonal E_{zi}. Uncertainties in E_{zi} were estimated using a Monte Carlo approach (supporting information). Transfer efficiencies of POC flux to a depth z_i below z_p, T_{zi}, were finally computed as the ratio E_{zi}:E_0 [Buesseler and Boyd, 2009].

5.3 RESULTS

During the 2 years of observations, the floats provided a repeated coverage of the Lofoten Basin (Figure 1). Particle backscattering and the estimated stocks of POC showed consistent seasonal patterns. Both quantities were minimal in January–February, began increasing within the upper productive layer at least a month before the shoaling of the mixed layer in spring, and reached maximum values following this shoaling (Figures 2, 3a, and 3b). In July, the POC stock in the productive layer began decreasing, likely due to the exhaustion of nutrients [Bagoien et al., 2012], and this decrease continued until the end of the summer.

In the mesopelagic layer, b_{bp} increased at deeper depths as the summer advanced (Figure 2). Integrated POC stocks ($iPOC_{zi}^{1000}$) began increasing as the mixed layer shoaled in the spring (Figures 3a and 3b). Accumulation of POC then continued throughout the summer until, in the fall, the deepening of the mixed layer marked the beginning of a relatively fast removal of POC from the entire water column. Despite the lower POC concentrations (i.e., b_{bp} values, Figure 2), $iPOC_{zi}^{1000}$ were comparable to or higher than in the upper productive layer during summer and fall (Figures 3a and 3b). At the end of the summer, up to 40% of the POC integrated between 0 and 1000 m was found deeper that 400 m below z_p (Figures 3a and 3b).

In July 2012, one of the two floats recorded a large peak in iPOC (Figure 3b) which corresponded to an almost doubling in the surface values of b_{bp}:chl with respect to 2011 (data not shown). Although no other in situ data are available, we hypothesize that this large peak was due to a bloom of coccolithophorids. This hypothesis is supported by MODIS estimates of particulate inorganic carbon (supporting information).

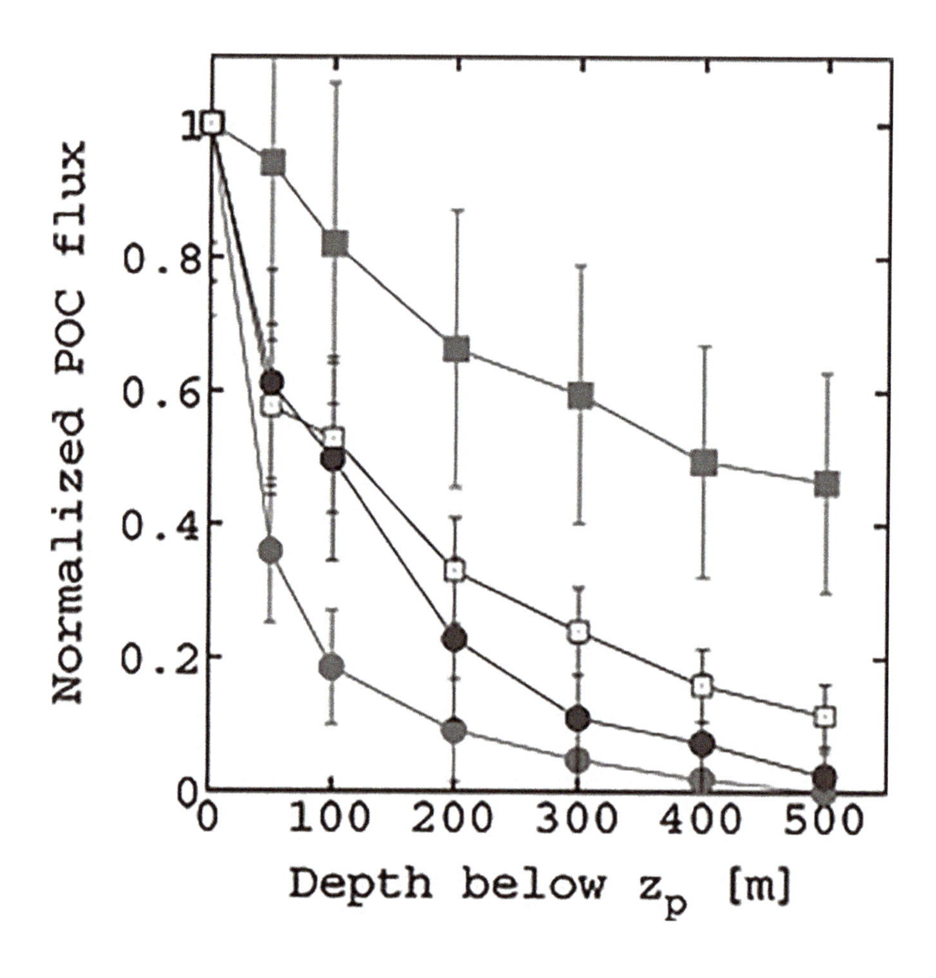

FIGURE 4: Examples of "instantaneous" estimates of transfer efficiency, i.e., POC flux normalized to the flux at the bottom of the upper productive layer (zp). Red and blue symbols refer to floats 6900798 and 6900799, respectively. Circles and squares refer to the first and second vertical dashed lines in Figure 3, respectively.

Instantaneous E_{zi} values showed a large positive peak due to the shoaling of the spring mixed layer, which transferred POC from the productive to the mesopelagic regions (Figures 3c and 3d). One or more positive peaks in E_{zi} were present during the summer (Figures 3c and 3d). E_{zi} ranged between 100 and 300 mg C m^{-2} d^{-1} with accumulation rates in the productive region (E_{zp}) peaking earlier (1–2 weeks, with the exception of the time before the shoaling of z_p) than in the mesopelagic. Note that E_{zp} is considerably smaller than net primary production, because it includes all particle losses (e.g., grazing and sinking). Instantaneous export production estimates were comparable to values reported in the literature for other ocean regions [e.g., Buesseler and Boyd, 2009]. Instantaneous transfer efficiencies 100 m below zp were variable and ranged between 20% and 80% (see examples in Figure 4).

Seasonal estimates of POC export flux just below (E0) and 500 m below (E500) the productive layer averaged around 6.7 and 1.5 g C m^{-2} yr^{-1}, respectively (Table 1). Average seasonal estimates of T_{100} and T_{500} were 60% and 23%, respectively (Table 1).

5.4 DISCUSSION

5.4.1 SMALL-PARTICLE DYNAMICS

Small particles can accumulate in the mesopelagic due to multiple mechanisms. One of the most important is the disaggregation of large sinking particles [Burd and Jackson, 2009]. Biological and physical forces can contribute to disaggregation, and both these processes are difficult to observe and to parametrize in models. Our estimates of small-particle fluxes could thus provide important insights regarding the fate of large particles in the mesopelagic. Lateral advection and/or sediment resuspension could also be responsible for the observed b_{bp} signals at depth [Peinert et al., 2001], although the latter is unlikely, because b_{bp} generally decreased as a function of depth (Figure 2). Finally, the observed deep b_{bp} values could have been due to small particles sinking from the surface [Alonso-Gonzalez et al., 2010; Riley et al., 2012].

TABLE 1: Estimates of Seasonal Export Production (g C m^{-2} yr^{-1}) Just Below and 500 m Below the Productive Layer (E_0 and E_{500}, Respectively) and Transfer Efficiencies 100 m and 500 m Below the Productive Layer (T_{100} and T_{500}, Respectively)[a]

	E0	E500	T100	T500
2011	6.6±0.5	2.2±0.4	0.53±0.07	0.34±0.06
	5.4±0.7	1.1±0.6	0.64±0.11	0.20±0.11
2012	5.8±0.8	1.5±0.2	0.68±0.14	0.27±0.05
	9.1±0.8	1.2±0.3	0.55±0.08	0.13±0.04

[a] *In a given year, the first and second rows correspond to floats 6900798 and 6900799, respectively. Errors represent 1 standard deviation.*

In this study, an additional process appears to have contributed to a significant export of small particles. Figures 3c and 3d show that the largest instantaneous fluxes of small particles were detected in correspondence of the spring shoaling of the mixed layer. Crucially, this shoaling occurred after particles had began accumulating in the upper water column (preshoaling bloom). The particle-laden layer was then isolated from the upper water column by a new shallower mixed layer, resulting in the export of these particles into the mesopelagic region. This physically mediated export process has been previously described and called the "mixed-layer" pump [Bishop et al., 1986; Ho and Marra, 1994; Gardner et al., 1995]. In the case of small particles in deeply mixed layers such as those observed in this study, the mixed-layer pump may be responsible for a significant fraction of the annual export. In addition, because particles are mixed relatively deep into the water column, they may escape remineralization. Thus, when mixed layers are deeper than the depth over which most of the remineralization takes place, the mixed-layer pump may be an efficient pathway for carbon export. Finally, because small particles behave similarly to dissolved organic matter (DOM, i.e., they do not sink), we hypothesize that the shoaling of the mixed layer in the spring could be an important mechanism also for exporting the DOM produced before the stratification period.

Independent estimates of export flux from moored sediment traps deployed in the Lofoten Basin are only available for the early 1990s [Peinert et al., 2001]; nevertheless, they provide a useful comparison for our results. In Peinert et al. [2001], the particle flux typically peaked between August and October and was in large part (i.e., 50%) composed of calcium carbonate ($CaCO_3$). The mean annual POC flux at 1000 m was 1.9±0.5 g C m^{-2} yr^{-1} [Peinert et al., 2001, Table 3]. Although our data set did not allow us to compute export at 1000 m, maximum particle concentrations inferred from bbp at about 900–1000 m were also found between August and November (Figure 2). Mean seasonal estimates of E500 from the floats were 1.6±0.7 and 1.3±0.4 g C m^{-2} yr^{-1} for 2011 and 2012, respectively (Table 1), suggesting that the flux of small particles could represent a considerable component of the POC flux at 500–1000 m. However, because $CaCO_3$ has a higher backscattering per unit of mass than POC [Gordon et al., 2001], our estimates could be overestimated by 44% if, as in Peinert et al. [2001], 50% of the flux was due to particles composed of calcite.

Our flux estimates did not include the export of large fast-sinking aggregates [Briggs et al., 2011] and should thus represent only a fraction of the total POC flux at depth. Our method further underestimates export flux, because it does not record a flux under steady state conditions (i.e., when concentrations do not vary, because input fluxes equal removal rates). Finally, our late summer export fluxes could be underestimated, because of a nonnegligible POC flux below 1000 m (see Figure 2). Thus, the estimates presented here should be considered as a lower bound of small-particle fluxes.

It is unclear if the small-particle flux recorded by the backscattering sensors would have been collected by sediment traps. If it was not collected, then our results would indicate that an important component is neglected from current carbon budgets [Burd et al., 2010]. More studies are needed to reconcile the results from these different methodologies.

Calculation of seasonal particle export by this method requires an assumption of spatial homogeneity. Although instantaneous estimates showed variations between the two floats (Figures 3c and 3d), seasonal estimates in the mesopelagic were, in general, consistent within each year (Table 1). However, in the global ocean floats are not confined to specific ocean regions. It is therefore useful to assess the typical spatial scales over which the b_{bp} signals would need to be integrated to achieve accurate carbon flux estimates. By propagating uncertainties in b_{bp} and POC:bbp ratios, we obtained conservative uncertainties in E_{zi} of about 40%, when we integrated signals over 30 days for a "high-export" scenario (i.e., 100 mg C m^{-2} d^{-1}; supporting information). If a float drifts at depth, in a straight line, and at an average speed of 5.4 cm/s [Ollitrault and Rannou, 2013], then during 30 days it would move by about 150 km. Thus, provided that our uncertainty estimates are reasonable, application of the technique presented here to data from a generic float would require an assumption of mesoscale homogeneity. Reducing uncertainties in the POC:b_{bp} ratio would allow us to relax this assumption.

Finally, although our data sets are limited to the upper 1000 m of the water column, the bbp signal propagated at the end of the summer to depths below the deepest winter mixing (Figure 2). Thus, the particle flux recorded in this study by the backscattering coefficient likely contributed to long-term carbon sequestration.

5.4.2 AUTONOMOUS PLATFORMS AND THE OCEAN CARBON CYCLE

An important step toward improving our understanding of the ocean biological carbon pump is to increase the number of carbon flux observations in the upper ocean. The cost and limited spatial coverage of ship-based measurements is a formidable obstacle to such an endeavor. Fortunately, autonomous platforms such as Bio-Argo floats have enormous potential for complementing current observations of carbon flux [e.g., Bishop and Wood, 2009].

Bio-optical proxies of particle concentration and characteristics (e.g., POC) carry, by nature, uncertainties due to the underlying empirical conversions. To reduce these uncertainties, more in situ data are needed, especially in the mesopelagic region. Nonetheless, few, if any, other measurements can parallel the spatiotemporal coverage that optical proxies afford when recorded by autonomous platforms in situ (or from space). These data can complement existing measurements and help resolve current method-method uncertainties in estimating carbon export.

Data from Bio-Argo floats have already demonstrated that multiyear, high-resolution, vertically resolved observations can transform the way we understand ocean ecosystems and biogeochemistry [Bishop et al., 2004; Riser and Johnson, 2008; Boss and Behrenfeld, 2010; Johnson et al., 2010; Estapa et al., 2013]. The results presented here provide another example of how a global fleet of Bio-Argo floats could allow us a quantum leap in understanding of the ocean carbon cycle.

5.5 CONCLUSIONS

1. Optical backscattering data from Bio-Argo floats unveiled the seasonal dynamics of small particles in the upper 1000 m of the Norwegian Sea.
2. In the summer and autumn the integrated POC stocks of small particles in the mesopelagic were comparable to or greater than in the upper productive layer.
3. The shoaling of the mixed layer in the spring was important for exporting small-particle carbon.
4. Estimates of small-particle POC export were comparable to literature values determined by sediment traps.
5. Small particles contributed to long-term carbon sequestration.

REFERENCES

1. Alonso-Gonzalez, I. J., J. Aristegui, C. Lee, A. Sanchez-Vidal, A. Calafat, J. Fabres, P. Sangra, P. Masque, A. Hernandez-Guerra, and V. Benitez-Barrios (2010), Role of slowly settling particles in the ocean carbon cycle, Geophys. Res. Lett., 37, L13608, doi:10.1029/2010GL043827.
2. Bagoien, E., W. Melle, and S. Kaartvedt (2012), Seasonal development of mixed layer depths, nutrients, chlorophyll and Calanus finmarchicus in the Norwegian Sea—A basin-scale habitat comparison, Prog. Oceanogr., 103, 58–79, doi:10.1016/j.pocean.2012.04.014.
3. Bishop, J.K.B., M.H. Conte, P.H. Wiebe, M.R. Roman, and C. Langdon (1986), Particulate matter production and consumption in deep mixed layers: Observations in a warm-core ring, Deep Sea Res. Part A, 33, 1813–1841.
4. Bishop, J.K.B., T.J. Wood, R.E. Davis, and J.T. Sherman (2004), Robotic observations of enhanced carbon biomass and export at 55 degrees S during SOFeX, Science, 304(5669), 417–420.

5. Bishop, J.K.B., and T.J. Wood (2009), Year-round observations of carbon biomass and flux variability in the Southern Ocean, Global Biogeochem. Cycles, 23, GB2019, doi:10.1029/2008GB003206.
6. Boss, E., and M. Behrenfeld (2010), In situ evaluation of the initiation of the North Atlantic phytoplankton bloom, Geophys. Res. Lett., 37, L18603, doi:10.1029/2010GL044174.
7. Briggs, N., M.J. Perry, I. Cetinic, C. Lee, E. D'Asaro, A.M. Gray, and E. Rehm (2011), High-resolution observations of aggregate flux during a sub-polar North Atlantic spring bloom, Deep Sea Res. Part I, 58(10), 1031–1039.
8. Buesseler, K.O., and P.W. Boyd (2009), Shedding light on processes that control particle export and flux attenuation in the twilight zone of the open oceans, Limnol. Oceanogr., 54(4), 1210–1232, doi:10.4319/lo.2009.54.4.1210.
9. Burd, A.B., and G.A. Jackson (2009), Particle aggregation, Annu. Rev. Mar. Sci., 1, 65–90, doi:10.1146/annurev.marine.010908.163904.
10. Burd, A. B., et al. (2010), Assessing the apparent imbalance between geochemical and biochemical indicators of meso- and bathypelagic biological activity: What the @$#! is wrong with present calculations of carbon budgets?, Deep Sea Res. Part II, 57, 1557–1571.
11. Cetinić, I., M.J. Perry, N.T. Briggs, E. Kallin, E.A. D'Asaro, and C.M. Lee (2012), Particulate organic carbon and inherent optical properties during 2008 North Atlantic Bloom Experiment, J. Geophys. Res., 117, C06028, doi:10.1029/2011JC007771.
12. Dall'Olmo, G., E. Boss, M. Behrenfeld, and T. Westberry (2012), Particulate optical scattering coefficients along an Atlantic meridional transect, Opt. Express, 20(19), 21,532–21,551.
13. Estapa, M.L., K. Buesseler, E. Boss, and G. Gerbi (2013), Autonomous, high-resolution observations of particle flux in the oligotrophic ocean, Biogeosciences, 10(8), 5517–5531.
14. Gardner, W.D., S.P. Chung, M.J. Richardson, and I.D. Walsh (1995), The oceanic mixed-layer pump, Deep Sea Res., 42(2–3), 1587–1590.
15. Gordon, H.R., G.C. Boynton, W.M. Balch, S.B. Groom, D.S. Harbour, and T.J. Smyth (2001), Retrieval of coccolithophore calcite concentration from SeaWiFS imagery, Geophys. Res. Lett., 28(8), 1587–1590.
16. Henson, S.A., R. Sanders, E. Madsen, P.J. Morris, F. LeMoigne, and G.D. Quartly (2011), A reduced estimate of the strength of the ocean's biological carbon pump, Geophys. Res. Lett., 38, L04606, doi:10.1029/2011GL046735.
17. Ho, C., and J. Marra (1994), Early-spring export of phytoplankton production in the northeast Atlantic Ocean, Mar. Ecol. Prog. Ser., 114, 197–202.
18. Johnson, K.S., S.C. Riser, and D.M. Karl (2010), Nitrate supply from deep to near-surface waters of the North Pacific subtropical gyre, Nature, 465(7301), 1062–1065, doi:10.1038/nature09170.
19. Laws, E.A., P.G. Falkowski, W.O. Smith, H. Ducklow, and J.J. McCarthy (2000), Temperature effects on export production in the open ocean, Global Biogeochem. Cycles, 14(4), 1231–1246.
20. Ollitrault, M., and J.P. Rannou (2013), ANDRO: An Argo-based deep displacement dataset, J. Atmos. Oceanic Technol., 30, 759–788.

21. Peinert, R., A. Antia, E. Bauerfeind, B. von Bodungen, O. Haupt, M. Krumbholz, I. Peeken, R. Ramseier, M. Voss, and B. Zietzschel (2001), Particle flux variability in the polar and Atlantic biogeochemical provinces of the Nordic Seas, in The Northern North Atlantic, edited by P. Schäfer et al., pp. 53–68, Springer, Berlin, Heidelberg.
22. Riley, J. S., R. Sanders, C. Marsay, F. A. C. Le Moigne, E. P. Achterberg, and A. J. Poulton (2012), The relative contribution of fast and slow sinking particles to ocean carbon export, Global Biogeochem. Cycles, 26, GB1026, doi:10.1029/2011GB004085.
23. Riser, S. C., and K. S. Johnson (2008), Net production of oxygen in the subtropical ocean, Nature, 451(7176), 323–325, doi:10.1038/nature06441.
24. Stramski, D., E. Boss, D. Bogucki, and K. J. Voss (2004), The role of seawater constituents in light backscattering in the ocean, Prog. Oceanogr., 61(1), 27–56.
25. Trull, T., S. Bray, K. Buesseler, C. Lamborg, S. Manganini, C. Moy, and J. Valdes (2008), In situ measurement of mesopelagic particle sinking rates and the control of carbon transfer to the ocean interior during the vertical flux in the global ocean (Vertigo) voyages in the North Pacific, Deep Sea Res. Part II, 55(14-15), 1684–1695.
26. Voet, G., D. Quadfasel, K. A. Mork, and H. Soiland (2010), The mid-depth circulation of the Nordic Seas derived from profiling float observations, Tellus Ser. A, 62(4), 516–529, doi:10.1111/j.1600-0870.2010.00444.x.

There are several supplemental files that are not available in this version of the article. To view this additional information, please use the citation on the first page of this chapter.

PART III

PHYTOPLANKTON AND OCEANIC CARBON CYCLE

CHAPTER 6

Ubiquitous Healthy Diatoms in the Deep Sea Confirm Deep Carbon Injection by the Biological Pump

S. AGUSTI, J. I. GONZÁLEZ-GORDILLO, D. VAQUÉ, M. ESTRADA, M. I. CEREZO, G. SALAZAR, J. M. GASOL, AND C. M. DUARTE

6.1 INTRODUCTION

The role of the ocean as a sink for anthropogenic CO_2 is critically dependent on the transport of carbon to depths below 1,000 m, where it is removed from ventilating back to the atmosphere over centennial timescales [1, 2]. Ocean plankton contributes to remove CO_2 through sinking of particles, transporting organic carbon at depth, the so-called biological pump [3, 4]. Sinking rates of individual phytoplankton cells are very slow, at ~1.5 m d^{-1} (ref. 5) expected for diatom cells [6]. However, Smayda [7] suggested that a range of physical and biological (aggregation, downwelling and density inversion currents, packaging of cells in faecal pellets) mechanisms can accelerate phytoplankton-sinking rates in situ [8, 9] beyond the rates expected for single cells.

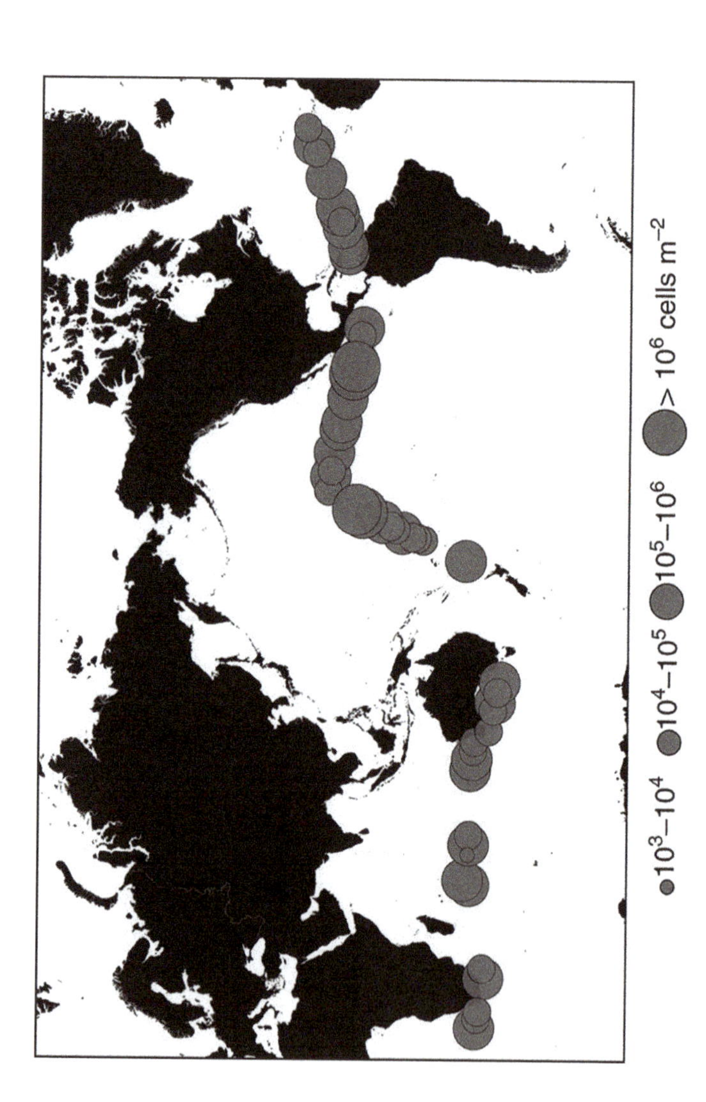

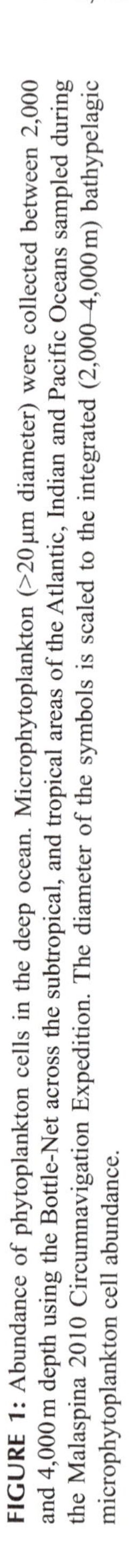

FIGURE 1: Abundance of phytoplankton cells in the deep ocean. Microphytoplankton (>20 µm diameter) were collected between 2,000 and 4,000 m depth using the Bottle-Net across the subtropical, and tropical areas of the Atlantic, Indian and Pacific Oceans sampled during the Malaspina 2010 Circumnavigation Expedition. The diameter of the symbols is scaled to the integrated (2,000–4,000 m) bathypelagic microphytoplankton cell abundance.

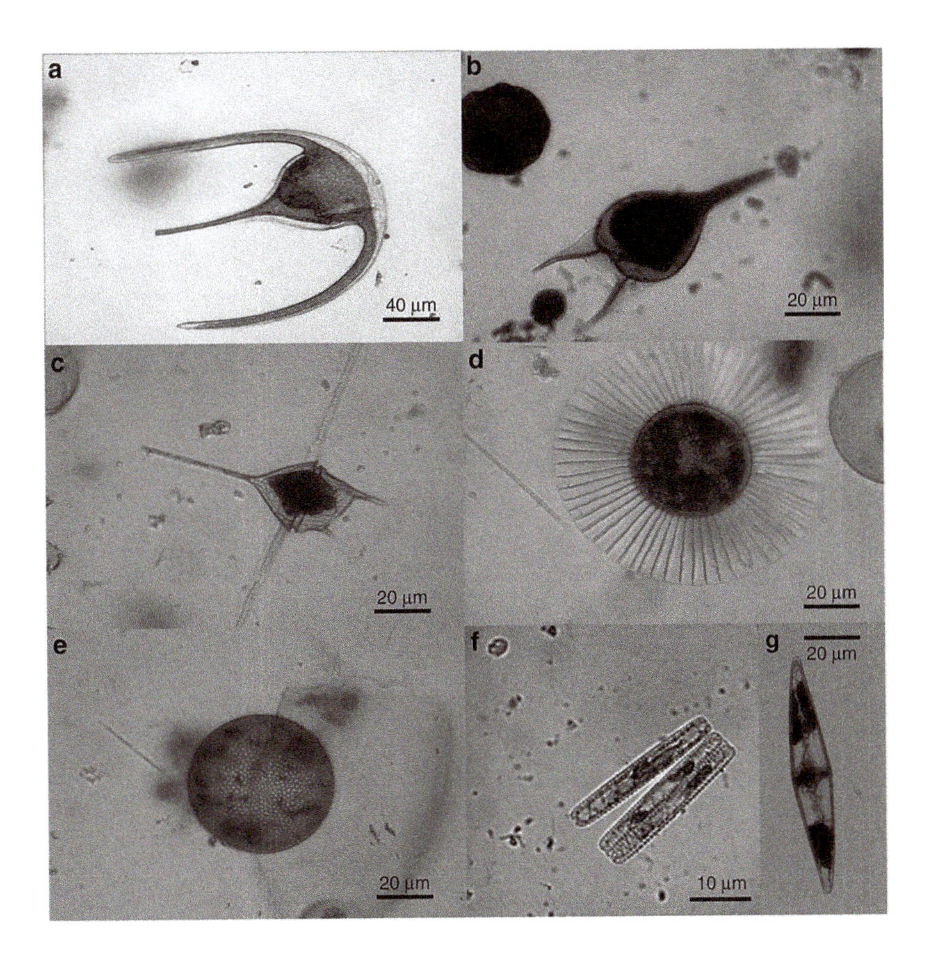

FIGURE 2: Intact phytoplankton cells in the dark deep ocean. Phytoplankton cells collected by the Bottle-Net between 2,000 and 4,000 m observed under the optical microscope included dinoflagellate (a–c) and diatom (d–g) genera. Most phytoplanktonic cells corresponded to vegetative cells.

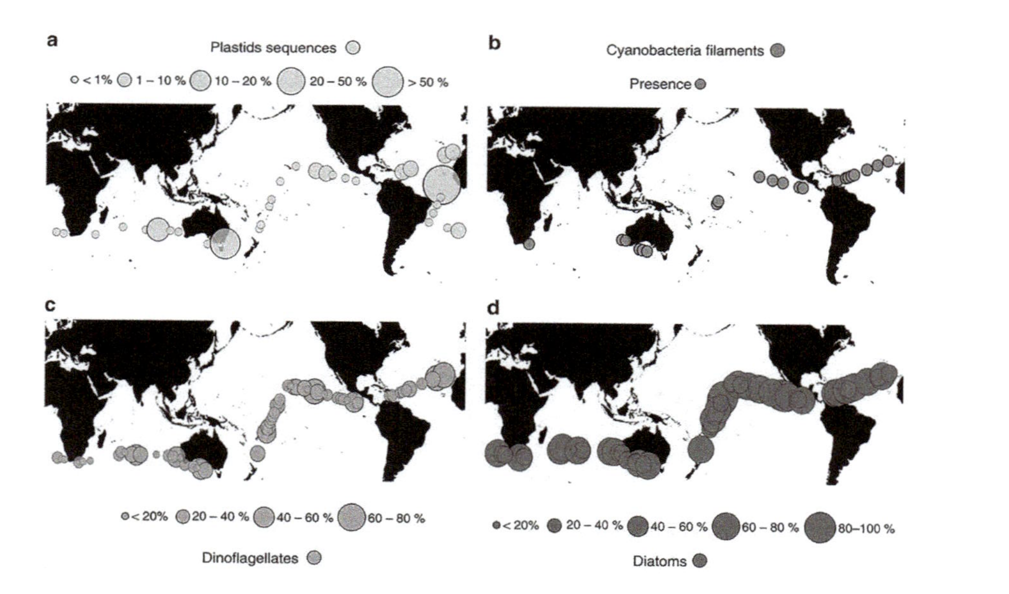

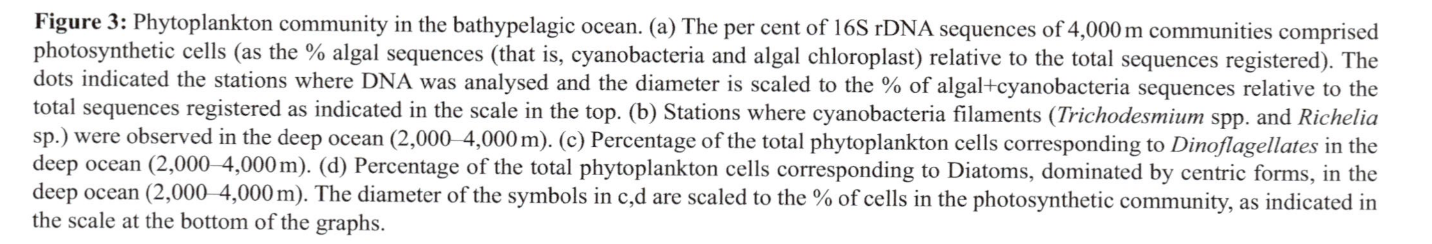

Figure 3: Phytoplankton community in the bathypelagic ocean. (a) The per cent of 16S rDNA sequences of 4,000 m communities comprised photosynthetic cells (as the % algal sequences (that is, cyanobacteria and algal chloroplast) relative to the total sequences registered). The dots indicated the stations where DNA was analysed and the diameter is scaled to the % of algal+cyanobacteria sequences relative to the total sequences registered as indicated in the scale in the top. (b) Stations where cyanobacteria filaments (*Trichodesmium* spp. and *Richelia* sp.) were observed in the deep ocean (2,000–4,000 m). (c) Percentage of the total phytoplankton cells corresponding to *Dinoflagellates* in the deep ocean (2,000–4,000 m). (d) Percentage of the total phytoplankton cells corresponding to Diatoms, dominated by centric forms, in the deep ocean (2,000–4,000 m). The diameter of the symbols in c,d are scaled to the % of cells in the photosynthetic community, as indicated in the scale at the bottom of the graphs.

TABLE 1: Mean (±s.e.) absolute and relative abundance of phototrophic and heterotrophic plankton in the surface and deep ocean.

	Cell abundance (cells m^{-2})	s.e.	N
Surface ocean (0–200 m)			
Microphytoplankton	9.78×10^7	3.63×10^7	11
Diatoms	6.73×10^7	3.08×10^7	11
Dinoflagellates	3.01×10^7	8.88×10^6	11
Others	7.61×10^5	4.49×10^6	11
Cyanobacteria	1.29×10^6	1.26×10^6	11
Ratio Diatoms/Dinoflag	2.5	0.9	11
Ciliates	9.15×10^6	2.83×10^6	11
Ratio Phytop/Ciliates	9.9	2.4	11
Picophytoplankton	2.72×10^{13}	1.10×10^{12}	226
Heterotrophic bacteria	1.33×10^{14}	1.24×10^{13}	123
Ratio Bact/Picophytop	4.9	11.3	123
Deep ocean (2,000–4,000 m)			
Microphytoplankton	2.52×10^5	5.29×10^4	58
Diatoms	2.02×10^5	5.08×10^4	58
Dinoflagellates	3.46×10^4	5.61×10^3	58
Others	1.12×10^3	3.04×10^3	58
Cyanobacteria	5.71×-10^2	4.21×10^1	58
Ratio Diatoms/Dinoflag	13.9	3.5	58
Ciliates	9.95×10^4	2.24×10^4	11
Ratio Phytop/Ciliates	8.9	2.9	11
Picophytoplankton	3.16×10^{12}	1.05×10^3	123
Heterotrophic bacteria	1.20×101^4	2.62×10^{14}	123
Ratio Bact/Picophytop	38.0	24.9	123

Bact/Picophytop, heterotrophic bacteria to picophytoplankton cell abundance ratio; Cyanobacteria, filamentous forms; Diatoms/Dinoflag, diatoms to dinoflagellates cell abundance ratio; N, number of samples; Others, flagellates and silicoflagellates; Phytop/Ciliates, phytoplankton to cilliates cell abundance ratio.

Subsequent laboratory experiments and observations, mostly derived from the epi- and mesopelagic layers of productive regions of the ocean, have documented phytoplankton cells within aggregates and faecal pellets to sink at rates of 10–1,000 m d^{-1} (refs 9, 10, 11, 12). These fast-sinking rates are consistent with estimates of sinking rates derived using geochemical tracers, such as $^{230, 234}Th$ isotopes [13, 14], reports of events of accelerated phytoplankton export [15, 16], and calculations suggesting high carbon use by heterotrophs in the deep ocean [17, 18]. Hence, that the transport of organic carbon at depth can be greatly accelerated by processes involving the packaging of phytoplankton cells within aggregates and faecal pellets is now well established [19, 20, 21]. However, whether these mechanisms delivering fresh organic carbon reach the deep sea and how prevalent they are across the oligotrophic regions of the ocean remains to be tested.

Research since the 1960–1970s reported the occasional presence of well-preserved phytoplankton cells in the deep sea [22, 23, 24, 25]; however, these observations, which could signal at rapid sinking rates [7], were considered anecdotal. Using new developments we tested the presence of healthy phytoplankton cells in the deep sea (2,000–4,000 m depth) along the Malaspina 2010 Circumnavigation Expedition, a global expedition sampling the bathypelagic Atlantic, Indian and Pacific Oceans [26, 27, 28]. In particular, we used a new microplankton sampling device, the Bottle–Net (Supplementary Fig. 1), 16S rDNA sequences, flow cytometric counts, vital stains and experiments to explore the abundance and health status of photosynthetic plankton cells in bathypelagic waters between 2,000 and 4,000 m depth along the circumnavigation track (Fig. 1).

6.2 RESULTS

6.2.1 PHYTOPLANKTON CELLS WERE UBIQUITOUS IN THE DEEP SEA

The samples retrieved in the Malaspina 2010 Circumnavigation Expedition revealed the ubiquitous presence in the deep, bathypelagic dark ocean of morphologically well-preserved (photosynthetic) microphytoplankton cells, characteristic of the lighted layers of the surface ocean (Fig. 2). The highest concentrations of phytoplankton in the deep sea were found in the Equatorial Pacific. The concentration of phytoplankton cells in the deep ocean was low, averaging 2.5 × 10^5 cells m^{-2} in the 4,000–2,000 m depth layer (range<50 to >500 cells m^{-3}, Fig. 1, Table 1). However, phytoplankton cells were more abundant than heterotrophic ciliates (Table 1, Supplementary Fig. 1), the organisms hitherto believed to dominate deep sea microplankton communities [29], and the ratio of photosynthetic to heterotrophic protist cells did not differ between surface and deep waters (Table 1).

Microphytoplankton communities in the deep sea were dominated by diatoms (81.5%), while photosynthetic dinoflagellates dominated the community in the Eastern Subtropical Atlantic (Fig. 3). Indeed, diatoms exerted a greater dominance in the deep-sea microphytoplankton community than in surface waters (Table 1). Nitrogen-fixing cyanobacteria were occasionally abundant, including *Trichodesmium* sp. trichomes in the deep waters of the Subtropical Atlantic and the Equatorial Pacific, and symbiotic N_2-fixing *Richelia* sp. within large *Rhizosolenia* (diatom) hosts in the Eastern Indian Ocean (Fig. 3; Table 1). Aggregates containing *Synechococcus* cells and other phytoplankton were also observed in deep waters. Single picophytoplankton cells were also detected using flow cytometry in the samples (Table 1; Supplementary Fig. 2), supporting previous reports [25]. Further, we also consistently observed the presence of significant numbers of 16S rDNA sequences affiliated to algae plastids and to cyanobacteria in samples collected at 4,000 m depth, where eukaryotic algae and cyanobacteria comprised up to 68.9% of the sequences in some microbial communities (Fig. 3).

Three observations pointed at the overlying waters, rather than lateral transport, as the source of the phytoplankton cells found in the deep sea: (1) the general correspondence between the community structure of microalgae in the surface and corresponding deep-sea samples across stations (Supplementary Table 1); (2) the occurrence in both surface waters and the corresponding bathypelagic waters of relatively rare taxa, as illustrated by the large *Rhizosolenia* diatom cells containing the symbiotic N_2-fixing cyanobacteria *Richelia* encountered in both epi- and bathypelagic waters of the Indian Ocean near Western Australia (Supplementary Table 1); and (3) that resting stages, which could be transported laterally within the deep sea, were not detected in the deep-sea microphytoplankton community for those taxa for which morphologically distinct resting stages have been reported, including dinoflagellates, and the diatom genera *Chaetoceros, Leptocylindrus* and *Rhizosolenia* [30].

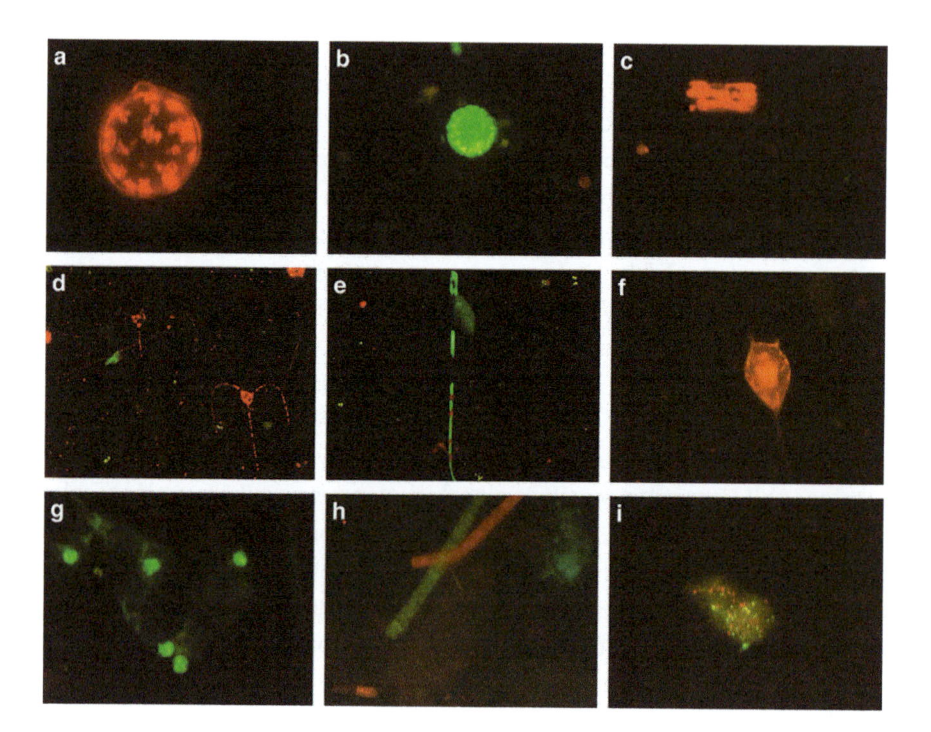

FIGURE 4: Testing the cell membrane permeability of bathypelagic phytoplankton. After stained with the double vital stain Bac-light Kit the living phytoplankton cells collected by the Bottle-Net fluoresce green and the dead cells fluoresce red when observed under the epifluorescence microscope. (a) Dead centric diatom; (b) alive centric diatom; (c) dead pennate diatom cells in a colony; (d) dead cells of the dinoflagellate *Ceratium* sp.; (e) alive cell of the dinoflagellate *Ceratium* fusus; (f) dead cell of *Ceratium* sp. red; (g) a group of alive cells aggregated in a particle; (h) dead (red) and alive (green) trichomes of the cyanobacteria *Trichodesmium*; (i) an aggregate of cells from the deep ocean containing picophytoplankton and bacteria. As this figure is to show differences in fluorescence colour there are no scale bars; the organisms and cells shown are all on the order of micrometres.

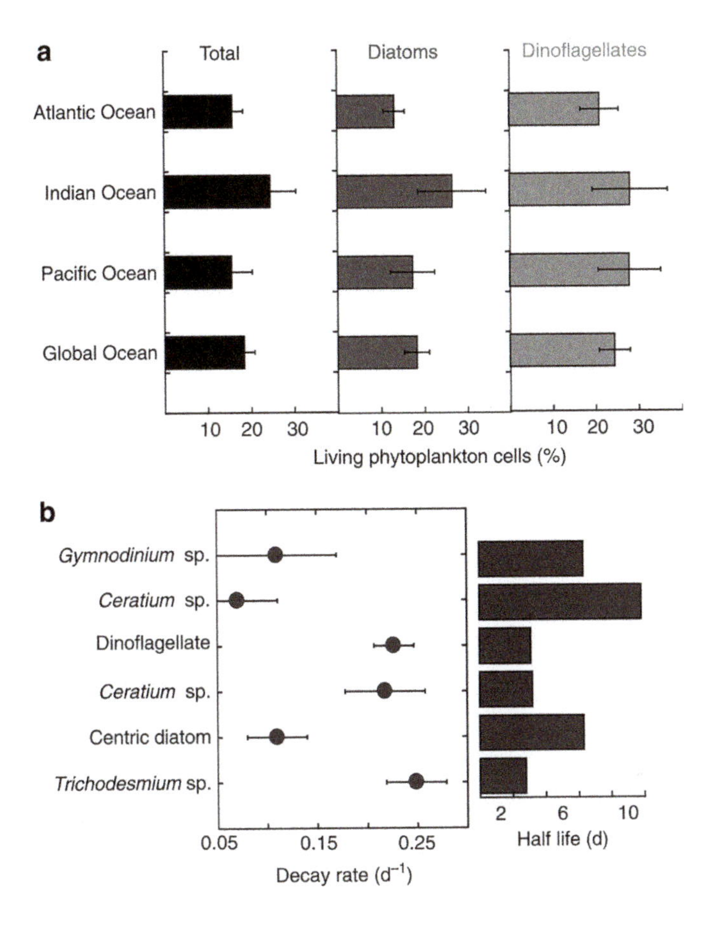

FIGURE 5: Living phytoplankton cells (>20 µm diameter) in the bathypelagic ocean and cell survival in the dark ocean. (a) The percentage (mean±s.e.) of living microphytoplankton cells found in the deep water column (2,000–4,000 m) of the Atlantic, Indian and Pacific Oceans, and for the global deep ocean sampled in the Malaspina 2010 Circumnavigation Expedition. Black columns represent the mean percentages of living cells for the total microphytoplankton cells encountered (±s.e.), and red and orange columns represent the mean percentages of living cells within diatom and dinoflagellate communities, respectively. Living and dead cells were identified by using a double vital stain to test cell membrane permeability. (b) Experimentally derived mean (±s.e.) phytoplankton living cell's decay rates (d−1) and half-lives of alive cells (days) for populations sampled at the photic surface layer and incubated in 4,000 m depth water and under deep ocean conditions (dark and cold temperature) for more than a month. Centric diatom: undetermined centric diatom. *Dinoflagellate*: undetermined dinoflagellate.

6.2.2 ALIVE PHYTOPLANKTON CELLS IN THE BATHYPELAGIC OCEAN

A proportion, 18.4±2.4% on average (Figs 4 and 5), of the phytoplankton cells present in the deep ocean had intact plasma membranes [31, 32], thereby demonstrating the presence of living phytoplankton cells in the deep ocean (Fig. 4). Experimental assessments of cell mortality of photic layer phytoplankton under the conditions of the dark deep ocean revealed high decay rates of living populations, with half-lives of living cells ranging from 3 to 10 days (Fig. 5; Supplementary Fig. 3). The time for living cells to decline to 18% of the community, the average cell viability in populations of photosynthetic microplankton retrieved from the deep ocean (Fig. 5), ranged between 6.8 and 24.2 days.

6.3 DISCUSSION

Microphytoplankton sinking as single cells at rates in the order of 1.5 m d^{-1} (refs 5, 6) would take 3.8–7.3 years to reach 2,000–4,000 m depth, respectively. Previous experiments demonstrated that phytoplankton populations may be preserved in the dark for over 2 months [33]; however, these experiments did not examine the rate of decay of viable cells in the dark. Our experimental results, which did not examine preservation but survival rates, show much shorter lifespans of phytoplankton cells in the conditions of the dark ocean. Our experiments imply that most cells of the phytoplankton communities we sampled in the deep ocean would be dead after 1 month and no living cells would be expected to remain in the population after transit times in excess of a year. Moreover, even most dead microphytoplankton cells in our bathypelagic samples were also morphologically well preserved, which would not be possible if they had left the photic layer months to years ago. Hence, the presence of healthy microphytoplankton cells between 2,000 and 4,000 m depth can only be explained by very fast-sinking rates.

On the basis of our experimental results we calculate that for 18% of phytoplankton cells to be still alive, populations sinking below the photic layer must take 6.8–24.2 days, at sinking rates of 124–435 m d^{-1}, to reach

3,000 m depth (midpoint between 2,000 and 4,000 m). These estimates of sinking rates are conservative, as these calculations assume that all cells sinking below the photic layer are alive, when available reports indicate that this is not the case [34]. Assuming that only 50% of the cells leaving the photic layer are alive, as observed in the epipelagic samples collected to initiate the decay experiments (mean±s.e.% living cells=48.5±3.8%), the sinking rates for the community to support 18% of living cells at 3,000 m depth ranges from 208 to 732 m d^{-1}.

The range of sinking rates of 124–732 m d^{-1} suggested by our experiments and observations of the fraction of living cells in the deep-sea communities is comparable to the sinking rates reported for fast-sinking particles such as aggregates [19, 21, 35] and faecal pellets [12, 36, 37]. Moreover, laboratory experiments have demonstrated that diatom cells can remain alive when packaged in faecal pellets [20]. Hence, we infer that the well-preserved phytoplankton cells we observed across the deep-sea must have been injected at depth embedded within fast-sinking aggregates, which may have disaggregated in situ or during the retrieval with the Bottle-Net instrument [38]. Hence, our results extend existing evidence of the penetration of fast-sinking particles [19, 22] to the deep, oligotrophic ocean.

Our results help constraint the fraction of the microphytoplankton community reaching the deep sea [39]. The pool of phytoplankton cells integrated over the 2,000- to 4,000-m water column represented 0.25% of that present in the photic layer (Table 1). This is a high fraction considering that only between 5 and 10% of microplankton production, which should be lower than the stock on a daily basis, sinks below the mixed layer in the warm waters sampled here [8, 40] and that this sinking flux is rapidly attenuated with depth in the oligotrophic ocean [41, 42, 43]. For comparison, the fraction of primary production expected with the sedimentary flux at 2,000–4,000 m depth would be 0.04–0.18% (central value 0.08%). The existence of a large, rapid and predictable seasonal pulse of particulate matter to the deep sea in the North Pacific Subtropical Gyre has been reported recently [16], again suggesting the operation of mechanisms accelerating the export of phytoplankton cells to the deep [35] sea. Indeed, multiple observations have converged to suggest sinking fluxes from 10 to 1,000 m d^{-1} (refs 11, 15, 22, 36), consistent with the sinking rates of 124–732 m d^{-1} derived here.

The greater dominance of diatoms in the deep-sea microphytoplankton community relative to that in surface waters (Table 1) suggested a role for opal ballasting in delivering the particles containing these cells to the deep sea [44, 45]. Whereas this observation does not preclude a role for calcite ballasting, coccolitophores (included within the 'other' microalgal category in Table 1) played a comparatively small role as members of the communities in both surface and deep-sea waters.

Identification in the past of well-preserved phytoplankton cells in the deep sea sampled with 3–5 l bottles was possible because of the phenomenal masses of cells occasionally found, such as 50,000 diatom cells per l at 900 m depth in the Pacific Ocean [23], which must derive from mass sinking events [16]. However, sampling limitations in the past suggested observations of well-preserved phytoplankton cells in the deep sea to be a rare event [22, 23, 24]. Use of the Bottle-Net instrument, capable of quantitatively sampling deep ocean microplankton communities by filtering tens of m^3 of deep-sea water, and sequencing of 16S rDNA concentrated from hundreds of litres of deep-sea water now demonstrate the presence of well-preserved phytoplankton at depth to be the norm [45].

Our results conclusively demonstrate that (1) the presence of healthy phytoplankton cells in the deep sea is ubiquitous at the global scale; (2) the abundance of phytoplankton in the deep sea is significant, and these communities show a higher dominance of diatoms relative to those in the upper ocean from which they originate; (3) identify the Equatorial Pacific as an area of high input of phytoplankton cells to the deep sea; and (4) provide evidence that these cells must have reached the bathypelagic ocean down to 4,000 m through fast-sinking mechanisms. Therefore, the results from the global survey presented here confirm the belief that fast-sinking processes should be able to inject fresh organic carbon down to the bathypelagic ocean [45, 46, 47]. Phytoplankton cells reaching the deep sea alive will eventually die, as supported by experimentally evidence provided here (Fig. 5) together with the 82% proportion, on average, of dead cells in the deep-sea phytoplankton community. Thus, the ubiquitous rapid transport of fresh phytoplankton cells to the deep ocean helps explain the high bacterial metabolic rates in the deep ocean [18, 29, 48], which are difficult to reconcile if the organic carbon flux was dominated by slowly sinking detritus [45, 47, 49]. Rapid sinking of phytoplankton has also been invoked as a likely explanation for the observation of phytodetritus 'fluff' layers in the sea bed [49, 50]. Whereas the development of these phytodetritus layers in the sea bed is believed to derive from episodic events [49, 50], our results provide evidence that healthy phytoplankton cells also reach the deep ocean outside of such events. Our confirmation that pathways for the rapid delivery of fresh organic carbon by the biological pump typically reach the deep ocean contributes, therefore, to understand the supply of carbon supporting bathypelagic microbial communities.

6.4 METHODS

6.4.1 MALASPINA 2010 CIRCUMNAVIGATION EXPEDITION

The Malaspina 2010 Circumnavigation Expedition sailed the oceans on board R/V Hésperides of the Spanish Navy from 15 December 2010 to 14 July 2011, surveying the Atlantic, Indian and Pacific Oceans [26, 27, 28]. The expedition left Spain sailing south via the Atlantic Ocean to enter the Indian Ocean south of Cape Town (South Africa); to enter the Pacific Ocean through the Bass Straight (Australia), and returning to the Northern Hemisphere in May 2011, crossing the Pacific to enter the Atlantic Ocean through the Panama Canal, sailing across the Atlantic to return to Spain. The expedition sampled mostly the subtropical, oligotrophic ocean (Fig. 1), with the sequence of the survey leading to sampling mostly during spring and summer. The expedition sampled the open ocean down to 4,000 m depth, or the maximum depth available if shallower than 4,000 m. Microphytoplankton samples integrated between 4,000 and 2,000 m depth were collected at 58 stations using a Rosette sampling system fitted with a CTD, 12-l Niskin Bottles, and a Bottle-Net sampling device (see below).

6.4.2 BOTTLE-NET DESCRIPTION

The Bottle-Net (Supplementary Fig. 4) is a new oceanographic device specifically developed for the Malaspina 2010 Circumnavigation Expedition (Pat: ES 2377070 B2), which consists of a conical plankton net housed in a cylindrical PVC pipe that acts as a case, with a net mouth arranged at the top. The case is opened at the bottom but presents a remote-closing cover on the top that hermetically closes the entrance of the net, with a mechanism identical to that of the Niskin bottles. The equipment is designed to be mounted on a standard rosette sampler of oceanographic bottles, where it replaces one 12-l bottle, and uses the same remote shooter as that operating regular oceanographic bottles. The Bottle-Net can also works autonomously when attached to a ballasted oceanographic cable. The Bottle-Net samples only during the ascension of the oceanographic rosette taking one sample per deployment. During sampling, the top cover of Bottle-Net remains opened while the oceanographic rosette descends; however, when the rosette reaches the maximum depth, 4,000 m in our survey, and starts the ascent, the Bottle-Net starts filtering water and collecting organisms. After rising to the target depth, 2,000 m in our survey, the remote shooter is activated and the cover hermetically closes the mouth of the Bottle-Net preventing the entrance of water during the final ascend of the device to the surface. On board, the net must be softly rinsed with filtered seawater in order to retrieve the sample from the collector. Sample volume is estimated as the product between the area of the mouth of the Bottle-Net and the vertical distance covered by the device from the start of the ascension to the closure of the mouth. The Bottle-Net presents an open area ratio of 4, displaying an efficiency of filtration of 96% for deep tows (2,000–4,000 m) and trawling velocities ~30 m min^{-1} (that is, standard rosette retrieval velocities).

6.4.3 BATHYPELAGIC MICROPLANKTON ABUNDANCE AND VIABILITY

A fraction of the Bottle-Net sample was stained fresh with the vital stain Back-light Kit (Molecular Probes) to identify living and dying phytoplankton cells. The Bac-light viability Kit (Molecular Probes Invitrogen) is a double-staining technique used to test cell membrane permeability by selective fluorescence signals [51], proven to be an effective method for phytoplankton [31, 32]. Living phytoplankton cells with intact membranes fluoresce green (Syto 9, nucleic acid stain) and dead phytoplankton cells with compromised membranes fluoresce red (Propidium Iodine, nucleic acid stain). The samples were filtered on black Nuclepore filters, stained with the Bac-light viability Kit, placed in slides and maintained frozen at −80 °C until examination under epifluorescence microscopy [31, 32]. The samples were examined under blue light, with those from the Atlantic, Indian and S. Pacific Oceans examined on board the research vessel under a Nikon epifluorescence microscope, and samples from the N. Pacific and the last leg on the Atlantic Ocean (Cartagena de Indias, Colombia, to Cartagena, Spain) examined at the IMEDEA laboratory under a Zeiss Axioplan Imaging and Leica epifluorescence microscopes. The fluorescence of the stained cells is well preserved at −80 °C for several months and samples transported to the laboratory were maintained at −80 °C during the transport.

Blank tows to test possible contamination of the Bottle-Net samples with surface plankton consisted of closing the Bottle-Net at 4,000 m and ascending closed to the surface, and were run in the S. Atlantic, the S. Pacific and the N. Atlantic Oceans, although in the last test the blank was taken from 500 m to surface because of operational reasons. The abundance of phytoplankton found in the blank samples was very small averaging 10.3 cells m^{-3} in the samples collected in the blanks, seven times lower than the abundance found in the parallel Bottle-Net tows at the same stations. The abundance in the blanks is equivalent to $5.97 \pm 5.1 \times 10^3$ cells m^{-2} if integrated to the water column sampled, equivalent to the minimum abundance value obtained in the Bottle-Net tows along the study.

Another fraction of the sample collected by the Bottle-Net was fixed with Lugol for further examination at the laboratory. For a comparison with the surface layer, phytoplankton, ciliates and radiolaria were counted in the samples of the Bottle-Net and compared with those in the surface plankton net (200 m) samples (as described below).

Subsamples were sedimented in 50-ml chambers for at least 24 h before enumeration at × 100–200 magnifications using an inverted microscope (AXIOVERT35, Zeiss). For each sample we enumerated microplankton cells at least in half of the sedimentation chamber. Ciliates were identified to the genus level when possible [52]. Radiolaria was quantified as a whole according to Haeckel [53]. Surface abundance was always higher than that for samples collected in the Bottle-Net, except for radiolaria in St 143.

6.4.4 DECAY RATES OF LIVING MICROPHYTOPLANKTON CELLS

The cell mortality rates of phytoplankton living cells in the dark and cold temperature conditions encountered in the bathypelagic ocean were examined with phytoplankton communities collected by vertical tows from the photic layer of the Indian, Pacific and Atlantic Oceans. An aliquot of the photic layer microphytoplankton sample was resuspended in 2 l of 4,000 m water and incubated in the dark at 4 °C for 1–2 months. The community was sampled at the onset of the experiment and at increasing time intervals (1, 2, 3, 4, 7, 14, 20, 30, 40, 60 days), and stained fresh with the vital stain Bac-light Kit, prepared and examined under epifluorescence microscope as described above to quantify the living cells in the community. The half-life (that is, the time for the number of living cells to decline to 50%) and the decay rate of the living cell population were then calculated from the decline in living cells over time.

6.4.5 PHYTOPLANKTON ABUNDANCE FROM THE SURFACE LAYER

Samples for micro- and nanophytoplankton enumeration were taken from the Niskin bottles at three levels, surface (3 m), the depth receiving 20% of photosynthetically active radiation (PAR) incident below the surface and the depth of the deep chlorophyll maximum, as established using the PAR and fluorescence sensors fitted in the CTD. Subsamples of 250 ml were introduced in amber glass bottles and fixed with 0.4% hexamine-buffered formaldehyde. For analysis, 100 ml of water were settled in composite chambers during 48 h. Subsequently, two or more transects of the chamber bottom were examined by means of an inverted microscope54, under 312 X magnification, to count the smaller cells. Additionally, the whole chamber bottom was scanned under 125 X to enumerate larger, less frequent forms. Organisms were classified to species when possible, but many taxa had to be pooled in categories such as 'small flagellates' and 'small dinoflagellates'.

Vertical net hauls were performed between 200 m depth and the surface, with a net of 28-cm mouth diameter and a mesh size of 47 μm. A rough estimate of the water volume filtered by the net was obtained from the expression $\pi r^2 d$, were r is the radius of the net and d the towing distance (200 m). The filtered plankton was washed into the collecting bucket, which was topped up to a suspension volume of 250 ml. As part of the Malaspina Expedition net phytoplankton collection, a subsample of 140 ml was placed into an amber glass bottle and fixed to a final concentration of 4% hexamine-buffered formaldehyde. For qualitative phytoplankton examination, 2 ml of this subsample were placed into a chamber and examined with an inverted microscope.

6.4.6 DNA ANALYSES

Sample collection: Samples for DNA at 4,000 m depth were collected from 31 stations along the Malaspina 2010 Circumnavigation Expedition and filtered on two size fractions (small 0.2–0.8 μm and large 0.8–20 μm) using 142-mm polycarbonate membrane filters. For each sample 120 l of seawater were first filtered through 200 and 20 μm mesh to remove large plankton. Further filtering was carried out by filtering water serially through 142-mm filters of 0.2 and 0.8 μm pore size with a peristaltic pump (Masterflex, EW-77410-10). Filters were then flash-frozen with N_2 and stored at −80 °C until DNA extraction.

Filters were cut into small pieces with sterile razor blades and half of every filter was resuspended in 3 ml of lysis buffer (40 mM EDTA, 50 mM Tris-HCl, 0.75 M sucrose). Lysozyme (1 mg ml^{-1} final concentration) was added and samples were incubated at 37 °C for 45 min with slight movement. Then, sodium dodecyl sulfate (1% final concentration) and proteinase K (0.2 mg ml^{-1} final concentration) were added and samples were incubated at 55 °C for 60 min under slight movement. The lysate was collected and processed with the standard phenol–chloroform extraction procedure. An equal volume of Phenol:$CHCl_3$:IAA (25:24:1, vol:vol:vol) was added to the lysate, carefully mixed and centrifuged 10 min at 3,000 r.p.m., the aqueous phase was recovered and the procedure was repeated. Finally, an equal volume of $CHCl_3$:IAA (24:1, vol:vol) was added to the recovered aqueous phase in order to remove residual phenol. The mixture was centrifuged and the aqueous phase was recovered for further purification. The aqueous phase was then concentrated by centrifugation with a Centricon concentrator (Millipore, Amicon Ultra-4 Centrifugal Filter Unit with Ultracel-100 membrane). Once the aqueous phase was concentrated, this step was repeated three times adding 2 ml of sterile MilliQ water each time in order to purify the DNA. After the third wash, between 100 and 200 µl of purified total genomic DNA product per sample was recovered. Extracted DNA was quantified using a Nanodrop ND-1000 spectrophotometer (NanoDrop Technologies Inc., Wilmington, DE, USA) and the Quant_it dsDNA HS Assay Kit with a Qubit fluorometer (Life Technologies, Paisley, UK).

Sequencing and sequence data processing: All library construction and sequencing were carried out at JGI (www.jgi.doe.gov) following ref. 55. Briefly, the variable region V4 of the 16S rDNA gene was targeted using F515/R806 primers (5′-GTGCCAGCMGCCGCGGTAA-3′/5′-GGACTACHVGGGTWTCTAAT-3′) and sequenced using Illumina MiSeq with 2 × 250 bp reads configuration. All the samples were run in a single lane using unique barcodes for every sample (that is, multiplexing). Before sequencing, a PhiX spike-in shottgun library reads were added to the amplicon pool for a final concentration of ~20–25% of the pair-end reads library.

Reads were first scanned for PhiX reads and contaminants (for example, Illumina adapter sequences) and all disrupted pair-end reads (every read pair for which one read has been lost because of the screening) were discarded. The remaining reads were trimmed to 165 bp and assembled using the FLASH software [56], and primer sequences were removed from the assembled reads. Assembled reads were trimmed from both 5′ and 3′ ends using a 20-bp sliding window (mean quality threshold >30). Trimmed reads with more than 5 or 10 nucleotides below quality 15 were discarded. Clustering was carried out using an in-house algorithm at JGI that consisted of clustering the filtered reads using USEARCH at 99% identity and clusters having abundances lower than 3 reads were discarded. A final cluster at 97% identity of the remaining clusters was carried out and the clusters obtained were considered as operational taxonomic units (OTUs). Finally, the obtained OTUs were checked for chimeric sequences using both the Chimera Slayer algorithm as implemented in the MOTHUR software [57] and the UCHIME reference-based algorithms [58]. OTUs detected as chimeric sequences by any of these methods were removed. Non-chimeric OTUs were taxonomically annotated using both the on-line RDP Naive Bayesan Classifier [59] and the BLAST-based classifier within the QIIME pipeline [60] using the SILVA database (release 108) as reference.

A total of 8,141,076 raw reads were obtained for the whole data set from which 42,254 reads corresponded to contaminant reads (that is, Illumina adapter sequences) and 1,526,330 to PhiX reads. That left a total of 6,572,492 non-contaminant non-PhiX reads from which 3,100,410 reads could be paired and assembled. From these, 2,584,926 reads passed the filtering process described before.

We considered all the sequences that were annotated by the two algorithms as belonging to cyanobacteria, and those annotated as belonging to algal chloroplasts, and expressed them as a percentage of the total recovered sequences. In the few cases in which the SILVA taxonomy did not agree with the RDP taxonomy at these large level, we followed the SILVA annotation.

6.4.7 FLOW CYTOMETRY

The abundance of pigment-containing prokaryotes in deep ocean samples were obtained using flow cytometry analyses of DNA-stained samples. Samples of 2 ml of seawater sampled from 4,000 m depth were fixed with 1% paraformaldehyde and 0.05% glutaraldehyde (immediately after collection, and after 15 min at room temperature in the dark they were deep-frozen in liquid nitrogen. A few days after sampling they were unfrozen, stained with SybrGreen I (Molecular Probes, Invitrogen) at a 1/10,000 dilution and run in a BectonDickinson FACSCalibur flow cytometer equipped with a blue 488-nm 15-mW Argon-ion laser as explained elsewhere [61]. At least 100,000 events were recorded and photosynthetic prokaryotes could be distinguished in a DNA-derived green fluorescence against a chlorophyll-derived red fluorescence scatter plot. Calibration of the machine for absolute counts was made daily by measuring the exact volume being analysed. Fluorescent beads (1 μm, Fluoresbrite carboxylate microspheres, Polysciences Inc., Warrington, PA) were added at a known density as internal standards.

REFERENCES

1. Sabine, C. L. et al. The oceanic sink for anthropogenic CO2. Science 305, 367–371 (2004).
2. England, M. H. The age of water and ventilation timescales in a global ocean model. J. Phys. Oceanogr. 25, 2756–2777 (1995).
3. Volk, T. & Hoffert, M. I. In: The Carbon Cycle and Atmospheric CO2: Natural Variations Archean to Present eds Sundquist E. T., Broecker W. S. American Geophysical Union (1985).
4. De La Rocha, C. L. In: Treatise on Geochemistry vol. 6, eds Heinrich D., Holland H. D., Turekian K. K. Pergamon Press (2006).
5. Bienfang, P. K. Phytoplankton sinking rates in oligotrophic waters off Hawaii, USA. Mar. Boil. 61, 69–77 (1980).
6. Sarthou, G., Timmermans, K. R., Blain, S. & Tréguer, P. Growth physiology and fate of diatoms in the ocean: a review. J. Sea Res. 53, 25–42 (2005).
7. Smayda, T. J. The suspension and sinking of phytoplankton in the sea. Oceanogr. Mar. Biol. 8, 353–414 (1970).
8. Martin, J. H., Knauer, G. A., Karl, D. M. & Broenkow, W. W. VERTEX: carbon cycling in the Northeast Pacific. Deep Sea Res 34, 267–285 (1987).
9. CASSmetacek, V. Role of sinking in diatom life-history cycles: ecological, evolutionary and geological significance. Mar. Biol. 84, 239–251 (1985).
10. Alldredge, A. L. & Gotschalk, C. C. Direct observations of the mass flocculation of diatom blooms: characteristics, settling velocities and formation of diatom aggregates. Deep Sea Res 36, 159–171 (1989).
11. CASTurner, J. T. Zooplankton fecal pellets, marine snow and sinking phytoplankton blooms. Aq. Microb. Ecol. 27, 57–102 (2002).
12. Ploug, H., Iversen, M. H. & Fischer, G. Ballast, sinking velocity, and apparent diffusivity within marine snow and zooplankton fecal pellets: Implications for substrate turnover by attached bacteria. Limnol. Oceanogr. 53, 1878–1886 (2008).
13. Bacon, M. P. & Anderson, R. F. Distribution of thorium isotopes between dissolved and particulate forms in the deep sea. J. Geophys. Res. Oceans 87, 2045–2056 (1982).
14. Buesseler, K. O., Cochran, J. K., Bacon, M. P. & Livingston, H. D. Carbon and nitrogen export during the JGOFS North Atlantic Bloom Experiment estimated from 234Th:238U disequilibria. Deep Sea Res 39, 1103–1114 (1992).
15. Smetacek, V. et al. Deep carbon export from a Southern Ocean iron-fertilized diatom bloom. Nature 487, 313–319 (2012).
16. Karl, D. M., Church, M. J., Dore, J. E., Letelier, R. M. & Mahaffey, C. Predictable and efficient carbon sequestration in the North Pacific Ocean supported by symbiotic nitrogen fixation. Proc. Natl Acad. Sci. USA 109, 1842–1849 (2012).

17. Arístegui, J. et al. Dissolved organic carbon support of respiration in the dark ocean. Science 298, 1967 (2002).
18. Arístegui, J., Gasol, J. M., Duarte, C. M. & Herndl, G. J. Microbial oceanography of the dark ocean's pelagic realm. Limnol. Oceanogr. 54, 1501–1529 (2009).
19. CASSimon, M., Grossart, H. P., Schweitzer, B. & Ploug, H. Microbial ecology of organic aggregates in aquatic ecosystems. Aquat. Microb. Ecol. 28, 175–211 (2002).
20. Jansen, S. & Bathmann, U. Algae viability within copepod faecal pellets: evidence from microscopic examinations. Mar. Ecol. Progr. Ser. 337, 145–153 (2007).
21. Ploug, H., Iversen, M., Koski, M. & Buitenhuis, E. T. Production, oxygen respiration rates, and sinking velocity of copepod fecal pellets: direct measurements of ballasting by opal and calcite. Limnol. Oceanogr. 53, 469–476 (2008).
22. Smayda, T. J. Normal and accelerated sinking of phytoplankton in the sea. Mar. Geol 11, 105–122 (1971).
23. Kimball, J., Corcoran, E. F. & Wood, F. E. Chlorophyll-containing microorganisms in the aphotic zone of the oceans. Bull. Mar. Sci. 13, 574–577 (1963).
24. Wiebe, P. H., Remsen, C. C. & Vaccaro, R. F. Halosphaera viridis in the Mediterranean sea: size range, vertical distribution, and potential energy source for deep-sea benthos. Deep Sea Res 8, 657–667 (1974).
25. Stukel, M. R. et al. The role of *Synechococcus* in vertical flux in the Costa Rica upwelling dome. Progr. Oceanogr. 112–113, 49–59 (2013).
26. Irigoien, X. et al. Large mesopelagic fish biomass and trophic efficiency in the open ocean. Nat. Commun. 5, 3271 (2014).
27. CASCózar, A. et al. Plastic debris in the open ocean. Proc. Natl Acad. Sci. USA 111, 10239–10244 (2014).
28. CASPernice, M. C. et al. Global abundance of planktonic heterotrophic protists in the deep ocean. ISME J. 9, 782–792 (2014).
29. CASdel Giorgio, P. A. & Duarte, C. M. Respiration in the open ocean. Nature 420, 379–384 (2002).
30. McQuoid, M. R. & Hobson, M. L. Diatom resting stages (Review). J. Phycol. 32, 889–902 (1996).
31. Agustí, S., Alou, E., Hoyer, M. V., Frazer, T. K. & Canfield, D. E. Cell death in lake phytoplankton communities. Freshwater Biol. 51, 1496–1506 (2006).
32. Llabrés, M. & Agustí, S. Extending the cell digestion assay to quantify dead phytoplankton cells in cold and polar Waters. Limnol. Oceanogr. Method 6, 659–666 (2008).
33. Peters, E. J. Prolonged darkness and diatom mortality: II. Marine temperate species. Exp. Mar. Biol. Ecol. 207, 43–58 (1996).
34. Alonso-Laita, P. & Agustí, S. Contrasting patterns of phytoplankton viability in the subtropical NE Atlantic Ocean. Aquat. Microb. Ecol. 43, 67–78 (2006).
35. Iversen, M. H. & Ploug, H. Ballast minerals and the sinking carbon flux in the ocean: carbon-specific respiration rates and sinking velocities of marine snow aggregates. Biogeosciences 7, 2613–2624 (2010).
36. Yoon, W., Kim, S. & Han, K. Morphology and sinking velocities of fecal pellets of copepod, molluscan, euphausiid, and salp taxa in the northeastern tropical Atlantic. Mar. Biol. 139, 923–928 (2001).

37. Komar, P. D., Morse, A. P., Small, L. F. & Fowler, S. W. An analysis of sinking rates of natural copepod and euphausiid fecal pellets. Limnol. Oceanogr. 26, 172–180 (1981).
38. Honjo, S., Doherty, K. W., Agrawal, Y. C. & Asper, V. L. Direct optical assessment of large amorphous aggregates (marine snow) in the deep ocean. Deep Sea Res. 31, 67–76 (1984).
39. Suess, E. Particulate organic carbon flux in the oceans-surface productivity and oxygen utilization. Nature 288, 260–263 (1980).
40. CASBuesseler, K. O. et al. Revisiting carbon flux through the ocean's twilight zone. Science 316, 567–570 (2007).
41. Francois, R., Honjo, S., Krishfield, R. & Manganini, S. Factors controlling the flux of organic carbon to the bathypelagic zone of the ocean. Global Biogeochem. Cycles 16, 34–41 (2002).
42. McDonell, A. M. P., Boyd, P. W. & Buesseler, K. O. Effects of sinking velocities andmicrobial respiration rates on the attenuation of particulate carbon fluxes through the mesopelagic zone. Global Biogeochem. Cycles 29, 175—193 (2015).
43. Marsay, C. M. et al. Attenuation of sinking particulate organic carbon flux through the mesopelagic ocean. Proc. Natl Acad. Sci. USA 112, 1089—1094 (2015).
44. CASFischer, G. & Karakas, G. Sinking rates and ballast composition of particles in the Atlantic Ocean: implications for the organic carbon fluxes to the deep ocean. Biogeosciences 6, 85–102 (2009).
45. Eloe, E. A. et al. Compositional differences in particle-associated and free-living microbial assemblages from an extreme deep-ocean environment. Environ. Microbial. Rep. 3, 449–458 (2010).
46. Berelson, W. M. Particle settling rates increase with depth in the ocean. Deep Sea Res. II 49, 237–251 (2001).
47. Ruff, S. E. et al. Indications for algae-degrading benthic microbial communities in deep-sea sediments along the Antarctic Polar Front. Deep Sea Res. II 108, 6–16 (2014).
48. Herndl, G. J. & Reinthaler, T. Microbial control of the dark end of the biological pump. Nat. Geosci. 6, 718–724 (2013).
49. Billet, D. S. M., Lampitt, R. S., Rice, A. L. & Mantoura, R. F. C. Seasonal sedimentation of phytoplankton to the deep-sea benthos. Nature 302, 520–522 (1983).
50. CASBeaulieu, S. E. Accumulation and fate of phytodetritus on the sea floor. Oceanogr. Mar. Biol. Annu. Rev. 40, 171–232 (2002).
51. Lee, D. Y. & Rhee, G. Y. Kinetics of cell death in the cyanobacterium Anabaena flos-aquae and the production of dissolved organic carbon. J. Phycol. 33, 991–998 (1997).
52. Lynn, D. H. & Small, E. B. in: Illustrated Guide to the Protozoa eds Lee J. J., Bradbury P. C., Leedale G. F. Lawrence (2000).
53. Haeckel, E. Art Forms from the Ocean: The Radiolarian Atlas of 1862 Prestel (2005).
54. Utermohl, H. Zur vervollkommung der quantitativen phytoplankton-methodik. Mitt. d. Internat 9, 1–39 (1958).
55. Caporaso, J. G. et al. Global patterns of 16S rRNA diversity at a depth of millions of sequences per sample. Proc. Natl Acad. Sci. USA 108, 4516–4522 (2011).

56. Magoč, T. & Salzberg, S. L. FLASH: fast length adjustment of short reads to improve genome assemblies. Bioinformatics 27, 2957–2963 (2011).
57. Schloss, P. D. et al. Introducing mothur: open-source, platform-independent, community-supported software for describing and comparing microbial communities. Appl. Environ. Microbiol. 75, 7537–7541 (2009).
58. Edgar, R. C., Haas, B. J., Clemente, J. C., Quince, C. & Knight, R. UCHIME improves sensitivity and speed of chimera detection. Bioinformatics 27, 2194–2200 (2011).
59. Wang, Q., Garrity, G. M., Tiedje, J. M. & Cole, J. R. Naive Bayesian classifier for rapid assignment of rRNA sequences into the new bacterial taxonomy. Appl. Environ. Microbiol. 73, 5261–5267 (2007).
60. Caporaso, J. G. et al. QIIME allows analysis of high-throughput community sequencing data. Nat. Methods 7, 335–336 (2010).
61. Gasol, J. M. & del Giorgio, P. A. Using flow cytometry for counting natural planktonic bacteria and understanding the structure of planktonic bacterial communities. Sci. Mar. 64, 197–224 (2000).

There are several supplemental files that are not available in this version of the article. To view this additional information, please use the citation on the first page of this chapter.

CHAPTER 7

Carbon Export Efficiency and Phytoplankton Community Composition in the Atlantic Sector of the Arctic Ocean

FRÉDÉRIC A. C. LE MOIGNE, ALEX J. POULTON, STEPHANIE A. HENSON, CHRIS J. DANIELS, GLAUCIA M. FRAGOSO, ELAINE MITCHELL, SOPHIE RICHIER, BENJAMIN C. RUSSELL, HELEN E. K. SMITH, GERAINT A. TARLING, JEREMY R. YOUNG, AND MIKE ZUBKOV

7.1 INTRODUCTION

Climate change is impacting the Arctic Ocean, with the most obvious manifestation of these changes being the reduction in summer sea-ice cover [Boé et al., 2009]. Modeling and observational studies imply that reduced sea-ice in the future will strengthen Arctic primary production (PP) [Arrigo et al., 2008; Pabi et al., 2008]. The biological carbon pump (BCP) is an important component of the global carbon (C) cycle, mainly driven by the sinking of organic material from the sunlit upper layer of the ocean [Boyd and Trull, 2007]. The fraction of PP that is exported below the euphotic zone (Ez) or below the surface layer (export/primary production, or export efficiency, ThE(Ez)-ratio [Buesseler, 1998; Buesseler and Boyd, 2009]) is a key determinant in how efficiently the BCP sequesters C to depth [Buesseler and Boyd, 2009]. Furthermore, the fraction of this material, which successfully transits deeper into the water column (upper mesopelagic zone) where the majority of remineralization of sinking particulate organic material occurs [Buesseler and Boyd, 2009; Giering et al., 2014], is indicative of BCP efficiency [Buesseler and Boyd, 2009]. The specific ecosystem related processes that drive the considerable variability observed in ThE(Ez)-ratios and further down into the upper mesopelagic remain largely unknown [Buesseler and Boyd, 2009; Giering et al., 2014; Henson et al., 2011] for much of the polar oceans.

Although a large body of literature reporting estimates of POC export fluxes using various techniques is available in the Arctic Ocean [Cai et al., 2010; Chen et al., 2003; Coppola et al., 2002; Lalande et al., 2011, 2007, 2008; Moran et al., 1997; Moran and Smith, 2000; Reigstad and Wassmann, 2007; Wassmann et al., 1990], little is known about the variability of the ThE(Ez)-ratio [Buesseler, 1998; Buesseler and Boyd, 2009]. Cai et al. [2010] and Chen et al. [2003] both report ratios of POC export flux to PP in the central Arctic Ocean, but this ratio has yet to be estimated in the deep ocean rather than over shelf waters, or over gradients of ice to ice-free conditions. Export efficiency from such regions and associated blooms has been estimated before, but with limited information on the plankton community composition [Cai et al., 2010; Chen et al., 2003] and/or with an export efficiency calculation methodology different from the ThE(Ez)-ratio approach [Reigstad et al., 2008; Smith et al., 1991], hence preventing direct comparisons. In cold waters, the coupling between surface carbon export and PP is known to be highly variable [Henson et al., 2011] and the specific processes that drive variability in ThE(Ez)-ratio in the Arctic Ocean remain elusive [Buesseler and Boyd, 2009]. One possible explanation for the variability in ThE(Ez)-ratio in cold high latitudes is the strong seasonality in phytoplankton bloom evolution and seasonal succession in phytoplankton functional types observed at high latitudes [Boyd and Newton, 1995, 1999].

Blooms of various phytoplankton groups have been reported in the Atlantic sector of the Arctic Ocean and Fram Strait, including diatoms [Gradinger and Baumann, 1991], coccolithophores [Smyth et al., 2004], and the colonial haptophyte *Phaeocystis* spp. [Smith et al., 1991; Wassmann, 1994]. However, it is unknown whether these phytoplankton groups result in similar ThE(Ez)-ratios when they are present and/or dominating the community composition. There are, therefore, grounds to hypothesize that some of the variability observed in ThE(Ez)-ratios may be explained by phytoplankton composition. Furthermore, it is not clear whether sinking material derived from different phytoplankton groups experience a similar amount of heterotrophic respiration once it has entered the upper mesopelagic zone [Buesseler and Boyd, 2009]. This aspect is critical if one wants to predict the future changes in the open Arctic BCP and, further, the influence of the Arctic Ocean on climate regulation.

To date, we know that future changes in temperature, CO_2, light regime, and macro(micro)-nutrient inputs may all induce important changes in both primary production and phytoplankton community composition in the Arctic [Bopp et al., 2005; Coello-Camba et al., 2014; Perrette et al., 2011; Popova et al., 2010, 2012; Vancoppenolle et al., 2013]. Unless clear links are established between phytoplankton community composition and carbon export metrics, such as the ThE(Ez)-ratio [Buesseler and Boyd, 2009], it will be difficult to predict how these changes will impact the efficiency of the Arctic BCP.

Here we surveyed variability in the ThE-ratio (as we defined and define the depth of integration for this study to be 100 m) [Buesseler, 1998] in the Atlantic sector of the Arctic Ocean and the Fram Strait where POC export flux estimates are rare [Le Moigne et al., 2013a], using the ^{234}Th technique (to estimate export of POC) and onboard ^{14}C incubations (to determine PP). These parameters were related to the major phytoplankton groups present in the water column during the survey period. Furthermore, oxygen saturation, bacterial production, and zooplankton oxygen demand are used to qualitatively assess to what extent this sinking material may be remineralized in the upper mesopelagic (<300 m). This represents the first attempt to estimate ThE-ratio and upper mesopelagic remineralization in the Atlantic sector of the Arctic Ocean.

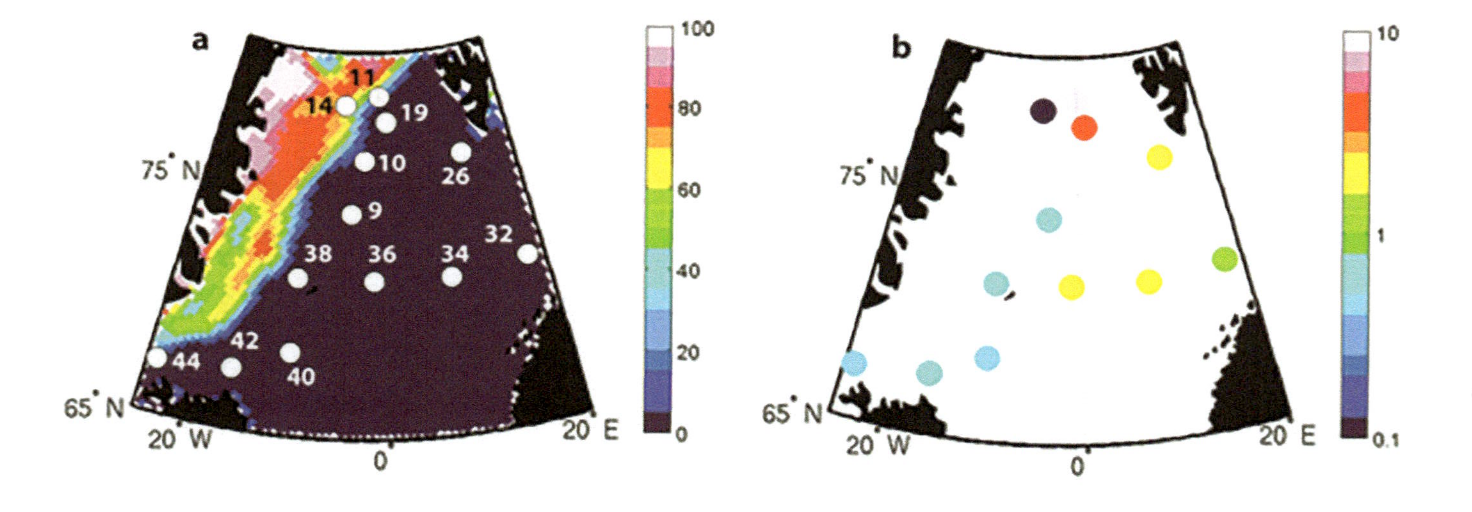

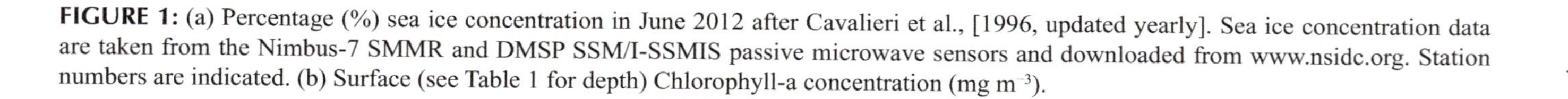

FIGURE 1: (a) Percentage (%) sea ice concentration in June 2012 after Cavalieri et al., [1996, updated yearly]. Sea ice concentration data are taken from the Nimbus-7 SMMR and DMSP SSM/I-SSMIS passive microwave sensors and downloaded from www.nsidc.org. Station numbers are indicated. (b) Surface (see Table 1 for depth) Chlorophyll-a concentration (mg m^{-3}).

TABLE 1: Biomass and Dominant Phytoplankton Species (or Genus) at Stations Sampled in the Mixed Layer During This Study[a]

Station	Date	Lat	Lon	Sampling Depth (m)	ML Depth (m)	Ez Depth (m)	Depth of the SCM (if Present) (m)	Chlorophyll-a (mg m^{-3})	Estimated C Phytoplankton (mmol C m^{-3})	Phytoplankton Biomass (mmol C m^{-3})			Dominant Species
										Diatom	*Phaeocystis*	Coccolithophores	
9	Jun 12 2012	74.1	−4.1	15	15	60	28	0.7	2.9	0.2		<0.1	*Thalassiosira* spp.
10	Jun 13 2012	76.1	−2.3	20	28	67	42	0.9	3.8	0.3		<0.1	Mixed diatoms
11	Jun 14 2012	78.4	0	10	20	21	10	8.4	35.0	0.2	31.3		*Phaeocystis pouchetii**
14	Jun 16 2012	77.5	1.1	10	70	63	23 (61)	0.1	0.4	<0.1	<0.1		*Thalassiosira* spp.
19	Jun 19 2012	71.7	17.9	16	15	28	14	3.5	14.6	0.3	4.6	<0.1	*Phaeocystis pouchetii*
26	Jun 22 2012	71.4	8.3	20	47	39		1.8	7.5	<0.1		0.1	*E. huxleyi*
32	Jun 25 2012	68.4	−10.3	13	15	62	12	1.1	4.6	0.3		<0.1	*Thalassiosira* spp.

TABLE 1: *Cont.*

Station	Date	Lat	Lon	Sampling Depth (m)	ML Depth (m)	Ez Depth (m)	Depth of the SCM (if Present) (m)	Chlorophyll-a (mg m^{-3})	Estimated C Phytoplankton (mmol C m^{-3})	Phytoplankton Biomass (mmol C m^{-3})			Dominant Species
										Diatom	*Phaeocystis*	Coccolithophores	
34	Jun 26 2012	67.5	−16.2	15	18	32		2.1	8.8	0.2		0.3	*E. huxleyi*
36	Jun 27 2012	71.4	−1.2	20	35	38	35	1.9	7.9	21.8	0.1	0.2	*Ephemera* sp.*
38	Jun 28 2012	71.4	−10.3	25	10	55	32	0.5	2.1	0.3		0.1	Mixed diatoms
40	Jun 29 2012	68.4	−10.3	20	20	43	30	3.6	15.0	50.9		0.4	Ephemera sp.*
42	Jun 30 2012	67.5	−16.3	20	15	48	33	0.7	2.9	0.1	<0.1	<0.1	Thalassiosira spp.

a *Note station 44 was not sampled for phytoplankton community structure. E. huxleyi stands for Emiliania huxleyi. Asterisks are monospecific blooms. ML stands for mixed layer, Ez for euphotic zone, and SCM for subsurface chlorophyll maximum.*

7.2 METHODS

7.2.1 SAMPLING AND ANALYSIS OF ANCILLARY DATA

The cruise took place onboard the R.R.S. James Clark Ross (British Antarctic Survey) in June 2012 as part of the UK's Ocean Acidification programme. Sampling took place from 12th June to the 1st July 2012 in the Greenland, Norwegian and Barents Seas (Figure 1; station positions are given in Table 1). Water samples were collected with Niskin bottles via deployment of a SeaBird CTD system. Dissolved oxygen (DO) profiles were taken from the CTD optode and calibrated against discrete O_2 measurements performed using a semiautomated whole bottle Winkler titration unit with spectrophotometric end point detection manufactured by SIS (http://www.sis-germany.com) following Dickson [1994] at all discrete depths. Samples (0.1–0.5 L) for chlorophyll-a (Chl-a) analysis were collected in the middle of the mixed layer (10–25 m, see Table 1) and filtered onto GF/F filters and then extracted into 90% acetone for 24 h in the dark before analysis with a fluorometer (Trilogy; Turner Designs). Samples for analysis of particulate organic carbon (POC) (0.5–1 L) were collected and analyzed following protocols described in Le Moigne et al. [2013b]. Raw fluorescence from the CTD fluorometer was cross-calibrated with discrete measurements of Chl-a.

7.2.2 PHYTOPLANKTON COMMUNITY COMPOSITION, ABUNDANCE, AND BIOMASS

Water samples (100–250 mL) were collected from the middle of the mixed layer (10–25 m, see Table 1) from each station and preserved with 2% final concentration acidic Lugol's solution for later analysis of microplankton community structure. Subsamples (10–50 mL) were settled overnight in 10–50 mL HydroBios settling chambers following [Uttermohl, 1958] and microplankton were enumerated using a SP-95-I inverted light microscope (Brunel Ltd). Biomass of each phytoplankton species was calculated following Poulton et al. [2010, 2007] and is presented in Table 1.

TABLE 2: Discrete Primary Production (mmol C m^{-3} d^{-1}) and Euphotic Zone Integrated Primary Production (mmol C m^{-2} d^{-1})[a]

Station	Sampling Depth (m)	Discrete Primary Production (mmol C m^{-3} d^{-1})	Integrated Primary Production (mmol C m^{-2} d^{-1})
9	15	0.9±0.1	11.3±0.8
10	20	2.2±0.1	32.4±1.2
11[b]	10	10.5±2.1	58.9±1.9
14[c]	10	0.8±0.5	20.6±12.4
19	16	5.2±0.5	46.7±5.0
26	20	1.8±0.4	26.7±5.3
32	13	1.5±0.2	23.3±2.9
34	15	2.3±1.2	42.5±21.6
36	20	3.9±0.2	63.5±3.9
38	25	1.1±0.1	15.1±1.3
40	20	1.6±0.3	47.5±8.1
42	20	1.3±0.1	53.1±6.0

[a] *Note station 44 was not sampled for primary production.*
[b] *Station 11 was sampled at the ice edge (see section 3.1).*
[c] *Station 14 was sampled in the ice (see section 3.1) and integrated PP may be overestimated in this case due to the deep Ez (see Table 1).*

7.2.3 PRIMARY PRODUCTION

Daily (dawn-to-dawn, 24 h) rates of primary production were determined following the methodology of Balch et al. [2000]. Four replicate water samples (70 mL, 3 lights, 1 formalin-killed) were collected at a depth approximating the middle of the mixed layer (10–25 m, Table 2), spiked with 30–40 μCi of ^{14}C-labeled sodium bicarbonate and incubated on deck at 40% of incident irradiance following the incubator setup described in Poulton et al. [2010]. Incubations were terminated by filtration through 25-mm 0.4-μm Nuclepore polycarbonate filters, with extensive rinsing with fresh filtered seawater to remove any labeled ^{14}C-DIC, and filters were placed in 20 mL glass vials, with organic (PP) and inorganic (CP) carbon fixation determined using the Micro-Diffusion technique [Poulton et al., 2014]. Ultima-Gold liquid scintillation cocktail was added to the vials and activity on the filters was then determined on a Tri-Carb 2100 low-level liquid scintillation counter, with counts converted to uptake rates using standard methodology. The ^{14}C counts from the formalin-killed samples were subtracted from the light bottles.

Single depth measurements of PP were scaled to Ez integrals by accounting for both variability in the depth distribution of biomass (Chl-a) and the vertical attenuation of irradiance. Subsurface chlorophyll maxima (SCM) are known to represent important sources of PP in the Arctic Ocean [Hill et al., 2013] and irradiance also exerts a strong control on the vertical distribution of PP. Ez depths were identified as the depth of penetration of 1% of surface irradiance based on a biospherical 2π Photosynthetically Available Radiation (PAR) sensor (Chelsea Technology Group Ltd) mounted on the CTD. The vertical diffuse attenuation coefficient of PAR (K_d) was then calculated based on the Ez depth. To integrate PP, we calculated the chlorophyll-specific rate of primary production (P^B; mg C (mg chl)$^{-1}$ d^{-1}) and used the calibrated fluorometer Chl-a profiles to estimate PP through the water column to the base of the Ez. Acknowledging that light availability will exert a strong vertical control on PP, we then used the K_d value to scale PP through the Ez and integrated these PP rates.

This method of estimating integrated PP assumes that phytoplankton photophysiology (e.g., P^B) is invariant with depth. If the phytoplankton community did show photophysiological variability with depth, for example, in an SCM, then our integration method would underestimate PP. However, as we did not make measurements of such physiological variability we are more confident in underestimating PP, although the potential effect on ThE-ratios should be noted (i.e., overestimates of ThE-ratio). A clear case where such physiological differences may be leading to an overestimate of Ez PP is station 14, where a deep Ez (70 m) is compensated by several SCM (Figure 2), leading to a high Ez PP relative to low rates of discrete PP at the sampling depth (Table 3). Hence, extrapolation of ThE-ratios should be treated with caution for this station.

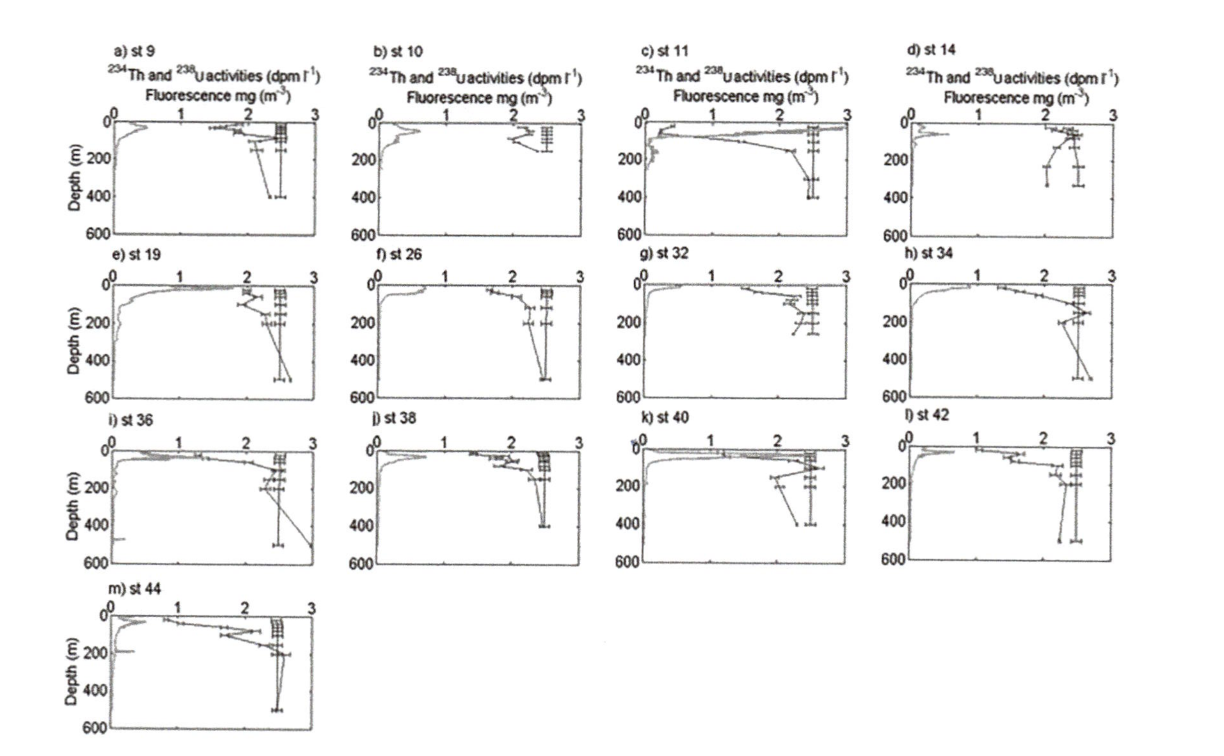

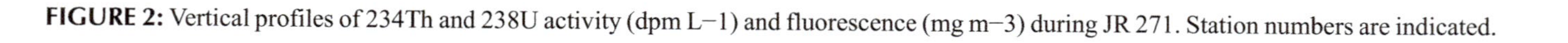

FIGURE 2: Vertical profiles of 234Th and 238U activity (dpm L−1) and fluorescence (mg m−3) during JR 271. Station numbers are indicated.

TABLE 3: Thorium Flux (dpm m^{-2} d^{-1}) Integrated at 100 m (See Section 2), C:Th Ratio (μmol dpm^{-1}), C:Th Ratio Sampling Depth (m), POC Export (mmol m^{-2} d^{-1}) From the ^{234}Th Technique During JR271[a]

Station	Thorium Export (dpm m^{-2} d^{-1})	C:Th Ratio (μmol dpm^{-1})	C:Th Ratio Sampling Depth (m)	POC Export (234Th) (mmol m^{-2} d^{-1})	ThE-Ratio
9	1342±382	4.6±0.5	50	6.2±1.9	0.5±0.2
10	862±204	4.0±0.4	40	3.4±0.9	0.10±0.03
11	4324±164	18.1±1.9	30	78.3±8.7	1.3±0.3
14	143±212	12.8±1.8	80	1.8±3.8	0.09±0.2
19	948±208	9.4±0.9	50	8.9±2.1	0.19±0.05
26	1267±215	11.5±1.5	60	14.6±3.1	0.5±0.2
32	1125±326	5.0±0.5	30	5.7±1.7	0.2±0.1
34	1374±199	4.2±0.5	30	5.8±1.1	0.1±0.1
36	1367±205	5.0±0.5	50	6.8±1.2	0.11±0.02
38	1576±374	4.1±0.5	60	6.4±1.7	0.4±0.1
40	1187±206	10.8±1.3	60	12.8±2.7	0.3±0.1
42	2023±302	8.6±1.0	60	17.4±3.3	0.3±0.1
44	2105±303	9.0±0.9	50	18.9±3.4	

[a] *ThE-ratio is also presented.*

7.2.4 BACTERIOPLANKTON AND PROTIST CONCENTRATIONS AND MICROBIAL METABOLIC ACTIVITY

Concentrations of bacterioplankton and aplastidic protists were determined using flow cytometry. Seawater subsamples of 1.6 mL were preserved with paraformaldehyde (PFA, 1% final concentration) in 2 mL polypropylene screw cap vials. The vials were then placed in a fridge and left for no longer than 12 h. Samples were stained with SYBR Green I nucleic acid dye and analyzed using a FACSort flow cytometer (BD, Oxford) with internal bead standards [Zubkov and Burkill, 2006; Zubkov and Tarran, 2008].

Bacterioplankton production was estimated as the microbial uptake rate of Leucine using ^{14}C-Leucine (Hartmann Analytic, Germany), added at a concentration of 20 nM, in samples from different depths from each morning CTD. Subsamples of 1.6 mL from each sample were dispensed into 2 mL polypropylene screw cap vials containing ^{14}C-Leucine [Zubkov et al., 2000]. Samples were fixed at each time point (20, 40, 60, and 80 min) by the addition of 80 μL 20% PFA (1% v/w final concentration). Fixed samples were filtered onto 0.2 μm polycarbonate membrane filters soaked in nonlabeled Leucine solution to reduce adsorption of radiotracer. Filtered samples were washed twice with 4 mL deionized water. Radioactivity of samples was measured as counts per minute (CPM) by liquid scintillation counting (Tri-Carb 3100, Perkin Elmer, UK). Microbial uptake of Leucine was computed using specific activity of the Leucine radiotracer.

7.2.5 CARBON EXPORT

Carbon export fluxes were calculated using the ^{234}Th "small-volume" technique with ICP-MS assessment of recoveries for ^{234}Th extraction [Pike et al., 2005]. Water samples were collected using GO-Flo bottles at 6–8 different depths within the upper 500 m with refined resolution in the top 150 m (Figure 2). All samples were checked for recovery of ^{234}Th extraction ($90.3 \pm 7.6\%$).Vertical profiles of ^{234}Th activity were converted to estimates of downward ^{234}Th flux using a one-dimensional steady state model [Buesseler et al., 1992] integrated to the depth of 100 m [Cai et al., 2010; Chen et al., 2003] as Ez integration [Buesseler and Boyd, 2009] was not suitable here. This is because at several stations the Ez was deeper than the depth of the mixed layer (Table 1 and Figure 2). We therefore report our export efficiency (see section 1) as ThE-ratios [Buesseler, 1998] and not Ez-ratios [Buesseler and Boyd, 2009]. We ignored the effect of horizontal and vertical advection and diffusion as these have been deemed to be insignificant in previous Arctic studies [Cai et al., 2010]. We did not use the nonsteady state approach as it only improves the flux estimates in cases where the sampling was conducted in a Lagrangian framework [Resplandy et al., 2012] which was not possible during this cruise. A triplicate of deep water (2000 m) samples was collected at station 9 in order to verify the calibration of the method. Averaged ^{234}Th:^{238}U ratio was 0.98 ± 0.02 (n=3). The ^{238}U activity was determined by using the ^{238}U-salinity relationship described in Chen et al. [1986].

These fluxes were then converted to estimates of downward organic carbon flux using the C:Th ratio in large (>53 μm) particles collected using an in situ Stand Alone Pumping System (SAPS) deployed for 1.5 h at a single depth beneath the mixed layer (depths given in Table 3). Particles were then rinsed off the screen using particulate thorium-free seawater (obtained from seawater filtered through a 0.4 μm polycarbonate filter), and the particle suspension evenly split into subsamples using a Folsom splitter. Splits were then analyzed for ^{234}Th and POC as described in Le Moigne et al. [2013b], with no replicates analyzed. Particulate ^{234}Th and C samples were analyzed from two distinct splits. POC flux at from the base of the mixed layer [Smith, 2014] (Table 1 and supporting information Table S1) was measured using the Marine Snow Catcher following methods described in Cavan et al. [2015] and Riley et al. [2012].

7.2.6 ZOOPLANKTON RESPIRATION RATE

Zooplankton community respiration rates were estimated from determination of the composition and abundance of the biomass-dominant species at each locality, subsequently converted to respiration rates as a function of species-weight and in situ sea surface temperature from the ship's underway system. Zooplankton were sampled between 0 and 200 m with a motion compensated Bongo net following sampling procedures described in Ward et al. [2012]. Specific respiration rate (μl O_2 ind^{-1} h^{-1}) of each biomass-dominant species at each station was determined from relationships detailed in Ikeda et al. [2001] through applying values for in situ temperature and species dry weight (supporting information Table S2) as described in equation (1):

$$\text{Specific respiration rate} = e^{(-0.399 + (0.801 \text{ x } (\ln DW)))} + (0.069 \text{ x SST}) \quad (1)$$

where DW is the dry weight of each individual species and stage (in mg) and SST is the sea surface temperature from the ship's underway system (in °C). Species respiration rate (μl O_2 ind^{-1} h^{-1}) was multiplied by species abundance (individuals m^{-2}) and then summed across all biomass-dominant species to determine community respiration rate ($\mu l\ O_2\ m^{-2}\ h^{-1}$) at each station as in equation (2):

$$\text{community respiration rate} = \Sigma \text{Specific respiration rate x Abundance}_{int} \quad (2)$$

where $Abundance_{int}$ is the integrated abundance of each species and stage over the top 200 m of the water column. We were not able to determine levels of error around our estimates of zooplankton community respiration since they are based on an extrapolation of the function developed by Ikeda et al. [2001] on temperature and weight specific zooplankton community respiration.

7.3 RESULTS

7.3.1 GENERAL HYDROGRAPHY

We sampled a large mix of oceanic habitats, including open water, ice-edge, and ice-covered waters (Figure 1a). In the Barents Sea (Stn. 32) and Norwegian Sea (Stns. 26, 34, and 36), surface waters were warmer than in the Greenland Sea (Stns. 9, 10, 38, 40, 42, and 44) reflecting the influence of Atlantic waters flowing northward. The Greenland Sea was under the influence of southward flowing polar waters. Three stations across the Fram Strait in close proximity to one another typified the local gradient in sea ice conditions, from under ice (more than 90% ice concentration, Stn. 14), ice-edge (Stn. 11) to ice-free waters (Stn. 19) (Figure 1a and supporting information Figure S1), with the ice station displaying a clear input of fresh water at the surface (supporting information Figures S2 and S3).

7.3.2 CHLOROPHYLL, PRIMARY PRODUCTION, AND PHYTOPLANKTON COMPOSITION

Chl-a concentration in the middle of the mixed layer (see sampling depth in Table 1) ranged from 0.1 to about 8.5 mg m^{-3} (Figure 1b). The full CTD calibrated fluorescence Chl-a is presented in Figure 2. The ice station (Stn. 14) had the lowest concentration (0.1 mg m^{-3}) while the two nearby stations (Stns. 11 and 19) had the highest (8.5 and 3.5 mg m^{-3}, respectively). Elsewhere, there was a clear divide between the Greenland Sea (average Chl-a=0.54±0.1 mg m^{-3}) and the Norwegian/Barents Seas (average Chl-a=1.7±0.4 mg m^{-3}). These trends were also true for mixed layer Chl-a concentrations (Table 1).

Clear subsurface chlorophyll maxima (SCM) were observed in the Greenland Sea (Figure 2). However, in the Barents/Norwegians Seas, vertical fluorescence profiles show that Chl-a concentrations were high within the mixed layer and decrease quasi-exponentially below it. At most stations, the Ez depth was deeper than the mixed layer depth (MLD) (Table 1), apart from at station 26. The presence of an SCM was clear at stations 9, 10, 14, 36, 38, 40, and 42 (Figure 2 and Table 1). SCM were generally located between the MLD and the Ez depth, with the exception of stations 14 and 36 where SCM, MLD, and Ez depths were all very close to each other (Table 1 and Figure 2).

The sampling depth for PP and phytoplankton composition ranged from 10 to 25 m rather than surface waters (i.e., <5 m), and was in most cases close to the depth of SCM or within 10–20 m of it (Table 1). For all but four of the sampling stations (Stns. 9, 10, 38, and 40), the SCM and sampling depth were shallower or near to the mixed layer depth, and hence we can have confidence that in most cases upper mixed layer (where we sampled PP and community structure) is representative of SCM (where a significant fraction of export can occur). Importantly, none of the key stations for our conclusions (Stns. 11, 14, and 19) suffer from significant differences in depths between sampling, SCM, and MLD.

Unlike Chl-a, integrated PP rates (Table 2) did not display any clear geographical pattern although the lowest PP was recorded at the ice stations (Figure 1a; Stn. 14, 20.6±12.4 mmol C m^{-2} d^{-1}) and the highest PP was recorded at the boundary between the Norwegian and Greenland Seas (Stn. 36, 63.5±3.9 mmol C m^{-2} d^{-1}) (Table 2). Other stations like station 11 (58.9±11.8 mmol C m^{-2} d^{-1}) located at the ice edge (Figure 1a) and station 42 (53.1±6.0 mmol C m^{-2} d^{-1}) located near Iceland, also had relatively high PP rates.

Three distinct plankton groups were dominant in the survey area during June 2012. Diatoms dominated the phytoplankton community at most of the open ocean stations with genera like *Thalassiosira* and *Ephemera* present (Table 1). At stations where *Ephemera* was observed, this species represented more than 90% of the total phytoplankton biomass (Table 1). The coccolithophore *Emiliania huxleyi* was observed in the Norwegian and the Barents Seas, as commonly observed in this region [Smyth et al., 2004]. In the high Chl-a area observed at the ice edge (Stns. 11 and 19, Figure 1), the colonial haptophyte *Phaeocystis pouchetii* was the major species in terms of biomass (Table 1). At station 11, *P. pouchetii* cells represented more than 90% of the phytoplankton carbon, while at station 19 the dominance of this species was much less pronounced (~30% of phytoplankton carbon, Table 1). In the ice covered region, diatoms and coccolithophores were observed but in small abundances which accounted for little biomass (Table 1).

7.3.3 POC EXPORT FLUXES

7.3.3.1 ^{234}TH ACTIVITY PROFILES AND INTEGRATED FLUXES

Vertical profiles of ^{234}Th and ^{238}U activities are presented alongside Chl-a fluorescence profiles in Figure 2. The lowest ^{234}Th activities (0.3–1 dpm L^{-1}) were observed in subsurface waters at station 11 and also in the Greenland Sea, near Iceland (Stns. 40, 42, and 44) suggesting a large removal of sinking particles at this stations. The highest ^{234}Th activity (3 dpm L^{-1}) was found at 500 m at station 36. While at this depth, the ^{234}Th is expected to be at equilibrium with the ^{238}U, this high value may reflect localized particle fragmentation (which does not necessarily means remineralization).

The most striking result is the ^{234}Th profile at station 14 (in ice) where the surface (0–60 m) ^{234}Th activities were low, and reach equilibrium approximately at the depth of the SCM (Table 1), though they decrease again below it. It is difficult to say whether this is due to particle export occurring deeper in the water column or some other process. Also, the ^{238}U activity at this station may not be accurate as the ^{238}U to salinity relationship we used here [Chen et al., 1986] does not cover the salinity range encountered at this station (32–35 for this station; see supporting information Figure S3 and 34.5–35.5 in Chen et al. [1986]). Lalande et al. [2008] observed a similar feature in the Barents Sea. This has, however, only a limited effect on the ^{234}Th fluxes as it occurs further down the water column relative to the depth the ^{234}Th fluxes were integrated to (i.e., 100 m).

The steady state 100 m integrated flux of ^{234}Th (see section 2) was large at station 11 and stations 42–44 (2023±302–4324±164 dpm m^{-2} d^{-1}) indicating one of the largest export of particles recorded worldwide [Le Moigne et al., 2013a]. At other stations, the deficits were moderate (from 862±204 to 1576±374 dpm m^{-2} d^{-1}) or low (e.g., Stn. 14, 143±212 dpm m^{-2} d^{-1}) (Table 3). Previous ^{234}Th deficits (integrated to a similar horizon depth as in this study) measured within *Phaeocystis* sp. blooms in the Arctic are consistent with our results [Lalande et al., 2007, 2008]. Furthermore, deficits recorded at stations 34, 36, 38, 40, and 42 are also consistent with previous ^{234}Th deficits in the presence of similar phytoplankton community structure ([Martin et al., 2011], 2200 dpm m^{-2} d^{-1}, diatom bloom) ([Sanders et al., 2010], 987–2700 dpm m^{-2} d^{-1}, coccolithophore rich waters). Under the ice, at station 14, the ^{234}Th deficit was low (143±212 dpm m^{-2} d^{-1}) (Table 3) consistent with other under ice studies [Cai et al., 2010; Chen et al., 2003].

7.3.3.2 C:TH RATIOS AND POC EXPORT FLUXES

^{234}Th fluxes were converted to estimates of downward particle flux using the POC:^{234}Th ratio (hereafter abbreviated as C:Th) in large (>53 μm) particles [Buesseler et al., 1992] (Table 3). C:Th ranged from 4 to 18 μmol C dpm^{-1}. The largest C:Th ratios were observed in the ice–edge and under the ice (Stns. 11 and 14), near the Barents Sea (Stn. 26) as well as in the southern sector of the Greenland Sea (Stns. 40–44) (Table 3).

Generally, POC export fluxes displayed a trend of decreasing flux from south to north with relatively large fluxes (12.8±2.7–18.9±3.4 mmol C m^{-2} d^{-1}) at stations close to Iceland (Stns. 40, 42, and 44) and low fluxes (1.8±3.8–6.2±1.97 mmol C m^{-2} d^{-1}) in the northernmost part of the survey and the Barents Sea (5.7±1.7 mmol C m^{-2} d^{-1}). The under ice station (Stn. 14) had the lowest flux (1.8±3.8 mmol C m^{-2} d^{-1}) (Table 3). An exception to this geographical pattern was station 11, located at the ice edge (Figure 1a) where the POC fluxes was surprisingly high (78.3±8.7 mmol C m^{-2} d^{-1}) (Table 3).

7.4 DISCUSSION

7.4.1 POC EXPORT

There are only a few studies that have looked at the export of carbon in the Atlantic sector of the Arctic Ocean [Le Moigne et al., 2013a]. Most of the effort in this sector of the Arctic Ocean has previously been concentrated on the shallow waters of the Barents Sea [Coppola et al., 2002; Lalande et al., 2008] and the ice covered Fram Strait [Rutgers van der Loeff et al., 2002]. Our data are thus unique for the Arctic, and are likely more comparable to what has been found in other deep-sea high latitude "ice-free" regions as described in section 3.3.1. POC export flux at station 32 in the Barents Sea (5.7±1.7 mmol C m^{-2} d^{-1}, Table 3) was on the lower range of what has been found previously [Coppola et al., 2002; Lalande et al., 2008]. In the open waters of Greenland and Norwegian Seas (Stns. 9, 10, 26, and 34–44), POC export flux ranged from 3.4±0.9 to 18.9±3.4 mmol C m^{-2} d^{-1} (Table 3). To our knowledge, this is the first attempt to measure the export of POC in this region [Le Moigne et al., 2013a], which makes comparison with POC fluxes from other oceanographic environments difficult. In the Fram Strait (Stns. 11–19), POC export flux showed the largest variability due to the presence of sea ice, ice-edge, and ice-free conditions. POC export fluxes in the Fram Strait (Table 3) are consistent with previous measurements over an ice covered/ice-free gradient in the Arctic

[Cai et al., 2010]. Station 11 located at the ice edge, however, displayed a surprisingly high POC export flux (78.3±8.7 mmol C m^{-2} d^{-1}), potentially due to the presence of a *P. pouchetii* bloom (Table 1 and Figure 1b).

C:Th ratios in blooms of *Phaeocystis* measured using large volume filtration units (as in our study) have previously been deemed to be overestimated relative to C:Th ratios measured in sediment traps due to the large proportion of dissolved organic carbon or mucilage released by *Phaeocystis* [Lalande et al., 2008]. C:Th measured in *Phaeocystis* blooms at different stages in the Barents Sea [Lalande et al., 2008, Table 3] with pumps were much higher (49±48 μmol dpm^{-1}) than those measured with traps (10±4 μmol dpm^{-1}), with the traps considered as providing more accurate C:Th ratio measurements. The authors [Lalande et al., 2008] concluded that the carbon mucilage released by blooming *Phaeocystis* may yield inaccurate (overestimated) C:Th ratios if collected using pumps. However, the C:Th measured at station 11 using pumps in our study (18 μmol dpm^{-1}) was almost identical to the value of Lalande et al. [2008] (16 μmol dpm^{-1}) in an ongoing *Phaeocystis* bloom. Hence, we are confident in our pump-derived C:Th ratio for this station where the *Phaeocystis* bloom was observed (Stn. 11). The high POC flux at station 11 was not only recorded by the ^{234}Th technique, but also using the Marine Snow Catcher (a modified settling chamber for instantaneous flux estimates) [Cavan et al., 2015; Riley et al., 2012], with full description in Smith [2014] (supporting information Table S1).

The instantaneous POC flux measured at 30 m (10 m below the MLD) using the Marine Snow Catcher yielded a flux of 156±51 mmol C m^{-2} d^{-1} for station 11 (supporting information Table S1) [Smith, 2014], which is roughly twice the value measured from 234Th (78.3±8.7 mmol C m^{-2} d^{-1}). For comparison, the average POC flux measured by the Marine Snow Catcher at similar depths across all the sampling sites was 32±16 mmol C m^{-2} d^{-1} (compared to 8±9 mmol C m^{-2} d^{-1} for the ^{234}Th-derived flux). The Marine Snow Catcher-derived POC fluxes are generally larger than the ^{234}Th-derived POC fluxes due to the fundamental difference in the time scale over which both approaches integrate the POC export fluxes and the export horizon considered for both approaches.

For instance, we know that in postbloom conditions, ^{234}Th-derived POC export is likely to overestimate the actual POC export relative to free-drifting sediment traps deployed for a couple of days [Le Moigne et al., 2013b]. In developing bloom conditions, like at some of the stations sampled in our study (e.g., Stn. 9, for example), it is possible that this discrepancy is limited. As mentioned above, the MSC flux at station 11 clearly stands out relative to the other stations. We therefore believe that both our ^{234}Th flux and C:Th ratio provide accurate measurements leading to the robust estimation of carbon export at station 11 (see data presented in supporting information Table S1).

7.4.2 EXPORT EFFICIENCY AND PHYTOPLANKTON COMMUNITY COMPOSITION

Overall, ThE-ratios ranged from 0.09±0.19 to 1.3±0.3 (Figure 3, Tables 3 and 4), which is similar to the range seen in open waters elsewhere [Buesseler, 1998; Buesseler and Boyd, 2009]. Only station 11 stands out as having a particularly high value (1.3±0.3). The three stations located in the Fram Strait are of a particular interest since, although levels of PP are similar (20.6±12.4–58.9±1.9 mmol C m^{-2} d^{-1}, Table 2, variance=117), ^{234}Th-derived carbon export fluxes were relatively variable (1.8±3.8–78±8.7 mmol C m^{-2} d^{-1}, Table 3, variance=1191), displaying higher variability than PP. This equates to ThE-ratios ranging from 0.09±0.19 to 0.19±0.05 under ice (Stn. 14) and in ice-free conditions (Stn. 19) to as high as 1.3±0.3 at the ice edge (Stn. 11). Such a high ThE-ratio at the ice edge implies that a substantial fraction of PP was exported out of the Ez. In the Barents Sea (Stn. 32) by comparison, the ThE-ratio was moderate (0.2±0.1). ThE-ratios ranged from 0.1±0.0 to 0.5±0.2 in the Greenland and Norwegian Seas (excluding stations in the Fram Strait). Station 10 also stood out because of its lower ThE-ratio (0.10±0.03), while consistently high ThE-ratios (0.3±0.1–0.4±0.1) were observed at stations located north of Iceland (Stns. 9, 38, 40, and 42).

On the global scale, phytoplankton community structure has been suggested to influence BCP efficiency [François et al., 2002], for example, through the supply of dense material, such as opal and calcite, to ballast sinking organic material. However, although global patterns of both ThE(Ez)-ratio [Siegel et al., 2014] and ballasted POC flux to a certain extent correlate [Le Moigne et al., 2014], other processes related to recycling and ecosystem structure may also greatly affect the ThE(Ez)-ratio [Henson et al., 2012; Lam et al., 2011; Le Moigne et al., 2014]. Therefore, whether there is a strong relationship between phytoplankton composition and ThE-ratios in the Arctic Ocean is very much an open question. We now examine ThE-ratios produced by the various phytoplankton groups we observed (Table 1) at each station undergoing a phytoplankton bloom (elevated biomass).

TABLE 4: Dominant Phytoplankton Community Structure (PCS) in Terms of Biomass (See Supporting Information Table S1), Euphotic Zone Integrated Primary Production, Carbon Export, and the Calculated Export Efficiency

Dominant PCS and Ice Conditionsa	Station	Integrated Primary Production (mmol C m^{-2} d^{-1})[b]	C Export (mmol C m^{-2} d^{-1})[c]	ThE-ratio
Phaeocystis sp. (ice edge)	11, 19	46.7±5.0–58.9±1.9	8.9±2.1–78.3±8.7	0.19±0.05–1.3±0.3
Ice	14	20.6±12.4	1.8±3.8	0.09±0.19
Coccolitho-phores (open)	26, 34	26.7±5.3–42.5±21.6	5.8±1.1–14.6±3.1	0.1±0.1–0.5±0.2
Diatoms (open)	9, 10, 32, 36, 38, 40, 42	11.3±0.8–63.5±3.9	3.4±0.9–17.4±3.3	0.10±0.03–0.5±0.2

[a] *See Table 1.*
[b] *See Table 2.*
[c] *See Table 3.*

At high latitudes, the POC ThE-ratio can vary seasonally [Baumann et al., 2013; Buesseler et al., 1992; Kawakami et al., 2007] to a greater extent than at low latitudes [Benitez-Nelson et al., 2001; Brix et al., 2006; Buesseler et al., 1995]. A recent modeling study confirmed that, in prebloom conditions, the ThE-ratio is low and then increases during the bloom to reach its pinnacle in postbloom conditions [Henson et al., 2015]. Therefore, to truly compare the effect of phytoplankton community structure on the ThE-ratio, one must compare stations where different phytoplankton communities are growing, at a similar point in their seasonal bloom cycle (e.g., sampled during prebloom, bloom, or postbloom conditions).

Figure 4 shows the time series of MODIS satellite-derived Chl-a (averaged over 100 × 100 km with 8 day resolution) for each station. We assume here that the blooming conditions correspond to the highest Chl-a satellite concentration recorded during the year. Stations 9, 10, and 19 were considered to be in a prebloom state (Figure 4). Although at stations 10 and 19, the PP was already high (Figure 1 and Table 2) suggesting that they were in an ascending bloom regime, Chl-a concentration had not yet reached its maximum. Export fluxes (Figure 1d) as well as ThE-ratios (0.10 ± 0.03–0.2 ± 0.0) (Figure 2) were consistently low at these stations. Conversely, stations 11, 26, 32, 34, 36, 38, 40, and 42 were sampled either at the beginning or during the main peak of bloom or shortly (less than a month) after the peak of it (Figure 4). At these stations, ThE-ratios ranged from 0.11 ± 0.02 to 1.3 ± 0.3. We hereafter refer to these stations as blooming stations. We cannot assess the bloom situation at station 14 (under ice) as no satellite Chl-a data were available at this station due to the presence of ice (Figure 1a).

Blooming stations near Iceland (Stns. 38, 40, and 42) were dominated by diatoms and had a consistently high ThE-ratio (0.3 ± 0.1–0.4 ± 0.1), comparable to previously reported values for a bloom in the North Atlantic [Buesseler and Boyd, 2009]. The blooms where *E. huxleyi* was present and comprised a significant fraction of the biomass without being the dominant species (Table 1) also had a fairly high ThE-ratio (0.1 ± 0.1–0.5 ± 0.2) (Table 4). However, the highest ThE-ratio (1.3 ± 0.3) (Figure 3) found in our study area was associated with high biomass (8.5 mg Chl-a m^{-3}), high integrated PP (58.9 mmol C m^{-2} d^{-1}) and dominance (>90% of biomass, Table 1) of the phytoplankton community by the haptophyte *P. pouchetii*. *P. pouchetii* was also present at the open ocean station (Stn. 19), but here it was not a mono-specific bloom as observed at station 11 (Table 1). Also, station 19 had not yet reached peak bloom at the time of sampling as revealed by the satellite Chl-a time series (Figure 3e) and had a lower ThE-ratio (0.19 ± 0.05). In contrast to the ice-edge station, this open ocean site had lower biomass, lower PP, and lower *P. pouchetii* biomass (Figure 1 and Table 1). This station is in a typical prebloom state (e.g., a station where

Chl-a and PP are increasing but export has yet to start, see Figures 1b, 1c, and 2) as described in Henson et al. [2015]. As expected from previous studies, the ThE-ratio was low (Figure 2) under the ice [Cai et al., 2010].

Buesseler and Boyd [2009] reported high latitude ThE-ratios from the north Atlantic, the north Pacific, and the Southern Ocean, all largely dominated by diatoms. To our knowledge, we present the first estimate of ThE-ratios associated with polar blooms in which significant biomass was associated with coccolithophores and *Phaeocystis* sp. Export efficiency from such regions has been estimated before but with limited information on the community composition [Cai et al., 2010; Chen et al., 2003] or with an export efficiency calculation methodology different from the ThE-ratio approach [Reigstad et al., 2008; Smith et al., 1991] used here, and thus prevents any direct comparison. Previously, only diatom dominated blooms were thought capable of producing high ThE-ratios with values reaching 0.45 during the diatom dominated North Atlantic bloom experiment [Buesseler and Boyd, 2009]. However, our observations from the ice edge of the Greenland Sea indicate that the presence of *P. pouchetii* can yield ThE-ratios at least as high as blooming diatoms [Buesseler and Boyd, 2009] (Figure 2 and Table 4).

7.4.3 REMINERALIZATION

Although we observed efficient export (i) in an ice edge associated *P. pouchetii* bloom (1.3 ± 0.3), (ii) where coccolithophores were present near the Barents Sea (0.1 ± 0.1–0.5 ± 0.2), and (iii) in diatom dominated conditions in the Greenland Sea near Iceland (0.3 ± 0.1–0.4 ± 0.1), this material may be highly remineralized just below the depth of the Ez and will not contribute greatly to long-term carbon sequestration. While a high ThE-ratio indicates that a significant portion of surface PP is exported out of the surface layer [Buesseler, 1998; Buesseler and Boyd, 2009], recent studies have suggested that significant remineralization within the upper part of the mesopelagic (<300 m depth) could prevent material from efficiently reaching deeper waters [Buesseler and Boyd, 2009; Giering et al., 2014] and thus contributing to C sequestration.

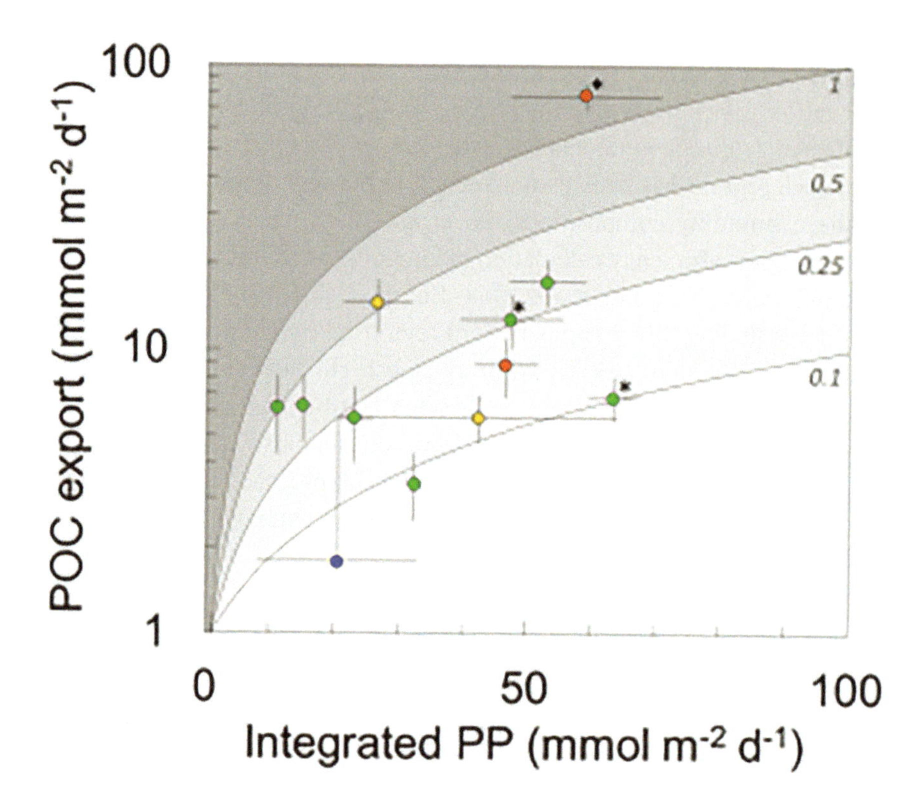

FIGURE 3: Euphotic zone integrated carbon export flux and primary production. Stations are colored according to the microphytoplankton functional type characterizing the community structure (green=diatom dominated stations, yellow=stations with coccolithophores present, red=Phaeocystis pouchetii dominated stations, and blue=under ice station). Stars indicate stations where the dominant phytoplankton type represented more than 90% of the microphytoplankton biomass. Error bars indicate standard deviation performed on replicates for integrated PP and the propagated analytical error for the POC export obtained from the 234Th technique. ThE-ratio values [Buesseler, 1998] are indicated and shaded. See Tables 2 and 3.

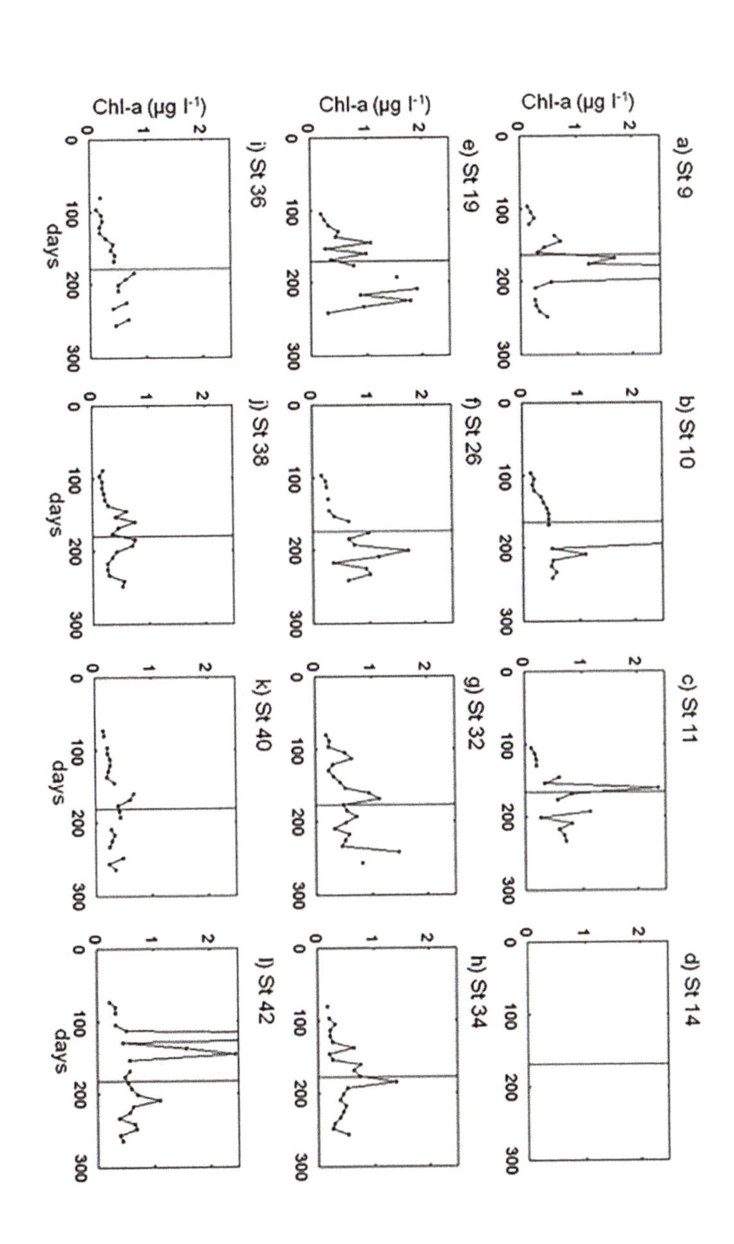

FIGURE 4: Time series of MODIS satellite Chl-a (averaged over a 100 × 100 km box centered on the station with 8 day temporal resolution) for each station. The blue vertical line represents the sampling date (note there is no satellite Chl-a data available for station 14 due to ice cover).

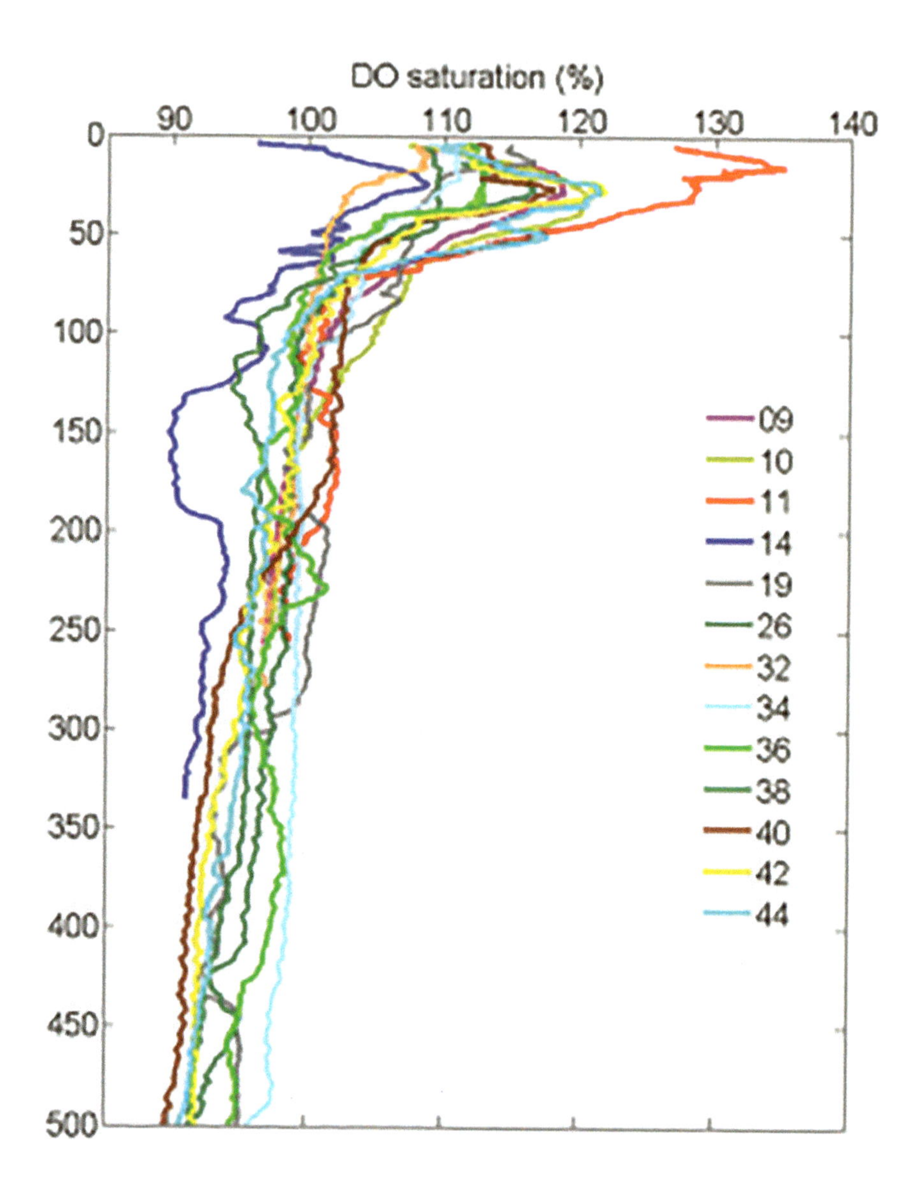

FIGURE 5: Dissolved oxygen (% saturation) at all stations (see section 4.1).

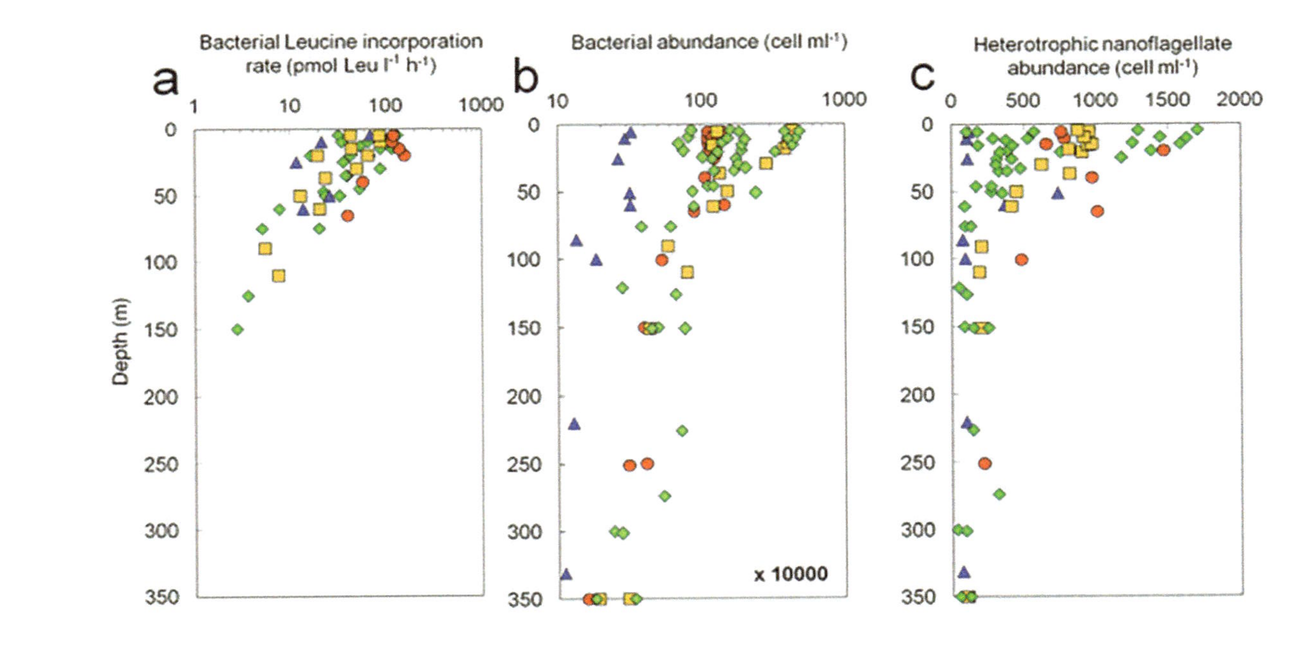

FIGURE 6: (a) Bacterial Leucine incorporation rate (pmol Leu L^{-1} h^{-1}). (b) Bacterial abundance (cell mL^{-1}). (c) Hetetrophic nanoflagellate abundance (cell mL^{-1}). Stations are colored according to the microphytoplankton functional type characterizing the community structure (green = diatom dominated stations, yellow = stations where coccolithophores were present, red = Phaeocystis pouchetii dominated stations, and blue = under ice station).

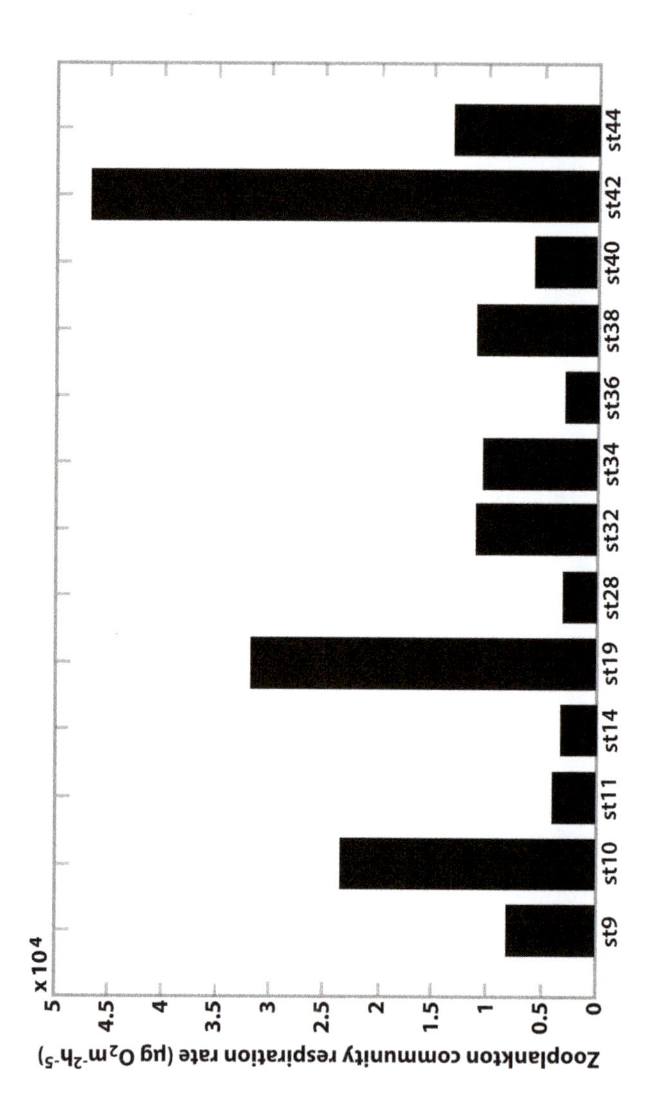

FIGURE 7: Zooplankton community respiration rate (μg O_2 L^{-1} h^{-1}) integrated over the top 200 m.

This hypothesis has not been examined previously in the Arctic. Hence, it has not previously been possible to assess the potential importance of Arctic blooms to the air-sea partitioning of atmospheric CO_2 in polar oceans. Here we do not attempt to estimate the T100 ratio [Buesseler and Boyd, 2009], as this metric requires two independent estimates of POC export flux: one below the Ez depth (which we have) (Figure 1d) and another 100 m below the Ez depth (which we do not have). This is because estimating POC flux using the ^{234}Th technique at 100 m below the Ez depth requires high resolution sampling in both the surface and deeper in the upper mesopelagic [Buesseler and Boyd, 2009], which was not possible during this cruise.

Instead, to examine whether substantially more remineralization (>100 m) of carbon occurred at certain stations relative to others, we compare several indirect but independent indicators of microbial activity rates. These are not actual carbon remineralization rates, but nevertheless provide relative information (integrated on various time scales) on the magnitude of the upper mesopelagic remineralization at each station. These include dissolved oxygen saturation ([DO] in %) (Figure 4), bacterial Leucine uptake, bacterial/heterotrophic flagellate abundance (Figure 5), and integrated zooplankton respiration rate (Figure 6).

We acknowledge that all these estimates and indicators integrate remineralization processes over different time scales. For instance, the bacterial Leucine uptake and the integrated zooplankton respiration rates look at instantaneous remineralization while the [DO] gives information on the seasonal time scale (several months). This has the great advantage in that encompasses the time scale over which our ThE-ratio (which couples monthly integrated estimates of POC export with instantaneous (daily) PP estimates) is integrated over.

7.4.3.1 OXYGEN SATURATION

Oxygen is produced during photosynthesis and consumed during the respiration (remineralization) of organic matter [Koeve, 2001]. Hence, in a water mass where the winter mixed layer is sufficiently deep that the water column is renewed at the end of the productive season, as in the Atlantic sector of the Arctic Ocean [de Boyer Montegut et al., 2004], vertical profiles show supersaturation of oxygen (>100%) in upper ocean productive waters and increasing under-saturation (<100%) with depth, due to the dominance of respiration over photosynthetic production [Koeve, 2001] (Figure 4a). This provides information on how large the upper mesopelagic respiration rate was from the end of the winter up to the sampling date. It is thus important at this stage to investigate whether the surface (surface to mixed layer depth) and subsurface water masses (mixed layer depth to 300 m) could potentially have a different origin and history, especially in the Fram Strait where the hydrography is complex due to the narrow topography and the influence of various waters masses [Woodgate, 2013].

We found station 14 (under the ice, see Figure 1) to be clearly influenced by cold and/or less saline (salinity<34) fresh water from ice melt [Woodgate, 2013] in the surface. However, the density profiles and temperature/salinity diagram (supporting information Figures S3 and S4) at station 11 displays an homogeneous water mass within the upper 250 m (salinity ~35; temperature 3–3.5°C) suggesting that the influence of Atlantic waters (salinity 34–5; temperature 0–2°C), occasionally observed in subsurface waters (>150 m) in the eastern sector of the Fram Strait [Woodgate, 2013] was limited at this station during our survey. Station 19 was influenced by fresher waters in the surface. Close to Svalbard, the surface waters may freshen slightly due to the nearby Barents Sea Polar Front separating the East Spitsbergen Current and West Spitsbergen Current, which is located approximately over the shelf break.

In the Greenland Sea (Stns. 9, 10, 38, 40, 42, and 44), the salinity and temperature range for this area are between 34.4–35 and −1.8–4°C, respectively [Hopkins, 1991]. There is a rather complex series of recirculation, atmospheric exchanges, interleaving, and mixing which gives rise to a broad variety of water masses in this region [Rudels et al., 2005]. Surface waters at stations 9, 10, 40, 42, and 44 originate from a single water mass but station 38 displayed unique features (supporting information Figure S4). This station is clearly influenced by fresher and colder waters. This is likely due to the position of this station, which is located slightly west of the East Greenland Front where the influence of the southward flowing East Greenland current is stronger. The water the Norwegian and Barents Seas (Stns. 26, 32, 34, and 36) has a strong influence from the North Atlantic Water, and has been defined by Loeng [1991] as salinity greater than 35 and temperature from 3.5 to 6.5°C. The temperature salinity diagram shows water sampled at stations 26, 34, and 36 originated from a single water mass. However, as for station 38, station 32 seems influenced by two different water masses (Figure S4). The presence of a surface coastal current of similar temperature but lower salinity is a possibility in the Norwegian Sea where station 32 was located.

For blooming stations in our study area, surface supersaturation (>105%) is evident in the upper 50 m (Figure 4), with [DO] reaching as high as 135% at the site of the *P. pouchetii* bloom. At depths down to 250 m, [DO] remains >95% at diatom dominated stations similar to stations in the Barents Sea (Figure 4). Below the *P. pouchetii* bloom (Stn. 11), [DO] also remains >95% indicating relatively similar levels of respiration of sinking organic matter as the other stations (Figure 4). In contrast, [DO] at the under ice station was <90% at depths of 150–200 m (Figure 4), indicating significant respiration of organic matter at depth. These subsurface waters have temperature-salinity characteristics similar to the central Arctic and so this [DO] under-saturation may represent long-term respiration (i.e., respiration of organic matter not generated during the preceding bloom season). It is therefore challenging to compare the upper mesopelagic remineralization between ice influenced and non-ice influenced stations.

7.4.3.2 HETEROTROPHIC RESPIRATION

With enhanced organic matter availability, it can be expected that bacteria, as the main drivers of remineralization [Giering et al., 2014], will increase in abundance and/or exhibit elevated metabolism. Within our study, examination of vertical profiles of bacterial activity (uptake of the amino-acid Leucine in this case; Figure 5a), bacterial abundance (Figure 5b), or heterotrophic flagellate abundance (Figure 5c) shows no clear enhancement below the 100 m depth horizon at any of the stations. In fact, the bacteria and heterotrophic flagellate abundance at depths of 100–300 m are within the same order of magnitude across all stations sampled (Figures 5b and 5c).

To a lesser extent, zooplankton also respire sinking carbon [Giering et al., 2014]. Zooplankton oxygen respiration rates (Figure 7) were estimated from allometric relationships and biomass measurements of zooplankton species collected from surface nets (down to 200 m; see section 2 and supporting information Table S1). In the *P. pouchetii* bloom, the zooplankton respiration rate was low (Figure 7) relative to other stations, indicating no intense remineralization by zooplankton took place at the ice-edge station. However, zooplankton respiration was high at diatom dominated stations 10 and 42 as well as at station 19 where *P. pouchetii* was also present but not dominating biomass. Station 19 had not yet reached peak bloom at the time of sampling as revealed by the satellite Chl-a time series (Figure 3e). This open ocean site had lower biomass, PP and *P. pouchetii* abundance (Figure 1, Tables 1, and 2) and is in a typical prebloom state (e.g., PP is increasing but export has yet to start, see Figure 1b and Table 2). When in colonial blooming form as observed at station 11, *Phaeocystis* can be a significant producer of dimethyl-sulfide (DMS) [DiTullio et al., 2000; van Leeuwen et al., 2007] and is repugnant to many marine grazers [Schoemann et al., 2005], which significantly increases its success and community dominance. The integrated (top 60 m) concentration of DMS at station 11 was almost twice as larger than at station 19 (523 and 331 nmol DMS m^{-2}, respectively) (F. Hopkins and J. Stephens, Plymouth Marine Laboratory, personal communication, 2012). This is a possible explanation for why at station 11 the zooplankton respiration rate is smaller than at station 19.

Although our approach only provides relative information and not quantitative rates of carbon remineralization at each station, several lines of evidence suggest that no enhanced remineralization of sinking material in the upper 300 m of the water column occurred below the *P. pouchetii* bloom (Stn. 11). Hence, the upper mesopelagic remineralization at the ice-edge bloom (Stn. 11) may have been similar to the diatom dominated stations and therefore, potentially as high as in diatom blooms in the North Atlantic that are amongst the highest observed globally [Buesseler and Boyd, 2009]. This trend also shows consistency in time as both bacterial/zooplankton remineralization rates and [DO] displays the same pattern. This suggests that the uniform remineralization (no matter the surface phytoplankton composition) we observed in the upper mesopelagic zone in the study region is not an artifact of decoupled integration time scales.

Extensive blooms of *Phaeocystis* spp. occur in the Southern Ocean [Arrigo et al., 1999; DiTullio et al., 2000; Salter et al., 2007] (P. antarctica) and in northern polar waters (*P. pouchetii*), such as the Barents Sea [Reigstad and Wassmann, 2007; Smith et al., 1991; Wassmann et al., 1990]. In the Ross Sea, observations from 300 to 500 m depth have previously suggested that P. antarctica is exported rapidly to depth with little remineralization through the upper water column [DiTullio et al., 2000]. Conversely, *P. pouchetii* blooms in the Barents Sea are thought to undergo more rapid remineralization in the upper water column than diatom blooms [Reigstad and Wassmann, 2007; Wassmann et al., 1990]. However, some recent observations suggest that ice-edge eddies can, on occasion, result in the export of *P. pouchetii* to depths greater than 340 m [Lalande et al., 2011].

Our results show that a high proportion of PP associated with a retreating ice-edge *P. pouchetii* bloom was exported below 100 m (high ThE-ratio), and, using [DO], bacterial/heterotrophic flagellate abundance/activity and zooplankton respiration rates, we infer that rates of remineralization of this material over the upper 100–350 m were similar to those found under diatom blooms. We only sampled one station (Stn. 11) where *P. pouchetii* was blooming; our results are therefore limited, but do highlight the potential for enhanced export related to the occurrence of Arctic *P. pouchetii*. We suggest that more effort be made to estimate accurately the carbon respiration rate under polar blooms of diatoms, coccolithophores and *Phaeocystis* spp. as none of the approaches cited above [Buesseler and Boyd, 2009; DiTullio et al., 2000; Reigstad and Wassmann, 2007] provide direct estimates of upper mesopelagic C remineralization rates in polar oceans.

7.5 CONCLUSIONS AND IMPLICATIONS

Large uncertainties remain about the fate of sinking organic matter in the Arctic. This prevents reliable predictions of the strength of the BCP in the future Arctic Ocean. In June 2012, we surveyed the efficiency with which the biomass produced in the surface ocean was exported out of the sunlit layer of the ocean (the ThE-ratio [Buesseler, 1998]) and link it to the dominant phytoplankton group present in the water column. Furthermore, we compared several qualitative indicators of sinking material remineralization in the upper mesopelagic (>100 m). Our main conclusions are that:

1. Several phytoplankton blooms were observed during our survey: a *P. pouchetii* bloom at the ice edge in the Fram Strait, a diatom bloom in the Greenland Sea, and a bloom, where coccolithophores were present in the Norwegian and Barents Seas (Table 1).
2. Primary production and POC export were very low under the Greenland ice pack. The highest rates of POC export were recorded in an ice-edge bloom of *P. pouchetii* (Fram Strait) and in the south of the Greenland Sea near Iceland (Figure 1).
3. The ThE-ratio [Buesseler, 1998] displayed large spatial variability. The common denominator in stations where high ThE-ratio were recorded was the time of sampling relative to the seasonal bloom dynamics (e.g., all sampled during the main Chl-a peak or no more than a month after the main bloom peak). The highest ThE-ratio was recorded in the mono-specific *P. pouchetii* bloom and in the diatom dominated stations near Iceland (Figure 3 and Table 3).
4. Our results suggest that, under certain circumstances, *Phaeocystis* blooms may experience little upper mesopelagic remineralization, as seen in diatom blooms.

Ultimately, the potential for phytoplankton bloom formation is driven by each species' ecological competitiveness [Schoemann et al., 2005], including its resistance to grazing. It is currently unclear whether the future Arctic Ocean PP is ultimately limited by light or nutrients; however, model predictions suggest that future Arctic primary production will be nutrient limited [Vancoppenolle et al., 2013]. Hence, although the behavior of grazers and the microbial community in the future Arctic is hard to forecast, predictions of a future reduction in nutrient concentrations in the Arctic [Codispoti et al., 2013], increased iron limitation due to reduction of ice cover [Measures, 1999] and possibly reduced stratification [Carmack and Wassmann, 2006], all suggest that there is potential for important changes in bloom composition in the future Arctic Ocean. In an increasingly ice-free Arctic Ocean [Boé et al., 2009; Stroeve et al., 2012], with a greater proportion of ice edge for bloom formation [Perrette et al., 2011], such fundamental processes will determine how climate change affects the strength of the Arctic carbon sink. We suggest that more effort is required to quantify the export ratio in the Arctic Ocean, and more importantly to assess the fate of sinking POC in the Arctic mesopelagic zone.

REFERENCES

1. Arrigo, K. R., G. van Dijken, and S. Pabi (2008), Impact of shrinking Arctic ice cover on marine primary production, Geophys. Res. Lett., 35, L19603, doi:10.1029/2008GL035028.
2. Arrigo, K. R., D. H. Robinson, D. L. Worthen, R. B. Dunbar, G. R. DiTullio, M. VanWoert, and M. P. Lizotte (1999), Phytoplankton community structure and the drawdown of nutrients and CO2 in the Southern Ocean, Science, 283(5400), 365–367.
3. Balch, W. M., D. T. Drapeau, and J. J. Fritz (2000), Monsoonal forcing of calcification in the Arabian Sea, Deep Sea Res., Part II, 47, 1301–1337.
4. Baumann, M. S., S. B. Moran, M. W. Lomas, R. P. Kelly, and D. W. Bell (2013), Seasonal decloupling of particulate organic carbon export ans net primary production in relation to sea-ice at the shelf break of the eastern Bering Sea: Implications for off-shelf carbon export, J. Geophys. Res. Oceans, 118, 5504–5522, doi:10.1002/jgrc.20366.
5. Benitez-Nelson, C., K. O. Buesseler, D. M. Karl, and J. Andrews (2001), A time-series study of particulate matter export in the North Pacific Subtropical Gyre based on Th-234:U-238 disequilibrium, Deep Sea Res., Part I, 48(12), 2595–2611.
6. Boé, J., A. Hall, and X. Qu (2009), September sea-ice cover in the Arctic Ocean projected to vanish by 2100, Nat. Geosci., 2, 341–343, doi:10.1038/NGEO467.
7. Bopp, L., O. Aumont, P. Cadule, S. Alvain, and M. Gehlen (2005), Response of diatoms distribution to global warning and potential implications: A global model study, Geophys. Res. Lett., 32, L19606, doi:10.1029/2005GL023653.
8. Boyd, P., and P. Newton (1995), Evidence of the potential influence of planktonic community structure on the interannual variability of particulate organic-carbon dlux, Deep Sea Res., Part I, 42(5), 619–639.
9. Boyd, P. W., and P. P. Newton (1999), Does planktonic community structure determine downward particulate organic carbon flux in different oceanic provinces?, Deep Sea Res., Part I, 46(1), 63–91.
10. Boyd, P. W., and T. W. Trull (2007), Understanding the export of biogenic particles in oceanic waters: Is there consensus?, Prog. Oceanogr., 72(4), 276–312, doi:10.1016/j.pocean.2006.10.007.
11. Brix, H., N. Gruber, D. M. Karl, and N. R. Bates (2006), On the relationships between primary, net community, and export production in subtropical gyres, Deep Sea Res., Part II, 53(5–7), 698–717.
12. Buesseler, K. O. (1998), The decoupling of production and particulate export in the surface ocean, Global Biogeochem. Cycles, 12(2), 297–310.

13. Buesseler, K. O., and P. W. Boyd (2009), Shedding light on processes that control particle export and flux attenuation in the twilight zone of the open ocean, Limnol. Oceanogr. Methods, 54(4), 1210–1232.
14. Buesseler, K. O., M. P. Bacon, J. K. Cochran, and H. D. Livingston (1992), Carbon and nitrogen export during the JGOFS North Atlantic bloom experiment estimated from 234Th:238U disequilibria, Deep Sea Res., Part A, 39(7–8), 1115–1137.
15. Buesseler, K. O., J. A. Andrews, M. C. Hartman, R. Belastock, and F. Chai (1995), Regional estimates of the export flux of particulate organic carbon derived from thorium-234 during the JGOFS EqPac program, Deep Sea Res., Part II, 42(2–3), 777–804.
16. Cai, P., M. Rutgers van der Loeff, I. Stimac, E. M. Nothig, K. Lepore, and S. B. Moran (2010), Low export flux of particulate organic carbon in the central Arctic Ocean as revealed by 234Th:238U disequilibrium, J. Geophys. Res., 115, C10037, doi:10.1029/2009JC005595.
17. Carmack, E., and P. Wassmann (2006), Food webs and physical-biological coupling on pan-Arctic shelves: Unifying concepts and comprehensive perspectives, Prog. Oceanogr., 71, 446–477, doi:10.1016/j.pocean.2006.10.004.
18. Cavalieri, D. J., C. L. Parkinson, P. Gloersen, and H. Zwally (1996, updated yearly), Sea Ice Concentrations From Nimbus-7 SMMR and DMSP SSM/I-SSMIS Passive Microwave Data, NASA DAAC at the Natl. Snow and Ice Data Cent., Boulder, Colo.
19. Cavan, E., F. A. C. Le Moigne, A. J. poulton, C. J. Daniels, G. Fragoso, and R. J. Sanders (2015), Zooplankton fecal pellets control the attenuation of particulate organic carbon flux in the Scotia Sea, Southern Ocean, Geophys. Res. Lett., 41, doi:10.1002/2014GL062744.
20. Chen, J. H., R. L. Edwards, and G. J. Wasserburg (1986), 238U, 234U and 232Th in seawater, Earth Planet. Sci. Lett., 80(3–4), 241–251.
21. Chen, M., Y. P. Huang, P. G. Cai, and L. D. Guo (2003), Particulate organic carbon export fluxes in the Canada Basin and Bering Sea as derived from Th-234/U-238 disequilibria, Arctic, 56(1), 32–44.
22. Codispoti, L. A., V. Kelly, A. Thessen, P. Matrai, S. Suttles, V. Hill, M. Steele, and B. Light (2013), Synthesis of primary production in the Arctic Ocean: III. Nitrate and phosphate based estimates of net community production, Prog. Oceanogr., 10(1), 126–150.
23. Coello-Camba, A., S. Agusti, J. Holding, J. M. Arrieta, and C. M. Duarte (2014), Interactive effect of temperature and CO2 increase in arctic phytoplankton, Front. Mar. Sci., 1(49), doi:10.3389/fmars.2014.00049.
24. Coppola, L., M. Roy-Barman, P. Wassmann, S. Mulsow, and C. Jeandel (2002), Calibration of sediment traps and particulate organic carbon using 234Th in the Barents Sea, Mar. Chem., 80(1), 11–26.
25. de Boyer Montegut, C., G. Madec, A. C. Fischer, A. Lazar, and D. Iudicone (2004), Mixed layer depth over the global ocean: An examination of profile data and a profile-based climatology, J. Geophys. Res., 109, C120003, doi:10.1029/2004JC002378.
26. Dickson, A. G. (1994), Determination of dissolved oxygen in sea water by Winkler titration, in WHP 91-1: WOCE Operations Manual, WOCE Hydrogr. Prog. Off., WOCE Operations Manual.

27. DiTullio, G. R., J. M. Grebmeier, K. R. Arrigo, M. P. Lizotte, D. H. Robinson, A. Leventer, J. B. Barry, M. L. VanWoert, and R. B. Dunbar (2000), Rapid and early export of Phaeocystis antarctica blooms in the Ross Sea, Antarctica, Nature, 404(6778), 595–598.
28. François, R., S. Honjo, R. Krishfield, and S. Manganini (2002), Factors controlling the flux of organic carbon to the bathypelagic zone of the ocean, Global Biogeochem. Cycles, 16(4), 1087, doi:10.1029/2001GB001722.26
29. Giering, S. L. C., et al. (2014), Reconciliation of the carbon budget in the ocean's twilight zone, Nature, 507(7493), 480–483.
30. Gradinger, R. R., and M. E. M. Baumann (1991), Distribution of phytoplankton communities in relation to the large-scale hydrographical regime in the Fram Strait, Mar. Biol., 111, 311–321.
31. Henson, S., R. Sanders, E. Madsen, P. Morris, F. A. C. Le Moigne, and G. Quartly (2011), A reduced estimate of the strength of the ocean's bioloical carbon pump, Geophys. Res. Lett., 38, L046006, doi:10.1029/2011GL046735.
32. Henson, S. A., R. J. Sanders, and E. Madsen (2012), Global patterns in efficiency of particulate organic carbon export and transfer to the deep ocean, Global Biogeochem. Cycles, 26, GB1028, doi:10.1029/2011GB004099.
33. Henson, S. A., A. Yool, and R. J. Sanders (2015), Variability in efficiency of particulate organic carbon export: A model study, Global Biogeochem. Cycles, 29, 33–45, doi:10.1002/2014GB004965.
34. Hill, V. J., P. A. Matrai, E. Olson, S. Suttles, M. Steele, L. A. Codispoti, and R. C. Zimmerman (2013), Synthesis of integrated primary production in the Arctic Ocean: II. In situ and remotely sensed estimates, Prog. Oceanogr., 110, 107–125.
35. Hopkins, T. S. (1991), The GIN Sea A synthesis of its physical oceanography and literature review 1972–1985, Earth Sci. Rev., 30, 175–318.
36. Ikeda, T., Y. Kanno, K. Ozaki, and A. Shimada (2001), Metabolic rates of epipelagic marine copepods as a function of body mass and temperature, Mar. Biol., 139, 587–596, doi:10.1007/S002270100608.
37. Kawakami, H., M. C. Honda, M. Wakita, and S. Watanabe (2007), Time-series observation of POC fluxes estimated from 234Th in the northwestern North Pacific, Deep Sea Res., Part I, 54, 1070–1090.ADS
38. Koeve, W. (2001), Wintertime nutrients in the North Atlantic—New approaches and implications for new production estimates, Mar. Chem., 74(4), 245–260.
39. Lalande, C., E. Bauerfeind, and E. M. Nothig (2011), Downward particulate organic carbon export at high temporal resolution in the eastern Fram Strait: Influence of Atlantic water on flux composition, Mar. Ecol. Prog. Ser., 440, 127–136, doi:10.3354/meps09385.
40. Lalande, C., K. Lepore, L. W. Cooper, J. M. Grebmeier, and S. B. Moran (2007), Export fluxes of particulate organic carbon in the Chukchi Sea: A comparative study using 234Th/238U disequilibria and drifting sediment traps, Mar. Chem., 103(1), 185–196.
41. Lalande, C., S. B. Moran, P. Wassmann, J. M. Grebmeier, and L. W. Cooper (2008), 234Th-derived particulate organic carbon fluxes in the northern Barents Sea with comparison to drifting sediment trap fluxes, J. Mar. Syst., 73(1–2), 103–113.

42. Lam, P. J., S. C. Doney, and J. K. B. Bishop (2011), The dynamic ocean biological pump: Insights from a global compilation of particulate organic carbon, CaCO3 and opal concentrations profiles from the mesopelagic, Global Biogeochem. Cycles, 25, GB3009, doi:10.1029/2010GB003868.
43. Le Moigne, F. A. C., S. A. Henson, R. J. Sanders, and E. Madsen (2013a), Global database of surface ocean particulate organic carbon export fluxes diagnosed from the 234Th technique, Earth Syst. Sci. Data, 5(2), 295–304, doi:10.5194/essd-5-295-2013.
44. Le Moigne, F. A. C., K. Pabortsava, C. L. J. Marcinko, P. Martin, and R. J. Sanders (2014), Where is mineral ballast important for surface export of particulate organic carbon in the ocean?, Geophys. Res. Lett., 41, 8460–8468, doi:10.1002/2014GL061678.
45. Le Moigne, F. A. C., M. Villa-Alfageme, R. J. Sanders, C. M. Marsay, S. Henson, and R. Garcia-Tenorio (2013b), Export of organic carbon and biominerals derived from 234Th and 210Po at the Porcupine Abyssal Plain, Deep Sea Res., Part I, 72, 88–101, doi:10.1016/j.dsr.2012.10.010.
46. Loeng, H. (1991), Features of the physical oceanographic conditions of the Barents Sea. Pp. S18 in Sakshaug, edited by E. Hopkins, C. C. E. and N. A. Britsland, Proceedings of the Pro Mare Symposium on Polar Marine Ecology, Trondheim, 12–16 May 1990, Polar Research 10/1.
47. Martin, P., R. Lampitt, M. J. Perry, R. Sanders, C. Lee, and E. D'asaro (2011), Export and mesopelagic particle flux during a North Atlantic spring diatom bloom, Deep Sea Res., Part I, 58(4), 338–349.
48. Measures, C. I. (1999), The role of entrained sediments in sea ice in the distribution of aluminium and iron in the surface waters of the Arctic Ocean, Mar. Chem., 68(1–2), 59–70.
49. Moran, S. B., K. M. Ellis, and J. N. Smith (1997), Th-234/U-238 disequilibrium in the central Arctic Ocean: Implications for particulate organic carbon export, Deep Sea Res., Part II, 44(8), 1593–1606.
50. Moran, S. B., and J. N. Smith (2000), Super(234)Th as a tracer of scavenging and particle export in the Beaufort Sea, Cont. Shelf Res., 20(2), 153–167.
51. Pabi, S., G. L. van dijken, and K. R. Arrigo (2008), Primary production in the Arctic Ocean, 1998–2006, J. Geophys. Res., 113, C08005, doi:10.1029/2007JC004578.
52. ADS
53. Perrette, M., A. Yool, G. D. Quartly, and E. E. Popova (2011), Near-ubiquity of ice-edge blooms in the Arctic, Biogeosciences, 8, 515–524, doi:10.5194/bg-8-515-2011.
54. Pike, S. M., K. O. Buesseler, J. Andrews, and N. Savoye (2005), Quantification of Th-234 recovery in small volume sea water samples by inductively coupled plasma-mass spectrometry, J. Radioanal. Nucl. Chem., 263(2), 355–360, doi:10.1007/s10967-005-0062-9.
55. Popova, E. E., A. Yool, A. C. Coward, Y. K. Aksenov, S. G. Alderson, B. A. de Cuevas, and T. R. Anderson (2010), Control of primary production in the Arctic by nutrients and light: Insights from a high resolution ocean general circulation model, Biogeosciences, 7(11), 3569–3591.
56. Popova, E. E., A. Yool, A. C. Coward, F. Dupont, C. Deal, S. Elliott, E. Hunke, M. Jin, M. Steele, and J. Zhang (2012), What controls primary production in the Arctic

Ocean? Results from an intercomparison of five general circulation models with biogeochemistry, J. Geophys. Res., 117, C00D12, doi:10.1029/2011JC007112.

57. Poulton, A., A. Charalampopoulou, J. R. Young, G. A. Tarran, M. I. Lucas, and G. D. Quartly (2010), Coccolithophore dynamics in non-bloom conditions during late summer in the central Iceland Basin (July-August 2007), Limnol. Oceanogr. Methods, 4(55), 1601–1613.
58. Poulton, A. J., C. M. Moore, S. Seeyave, M. I. Lucas, S. Fielding, and P. Ward (2007), Phytoplankton community composition arounf the Crozet Plateau, with emphasis on diatoms and Phaeocystis, Deep Sea Res., Part II, 54, 2085–2105.
59. Poulton, A. J., M. C. Stinchcombe, E. P. Achterberg, D. C. E. Bakker, C. Dumousseaud, H. E. Lawson, G. A. Lee, S. Richier, D. J. Suggett, and J. R. Young (2014), Coccolithophores on the north-west European shelf: Calcification rates and environmental controls, Biogeosci. Discuss, 11, 2685–2733.
60. Reigstad, M., and P. Wassmann (2007), Does Phaeocystis spp. contribute significantly to verticalexport of organic carbon?, Biogeochemistry, 83, 217–234, doi:10.1007/s10533-007-9093-3.
61. Reigstad, M., C. Wexels Riser, P. Wassmann, and T. Ratkova (2008), Vertical export of particulate organic carbon: Attenuation, composition and loss rates in the northern Barents Sea, Deep Sea Res., Part II, 55, 2308–2319.
62. Resplandy, L., A. P. Martin, F. A. C. Le Moigne, P. Martin, A. Aquilina, L. Mémery, M. Lévy, and R. Sanders (2012), Impact of dynimical spatial variability on estimates of organic material export to the deep ocean, Deep Sea Res., Part I, 68, 24–45.
63. Riley, J., R. Sanders, C. Marsay, F. A. C. Le Moigne, E. Achterberg, and A. Poulton (2012), The relative contribution of fast and slow sinking particles to ocean carbon export, Global Biogeochem. Cycles, 26, GB1026, doi:10.1029/2011GB004085.
64. Rudels, B., G. Bjork, J. Nilsson, P. Winsor, I. Lake, and C. Nohr (2005), The interaction between waters from the Arctic Ocean and the Nordic Seas north of Fram Strait and along the East Greenland Current: Results from the Arctic Ocean-02 Oden expedition, J. Mar. Syst., 55, 1–30.
65. Rutgers van der Loeff, M. M., R. Meyer, B. Rudels, and E. Rachor (2002), Resuspension and particle transport in the benthic nepheloid layer in and near Fram Strait in relation to faunal abundances and Th-234 depletion, Deep Sea Res., Part I, 49(11), 1941–1958.
66. Salter, I., R. S. Lampitt, R. Sanders, A. Poulton, A. E. S. Kemp, B. Boorman, K. Saw, and R. Pearce (2007), Estimating carbon, silica and diatom export from a naturally fertilised phytoplankton bloom in the Southern Ocean using PELAGRA: A novel drifting sediment trap, Deep Sea Res., Part I, 54(18–20), 2233–2259, doi:10.1016/j.dsr2.2007.06.008.
67. Sanders, R., P. J. Morris, A. J. Poulton, M. C. Stinchcombe, A. Charalampopoulou, M. I. Lucas, and S. J. Thomalla (2010), Does a ballast effect occur in the surface ocean?, Geophys. Res. Lett., 37, L08602, doi:10.1029/2010GL042574.
68. Schoemann, V., S. Becquevort, J. Stefels, W. Rousseau, and C. Lancelot (2005), Phaeocystis blooms in the global ocean and their controlling mechanisms: A review, J. Sea Res., 53(1–2), 43–66, doi:10.1016/j.seares.2004.01.008.
69. Siegel, D. A., K. O. Buesseler, S. C. Doney, S. F. Sailley, M. J. Behrenfeld, and P. W. Boyd (2014), Global assessment of ocean carbon export by combining satel-

lite observations and food-web models, Global Biogeochem. Cycles, 28, 181–196, doi:10.1002/2013GB004743.
70. Smith, H. E. K. (2014), The contribution of mineralising phytoplankton to the biological carbon pump in high latitudes, PhD thesis, Univ. of Southampton, Southampton, U. K.
71. Smith, W. O., L. A. Codispoti, D. M. Nelson, T. Manley, E. J. Buskey, H. J. Niebauer, and G. F. Cota (1991), Importance of Phaeocystis blooms in the high latitude ocean carbon cycle, Nature, 352, 514–516.
72. Smyth, T. J., T. Tyrrell, and B. Tarrant (2004), Time series of coccolithophore activity in the Barents Sea, from twenty years of satellite imagery, Geophys. Res. Lett., 31, L11302, doi:10.1029/2004GL019735.
73. Stroeve, J. C., V. Kattsov, A. Barrett, M. Serreze, T. Pavlova, M. Holland, and W. N. Meier (2012), Trends in Arctic sea ice extent from CMIP5, CMIP3 and observations, Geophys. Res. Lett., 39, L16502, doi:10.1029/2012GL052676.
74. Uttermohl, H. (1958), Zur vervollkommung der quantitativen phytoplankton methodik, Int. Ver. Theor. Angew. Limnol., 9, 38.
75. van Leeuwen, M. A., J. Stefels, S. Belviso, C. Lancelot, P. G. Verity, and W. W. C. Gieskes (2007), Phaeocystis, major link in the biogeochemical cycling of climate-relevant elements, Biogeochemistry, 83, 1–3.
76. Vancoppenolle, M., L. Bopp, G. Madec, J. P. Dunne, T. Ilyina, P. R. Halloran, and N. Steiner (2013), Future Arctic Ocean primary productivity from CMIP5 simulations: Uncertain outcome, but consistent mechanisms, Global Biogeochem. Cycles, 27, 605–619, doi:10.1002/gbc.20055.
77. Ward, P., A. Atkinson, and G. Tarling (2012), Mesozzoplankton community structure and variability in the Scotia Sea: A seasonal comparison, Deep Sea Res., Part II, 59–60.
78. Wassmann, P. (1994), Significance of sedimentation for the termination of Phaeocystis blooms, J. Mar. Syst., 5, 81–100.
79. Wassmann, P., M. Vernet, B. G. Mitchell, and F. Rey (1990), Mass sedimentation of Phaeocystis pouchetii in the Barents Sea, Mar. Ecol. Prog. Ser., 66, 183–195.
80. Woodgate, R. (2013), Arctic Ocean circulation: Going around at the top of the world, Nat. Educ. Knowledge, 4(8), 8.
81. Zubkov, M. V., and P. H. Burkill (2006), Syringe pumped high speed flow cytometry of oceanic phytoplankton, Cytometry Part A, 69, 1010–1019.
82. Zubkov, M. V., and G. A. Tarran (2008), High bacterivory by the smallest phytoplankton in the North Atlantic Ocean, Nature, 455(7210), 224–226, doi:10.1038/nature07236.
83. Zubkov, M. V., M. A. Sleigh, P. H. Burkill, and R. J. G. Leakey (2000), Bacterial growth and grazing loss in contrasting areas of North and South Atlantic, J. Plankton Res., 22, 685–711.

There are several supplemental files that are not available in this version of the article. To view this additional information, please use the citation on the first page of this chapter.

PART IV

OCEAN ACIDIFICATION

CHAPTER 8

Ocean Warming–Acidification Synergism Undermines Dissolved Organic Matter Assembly

CHI-SHUO CHEN, JESSE M. ANAYA , ERIC Y-T CHEN, ERIK FARR, AND WEI-CHUN CHIN

8.1 INTRODUCTION

Existing as part of an organic matter continuum, the ability of dissolved organic matter (DOM) polymers to spontaneously assemble into a more bioactive microgels represents a 70 Gt carbon flux [1,2], out of a total DOM budget of 700 Gt carbon. The DOM–particulate organic matter (POM) shunt plays many roles: it redirects organic carbon flow in marine microbial communities [3–5]; reshapes trophic cycling [6–8]; and even serves as cloud condensation nuclei [9]. Conventionally, DOM has been considered a refractory macromolecule, revealing complex chemical compositions and structures [1,10]. However, notwithstanding their broad significance, it remains unknown whether the macromolecular nature of DOM is susceptible to multiple environmental fluctuations—fluctuations realistic under future climate scenarios [11,12]. Considering the critical nature of the DOM–POM shunt, minute perturbations to DOM assembly, induced by moderate temperature or pH changes, would have effects on the ocean carbon flux and marine ecosystems [13].

8.2 MATERIALS AND METHODS

8.2.1 EXPERIMENTAL DESIGN

We conducted five independent experiments to investigate the influence of temperature and pH on marine gel assembly. Our hypotheses were: 1) there is a particular point for both temperature and pH beyond which marine gels cannot assemble and existing gels are dispersed; and 2) there is a warming–acidification synergism to microgel assembly. The first experiment investigated an upper bound for temperatures beyond which marine gels could not assemble. The second experiment investigated the potential dispersion of microgels to temperatures above 30°C. The third experiment investigated an upper bound for pH beyond which marine gels could not assemble. The fourth experiment investigated a potential warming–acidification synergism on an upper bound beyond which marine gels could not assemble. The fifth experiment evaluated Ca^{2+}, a DOM polymer cross-linker, and marine gel hydrophobicity, an additional DOM polymer cross-linking mechanism [1,14], to assess which mechanism might contribute to the assembly/dispersion. All data are reported with the mean and standard deviation.

8.2.2 WATER SAMPLING AND FILTRATION

Seawater samples collected at Puget Sound (WA, USA) near Friday Harbor Marine Laboratories in August 2009 were filtered through a 0.22-μm membrane (Millipore polyvinylidene fluoride low protein binding filter, prewashed with 0.1 N HCl), treated with 0.02% sodium azide—a microbial biocide—and stored in sterile, Parafilm-sealed bottles in the dark at 4°C until further processing. Specific permission was not required to collect seawater sample (<5 L) at this site (Friday Harbor, WA, USA, 48.54619, -123.00761). No endangered or protected species were involved. DOC (dissolved organic carbon) concentration (2.566 mg L^{-1}) was measured using a Shimadzu TOC-Vcsh Total Organic Carbon Analyzer [15].

8.2.3 GROWTH KINETICS

Microgel assembly was monitored with dynamic laser scattering as described previously [4]. Seawater aliquots (10 mL) were syringe-filtered through a 0.22-μm membrane (low-protein binding Durapore, Millipore) directly into scintillation vials (pre-rinsed with Milli-Q Millipore DI water). Scattering cells were placed in the goniometer of a Brookhaven laser spectrometer (Brookhaven Instruments, NY), where the scattering fluctuation signals were detected at a 45° angle. The autocorrelation function of scattering intensity fluctuations was averaged over a 12-minute sampling time. Hydrodynamic diameters of polymer gels were analyzed by the CONTIN method [1,4,16]. Seawater samples were measured every 24 or 72 hrs by dynamic laser scattering to monitor size changes.

8.2.4 TEMPERATURE-PH ADJUSTMENTS

Based on climate changes models [17,18], we set up the ranges of experimental temperature and pH. We understand only specific regions may experience wide-range temperature/pH changes; however, we aimed to explore potential impacts and possible consequences under extreme environmental conditions. In temperature-dependent experiments (Fig. 1), samples in airtight, sealed scintillation vials were grouped and incubated at 22, 32 and 35°C for 24 hrs and stored/monitored for 15 d at 22°C. For microgel dispersion-temperature experiments (Fig. 2), 10 mL of seawater were syringe-filtered into vials and incubated in dark for 10 d at 22°C.

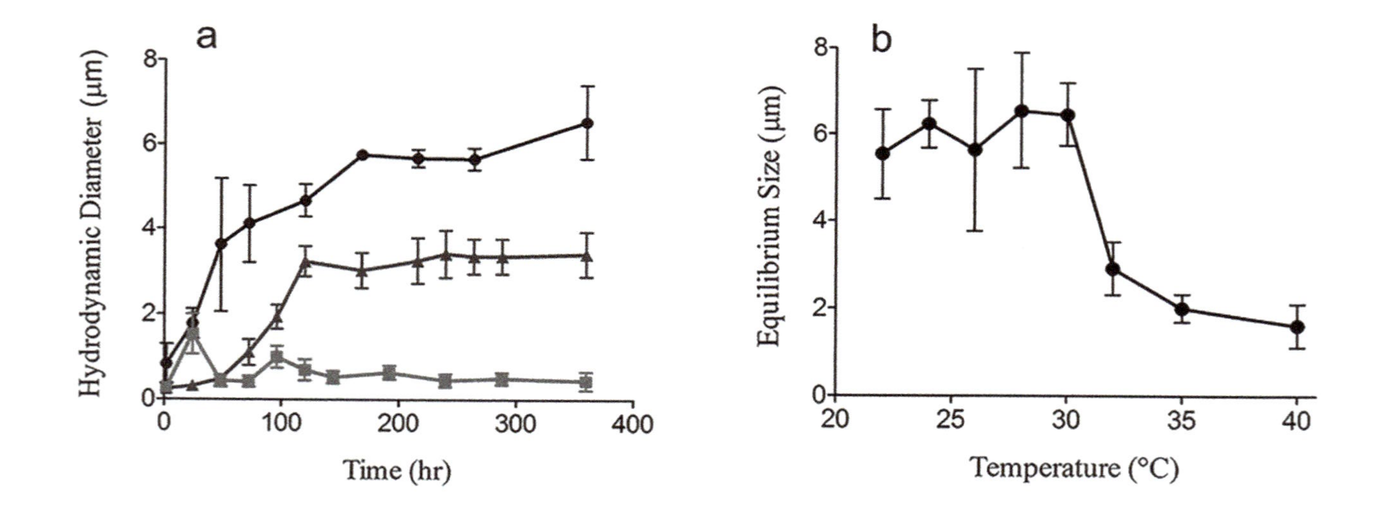

FIGURE 1: Microgel assembly / dispersion are temperature dependent. 1-a. Microgel assembly rate and equilibrium size decreases with increased temperature. Samples were incubated at 22˚C (black circles), 32˚C (blue triangles) and 35˚C (red squares) for 24 hours, then stored in the dark at 22˚C for the remainder of the experiment. Assembly was measured using dynamic laser scattering at 22˚C. Each data point represents (mean ± SD) of six measurements made in each of six replicate samples. Data highlight that short-term temperature exposure above 35˚C confers significant DOM assembly loss with no obvious recovery. 1-b. Microgel dispersion depends on temperature variation. Self-assembled microgels (size ~ 6 μm) were incubated at various temperatures (from 22˚C to 40 ˚C) for 24 hours. The equilibrium microgel sizes were monitored with dynamic laser scattering spectroscopy. Each data point represents six replicate samples. Non-linear temperature responses of microgels were observed—particularly for microgels incubated at temperatures above 32˚C, which showed a marked size decrease.

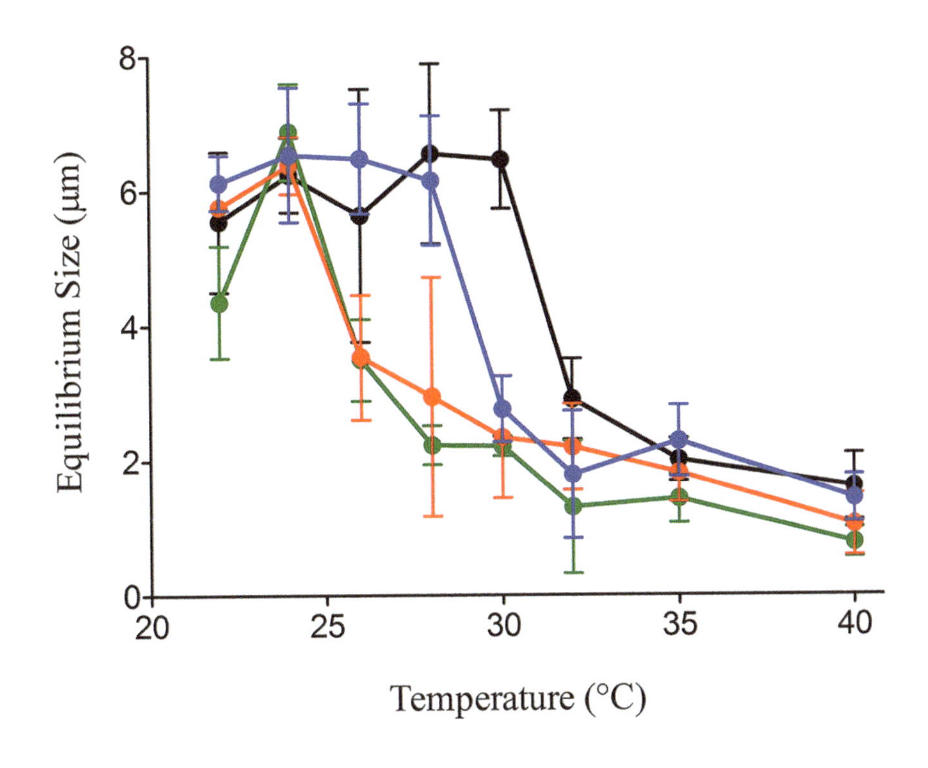

FIGURE 2: Dispersion-temperature of microgels decreases at lower pHs. As pH is reduced from 8.0 (black) to 7.7 (blue) to 7.5 (red) to 7.3 (green), non-linear microgel dispersion-temperature changes were observed while pH decreases. Dispersion temperature dropped ~2°C with a 0.2 pH decrease. Each data point represents mean (+/−SD) of six measurements made in each of six replicate samples.

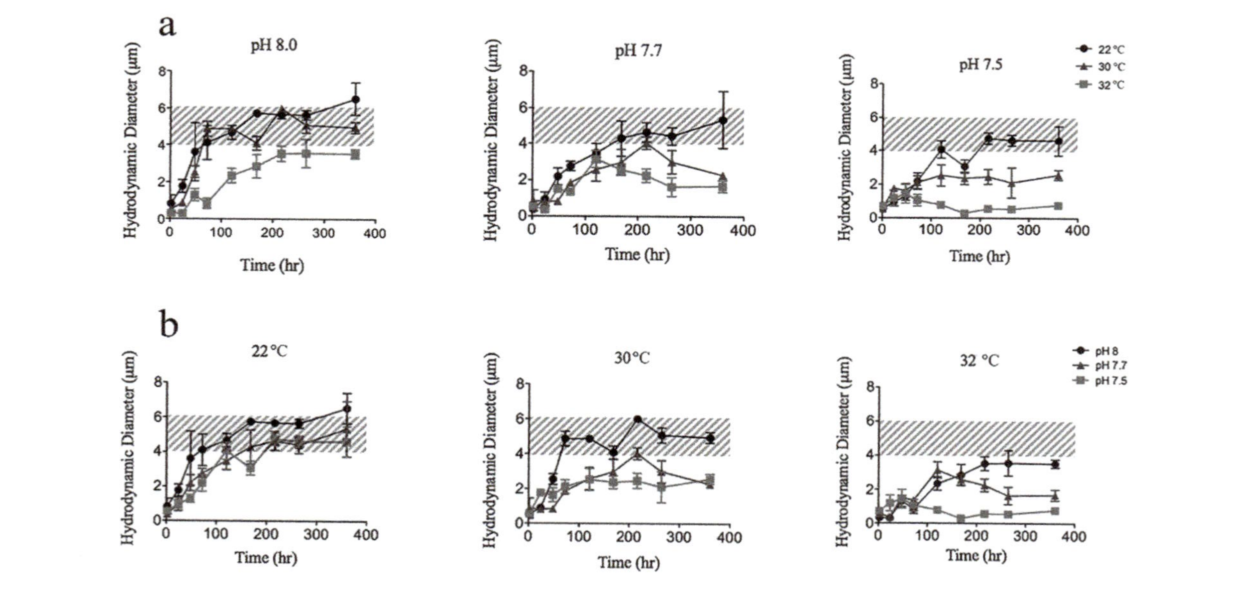

FIGURE 3: DOM assembly monitored with temperature and pH reveals that, as either pH decreases or temperature increases, microgel equilibrium size and assembly rates decrease at a non-linear rate. a. DOM assembly at three temperatures—22°C (black circles), 30°C (blue triangles), 32°C (red squares)—over time at three pH units. Each data point represents (mean ± SD) of six measurements made in each of six replicate samples. b. DOM assembly at three pHs—8.0 (black circles), 7.7 (blue triangles), 7.5 (red squares)—over time at three constant temperature incubations. Microgels assembled in identical pH conditions showed equilibrium size reduction and decelerated non-linear assembly rates when exposed to increased temperature. Each data point represents the mean (+/− SD) of six measurements made in each of six replicate samples. Shaded windows represent an average microgel equilibrium size range (4–6 μm) at 22°C and pH 8.

After confirming microgel equilibrium sizes by dynamic laser scattering, seawater sample pHs were adjusted with 0.1N HCl without an extra buffer system, and seawater samples were incubated at desired experimental temperatures for 24 hrs. Microgel sizes were measured with dynamic laser scattering spectroscopy immediately after pH/temperature adjustments. For the combined pH-temperature impact on microgels experiment, DOM assembly was monitored under pHs ranging from 7.3 to 8.0 and temperatures ranging from 22 to 32°C (Figs. 2 and 3). 0.1N HCl was used to adjust seawater samples to desired pH, and scintillation vials were incubated and measured at each experimental temperature over 15 d.

8.3 RESULTS AND DISCUSSION

8.3.1 INCREASED TEMPERATURE HINDERS DOM-MICROGEL ASSEMBLY

One of the most salient outcomes of climate change is surface warming, projected to increase by 3°C, relative to the 1961–1990 mean, in certain ocean regions by the end of the century [19]. These projections involve temperatures above 30°C [14,17], a previously studied temperature limit. To investigate increasing temperature effects on DOM, we monitored microgel assembly as a function of time using dynamic laser scattering spectroscopy as described previously [4]. DOM in seawater (Friday Harbor, WA, USA), passed through a 0.22 μm-filter, was incubated (at 22, 32 and 35°C) for 24 hrs before assembly was monitored. Our control group (22°C) showed DOM polymers can spontaneously assemble into gels with sizes ranging from 200 nm to 1 μm within 30 minutes. Microgels continued to grow following a nonlinear assembly process to reach equilibrium sizes (~5 μm) within 100 hrs. Unexpectedly, microgel assembly was hindered after 24-hr incubation at 35°C (Fig. 1a). Our results show that increasing seawater temperature progressively hinders microgel assembly. DOM exposed to 35°C can undergo long lasting inhibition of multi-micrometer gel (microgel) formation, yielding only nanogels of less than a micrometer in dimension. No subsequent micron-scale assembly following initial dispersion was observed after 10 days of monitoring.

DOM polymers can spontaneously assemble to form microgels [11]. In order to further test the impacts of increased temperature on DOM self-assembly, we performed another experiment (Fig. 1b). We measured DOM that had reached its equilibrium size (120 hrs at 22°C), first confirming the equilibrium size of microgel (~6 μm) using dynamic laser scattering spectroscopy. Our results from dynamic laser scattering showed that microgel sizes reduce considerably—from 6 μm to 2 μm—after incubation at temperatures above 32°C for 24 hrs (Fig. 1b). Contrary to the previously reported reversible volume transition through pH/temperature titration within a short exposure time [4,11], the significant microgel size reduction observed here appears irreversible (Fig. 1a). The instability (size changes) of DOM microgels above 32°C is consistently evident in Figs 1a and 1b, implying potential impacts of higher temperatures on organic matter.

The non-linear temperature dependency demonstrates one of the polymer characters of microgel kinetics—but implies even short-term environmental perturbation may cause long-term impacts. If our experimental findings in the lab can be extrapolated to complex marine environments, our results indicate that increasing temperatures may impede DOM's ability to form microgels, thereby potentially disrupting the marine colloidal pump—a major carbon cycling driving mechanism, and leading to decreasing downward carbon flux from the surface ocean to the deep ocean [11,20].

8.3.2 POSSIBLE MECHANISM FOR MICRO/NANO MARINE GEL TRANSITION

To probe the mechanism of gel assembly changes, we studied microgels using spectrofluorophotometry. Two major mechanisms of DOM cross-linking have been demonstrated: Ca^{2+} cross-linking and hydrophobic binding [1]. Divalent ion (Ca^{2+}) cross-linking has been identified as a major driving mechanism for DOM microgel formation [1,4]. We used chlortetracycline (CTC) to quantify bound Ca^{2+} on DOM polymers to assess the relative contribution of Ca^{2+} cross-linking [4,21,22]. The CTC fluorescence intensity of heated DOM decreased ~50%, which indicated fewer bound Ca^{2+} ions on polymer surfaces (Fig. 4). Thus, decreased levels of bound Ca^{2+} on DOM polymers associated with increased temperatures would explain smaller equilibrium sizes of forming gels.

Additionally, DOM can be cross-linked by changes in hydrophobic binding [1]. We used Nile red staining to evaluate relative hydrophilic/hydrophobic changes for DOM polymers [23]. Compared to native DOM—with low hydrophobicity (Fig. 4)—our Nile red data revealed increased hydrophobic ratios on heated DOM polymers.

Increasing hydrophobicity leads to an increase in DOM polymer association [14], either via intermolecular or intramolecular interactions. Within the moderate temperature range studied here, the observed nanogel formation (microgel size decrease), DOM Ca^{2+} binding and associated hydrophobic/hydrophilic differences strongly suggest involvement of polymer conformational changes. We speculate different polymer conformations favor the stable nanogel formation, rather than polymer chain entangling, which serves as a major driving force for the observed gel size decrease. From the perspective of polymer assembly, in the absence of covalent cross-linkers, an increase of polymer reptation concomitants a decrease of entangle frictions [24]. In contrast to extended polyionic coils, polymer globule conformation resulting from increased hydrophobicity decreases the intermolecular polymer tangle interactions, which lead to the change of assembly dynamics and final gel equilibrium size. Due to abundant anionic charges and functional groups on DOM polymers, as H^+ concentration increases (owing to a pH decrease), H^+ ions can protonate certain functional groups on DOM polymers, causing fewer available binding sites for divalent Ca^{2+} cross-linking [1,4]; fewer cross-linked polymer chains would result in longer correlation lengths and decreased entangle friction [24]. Thus, low pH situations would lead to smaller equilibrium microgel sizes, as observed in our results (Fig. 3). To verify the influences of low pHs on the stability of microgel matrices, we monitored the impacts of temperatures on microgels at different pHs. Given fewer available cross-linking sites, as hypothesized, the critical temperature decreased with lower pHs: to 30°C at pH 7.7 and again to 28°C at pH 7.5 (Fig. 3). The model of polymer conformation was supported by our current results. However, without direct DOM conformation measurement, future investigation will be needed to further verify this proposed model.

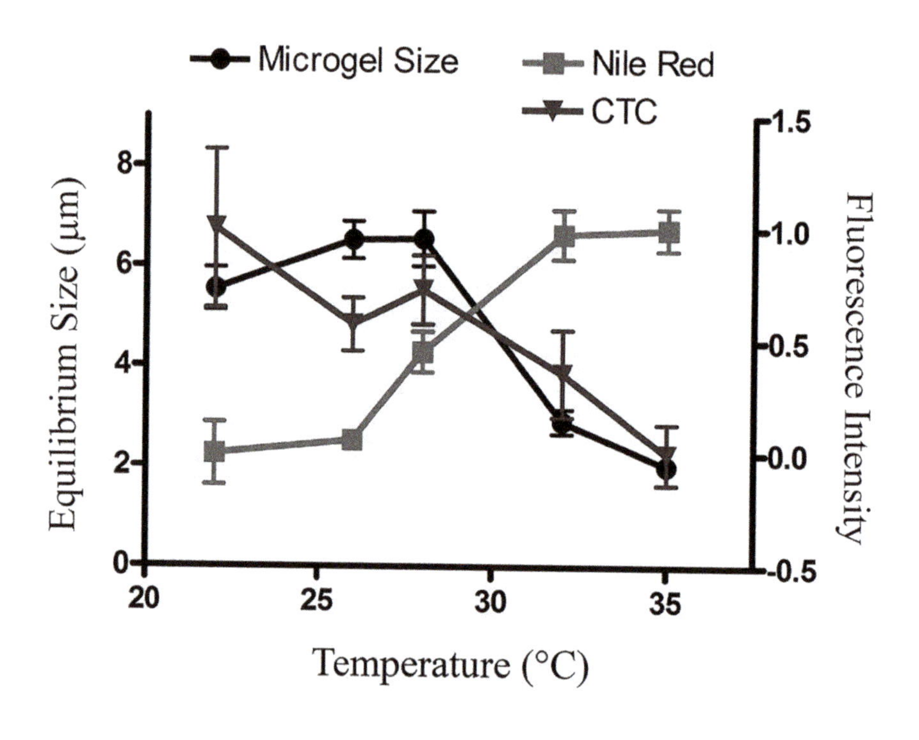

FIGURE 4: Decreased microgel equilibrium size (black circles) and bound Ca^{2+} (blue triangles) with concomitant increase in hydrophobicity (red squares). The non-linear rate of declining microgel size with increased temperature indicates potential cooperativity; around 32°C all three parameters experienced the most pronounced associative effect—a major drop in microgel size and bound Ca^{2+}, with a concomitant rise in hydrophobicity.

8.3.3 MICROGEL/NANOGEL TRANSITION TEMPERATURE LOWERED WITH INCREASED ACIDIFICATION

Microgel assembly is dynamic, stable and reversible in environments up to 30°C [11]. However, because recent ocean warming trends and climate model predictions point to temperatures above this threshold [17,18], we investigated microgel stability in higher temperature environments. In addition to warming, ocean pH is projected to decrease by as much as 0.5 pH units by 2300 [25,26]. To accommodate the more realistic scenario of simultaneous temperature and pH fluctuations, DOM assembly was investigated at 3 pH values (7.5, 7.7, 8.0) coupled with 3 temperature conditions (22, 30, 32°C). We found the equilibrium microgel size decreased with increased temperature, although microgel size at pH 8.0 remained ~4 μm, even at 32°C (Fig. 3). Dropping pH from 8.0 to 7.5 at 22°C, DOM assembly rates decreased (from 120 to 200 hrs), as well as equilibrium sizes (from ~6 to ~4 μm). At pH 7.5, for temperature >30°C, microgel equilibrium sizes dramatically decreased to <2 μm. In addition, the transition temperature of microgels to nanogels was investigated under different pHs (Fig. 2). In general, we found temperature impacts on DOM assembly were amplified with a concurrent pH decrease: at pH 7.5, the microgel size reached only ~0.5 μm at 32°C. This relationship highlights an unexpected synergism—i.e., increasing temperature and acidification—on microgel formation to disturb DOM-microgel exchange and potentially impact the marine carbon cycle.

8.3.4 POTENTIAL FAR-REACHING INFLUENCES ON MARINE PROCESSES

Our results that demonstrate DOM susceptibility to moderate environmental changes provide alternative scenarios for DOM utilization by microbial communities. Microbial respiration depends on the bioavailability of organic carbon in the water column, which is lower for DOM than for microgels [3,27]. In addition, several mechanistic pathways have also been proposed for refractory DOM formation; these include ectoenzyme conversion and exudation during bacterial production [28] as well as hydrophobicity and micelle-like macromolecular structures—proposed to contribute to DOM degradation resistance [29]. The hydrophobicity increase induced by environmental changes observed here permits an understanding of an alternative abiotic mechanism for refractory DOM formation.

The ocean colloidal pump plays an important role in the carbon cycle [30], downward nutrient transport, and organic particle dynamics in the ocean [1,11]. It also has been reported that small size particles (<10 μm) play an important role for carrying organic carbon out of the euphotic zone [31]. The sensitivity of the DOM/microgels shunt to temperature and pH fluctuations reveals the vulnerability of this critical driving force (i.e., the colloidal pump) in future climate change scenarios. Ocean acidification changes the algal metabolism [32]; one mesocosm study showed warming can increase the extracellular release of organic materials from phytoplankton [33]. Environmental impacts resulting from abiotic micro/nanogel transitions warrant further investigations to better understand this complicated system. First, a possible reduction of downward organic carbon fluxes driven by the colloidal pump may be expected. The decreased burial of organic carbon on the seafloor could dampen the capacity of the oceanic carbon sink. In addition, a decrease in microgel downward flux can partially reduce nutrient transport that sustains microbial communities in the twilight (200–1000 m) and deep-sea (>1000 m) regions [5,6,34]. As self-assembled gels are found over a broad range, from the surface to 4000 m depth, this shift of organic carbon supply to the microbial communities would stress biodiversity in the deeper oceans [35], though the magnitude of this disturbance—or any other—remains uncertain. In the surface ocean, reduced microgel assembly could restructure nutrition availability for colonized bacterial communities [1,11,36]. Furthermore, it has been shown that DOM assembly may affect trace metal cycling and the marine biological pump [1], which mediates carbon fixation through the photosynthesis of surface ocean phytoplankton [6,34]. The observed micro/nanogels transition not only indicates the changes of surface area/volume ratio of marine gels, but the different surface properties observed in our study. Studies of the interaction between trace elements and nanogels will significantly contribute to our understanding of the capacity of biological carbon pump in future warmer, more acidic ocean environments [26].

Assessing how climate changes will affect global carbon cycling is one of Earth sciences' great challenges. Our findings suggest ocean acidification and warming, both the result of unabated CO_2 levels, would significantly, and irreversibly, alter marine DOM and DOM/microgel dynamics. These effects reshape the picture of DOM as chemically stable and refractory. Instead, they reveal a more nuanced view of the DOM/microgel transition—one that reveals a critical point sensitive to pH and temperature. From the surface ocean to the 'dark side'—depths past the euphotic zone, through DOM and its associations within the carbon cycle—we propose this synergism of ocean warming and acidification carries potential to be far-reaching, likely affecting major carbon fluxes and microbial communities. Our findings imply a novel potential synergism from seemingly independent changes (e.g., increased temperature and ocean acidification) may hold significant impacts on carbon cycle dynamics and stress the urgency to study marine processes as parts of an integrated system.

REFERENCES

1. Verdugo P, Alldredge AL, Azam F, Kirchman DL, Passow U, Santschi P. The oceanic gel phase: a bridge in the DOM-POM continuum. Marine Chemistry. 2004; 92: 67–85. doi: 10.1016/j.marchem.2004.06.017
2. Verdugo P. Marine microgels. Annual Review of Marine Science. 2012; 4: 1–25. pmid:22457966 doi: 10.1146/annurev-marine-120709-142759
3. Azam F, Long RA. Sea snow microcosms. Nature. 2001; 414: 495, 497–498. pmid:11734832 doi: 10.1038/35107174
4. Chin WC, Orellana MV, Verdugo P. Spontaneous assembly of marine dissolved organic matter into polymer gels. Nature. 1998; 391: 568–572. doi: 10.1038/35345
5. Suess E. Particulate organic-carbon flux in the oceans—surface productivity and oxygen utilization. Nature. 1980; 288: 260–263. doi: 10.1038/288260a0
6. Azam F, Malfatti F. Microbial structuring of marine ecosystems. Nature Reviews Microbiology. 2007; 5: 782–791. pmid:17853906 doi: 10.1038/nrmicro1747
7. Kiorboe T. Formation and fate of marine snow: small-scale processes with large-scale implications. Scientia Marina. 2001; 65: 57–71. doi: 10.3989/scimar.2001.65s257
8. Kiorboe T, Jackson GA. Marine snow, organic solute plumes, and optimal chemosensory behavior of bacteria. Limnology and Oceanography. 2001; 46: 1309–1318. doi: 10.4319/lo.2001.46.6.1309
9. Orellana MV, Matrai PA, Leck C, Rauschenberg CD, Lee AM, Coz E. Marine microgels as a source of cloud condensation nuclei in the high Arctic. Proc Natl Acad Sci U S A. 2011; 108: 13612–13617. doi: 10.1073/pnas.1102457108. pmid:21825118

10. Benner R, Pakulski JD, McCarthy M, Hedges JI, Hatcher PG. Bulk chemical characteristics of dissolved organic-matter in the ocean. Science. 1992; 255: 1561–1564. pmid:17820170 doi: 10.1126/science.255.5051.1561
11. Verdugo P. Marine microgels. Ann Rev Mar Sci. 2012; 4: 375–400. pmid:22457980 doi: 10.1146/annurev-marine-120709-142759
12. Ćosović B, Kozarac Z. Temperature and pressure effects upon hydrophobic interactions in natural waters. Marine Chemistry. 1993; 42: 1–10. doi: 10.1016/0304-4203(93)90245-j
13. David MK. Nutrient dynamics in the deep blue sea. Trends in Microbiology. 2002; 10: 410–418. pmid:12217506 doi: 10.1016/s0966-842x(02)02430-7
14. Ding Y- X, Chin W- C, Rodriguez A, Hung C- C, Santschi PH, Verdugo P. Amphiphilicexopolymers from Sagittula stellata induce DOM self-assembly and formation of marine microgels. Marine Chemistry. 2008; 112: 11–19. doi: 10.1016/j.marchem.2008.05.003
15. Zhang S, Ho Y- F, Creeley D, Roberts KA, Xu C, Li HP, et al. Temporal variation of Iodine concentration and speciation (127I and 129I) in wetland groundwater from the Savannah River Site, USA. Environmental Science & Technology. 2014; 48: 11218–11226. doi: 10.1016/j.marenvres.2015.01.004. pmid:25636164
16. Provencher SW. Contin—a general-purpose constrained regularization program for inverting noisy linear algebraic and integral-equations. Computer Physics Communications. 1982 27: 229–242. doi: 10.1016/0010-4655(82)90174-6
17. Sheppard CRC. Predicted recurrences of mass coral mortality in the Indian Ocean. Nature. 2003; 425: 294–297. pmid:13679917 doi: 10.1038/nature01987
18. Solomon S, Qin D, Manning M, Chen Z, Marquis M, Averyt KB, et al. (eds) Climate Change 2007: The physical science basis. Contribution of Working Group I to the Fourth Assessment Report of the Intergovernmental Panel on Climate Change: Cambridge Univ Press, Cambridge, UK; 2007.
19. 19. Johnson NC, Xie SP. Changes in the sea surface temperature threshold for tropical convection. Nature Geoscience. 2010; 3: 842–845. doi: 10.1038/ngeo1008
20. Verdugo P, Santschi PH. Polymer dynamics of DOC networks and gel formation in seawater. Deep Sea Research Part II: Topical Studies in Oceanography. 2010; 57: 1486–1493. doi: 10.1016/j.dsr2.2010.03.002
21. Orellana MV, Petersen TW, Diercks AH, Donohoe S, Verdugo P, van den Engh G. Marine microgels: Optical and proteomic fingerprints. Marine Chemistry. 2007; 105: 229–239. doi: 10.1016/j.marchem.2007.02.002
22. Oliver AE, Baker GA, Fugate RD, Tablin F, Crowe JH. Effects of temperature on calcium-sensitive fluorescent probes. Biophys J. 2000; 78: 2116–2126. pmid:10733989 doi: 10.1016/s0006-3495(00)76758-0
23. Greenspan P, Mayer EP, Fowler SD. Nile red: a selective fluorescent stain for intracellular lipid droplets. J Cell Biol. 1985; 100: 965–973. pmid:3972906 doi: 10.1083/jcb.100.3.965
24. Gennes PGD. Scaling concepts in polymer physics: Cornell University Press; 1979.
25. 25. Caldeira K, Wickett ME. Oceanography: anthropogenic carbon and ocean pH. Nature. 2003; 425: 365. pmid:14508477 doi: 10.1038/425365a
26. Sunda WG. Oceans. Iron and the carbon pump. Science. 2010; 327: 654–655. doi: 10.1126/science.1186151. pmid:20133563

27. Kiorboe T, Grossart HP, Ploug H, Tang K. Mechanisms and rates of bacterial colonization of sinking aggregates. Applied and Environmental Microbiology. 2002; 68: 3996–4006. pmid:12147501 doi: 10.1128/aem.68.8.3996-4006.2002
28. Jiao N, Herndl GJ, Hansell DA, Benner R, Kattner G, Wilhelm SW, et al. Microbial production of recalcitrant dissolved organic matter: long-term carbon storage in the global ocean. Nat Rev Micro. 2010; 8: 593–599. doi: 10.1038/nrmicro2386
29. Dittmar T, Kattner G. Recalcitrant dissolved organic matter in the ocean: major contribution of small amphiphilics. Marine Chemistry. 2003; 82: 115–123. doi: 10.1016/s0304-4203(03)00068-9
30. Wells ML, Goldberg ED. Colloid Aggregation in Seawater. Marine Chemistry. 1993; 41: 353–358 doi: 10.1016/0304-4203(93)90267-r
31. Hung C- C, Gong G- C, Santschi PH. 234Th in different size classes of sediment trap collected particles from the Northwestern Pacific Ocean. Geochimica et Cosmochimica Acta. 2012; 91: 60–74. doi: 10.1016/j.gca.2012.05.017
32. Hurd CL, Hepburn CD, Currie KI, Raven JA, Hunter KA. Testing the Effects of Ocean Acidification on Algal Metabolism: Considerations for Experimental Designs. Journal of Phycology. 2009; 45: 1236–1251. doi: 10.1111/j.1529-8817.2009.00768.x
33. Engel A, Handel N, Wohlers J, Lunau M, Grossart HP, Sommer U, et al. Effects of sea surface warming on the production and composition of dissolved organic matter during phytoplankton blooms: results from a mesocosm study. Journal of Plankton Research. 2011; 33: 357–372. doi: 10.1093/plankt/fbq122
34. Azam F. Microbial control of oceanic carbon flux: The plot thickens. Science. 1998; 280: 694–696. doi: 10.1126/science.280.5364.694
35. Smith KL, Ruhl HA, Bett BJ, Billett DSM, Lampitt RS, Kaufmann RS, et al. Climate, carbon cycling, and deep-ocean ecosystems. Proc Natl Acad Sci U S A. 2009; 106: 19211–19218. doi: 10.1073/pnas.0908322106. pmid:19901326
36. Passow U. The abiotic formation of TEP under different ocean acidification scenarios. Marine Chemistry. 2011; 128: 72–80. doi: 10.1016/j.marchem.2011.10.004

CHAPTER 9

Ocean Acidification with (De)eutrophication will Alter Future Phytoplankton Growth and Succession

KEVIN J. FLYNN, DARREN R. CLARK, ADITEE MITRA, HEINER FABIAN, PER J. HANSEN, PATRICIA M. GLIBERT, GLEN L. WHEELER, DIANE K. STOECKER, JERRY C. BLACKFORD, AND COLIN BROWNLEE

9.1 INTRODUCTION

Effects of ocean acidification (OA), and specifically the impacts of higher pCO_2 and a lower pH (higher $[H^+]$) on marine life are subjects of much research. Taken together with other climate change events, notably changes in temperature and water-column stability, OA has the potential for various impacts upon marine plankton communities and production [1–9]. The removal of CO_2 by primary production leads to seawater basification (increase in pH), and the tendency towards this event is enhanced by increased nutrient availability. Indeed, in coastal waters subjected to eutrophication by addition of inorganic nutrients that support large algal blooms, the increase in pH can be highly significant and can override any signal from OA [10]. On the contrary, when organic eutrophication promotes increased system respiration (net addition of CO_2), then acidification is increased [11]. Events can also run in series, with inorganic eutrophication promoting first primary production and then respiration during decay of the bloom biomass. Either way, the system pH displays variations that may affect ecology, and these events will be affected with OA by commencement at a lower pH. Such transients are of most consequence in the highly dynamic coastal zones which contain elevated nutrient loadings in comparison with the relatively stable and low nutrient oligotrophic oceans.

Significant basification during bloom development directly influences phytoplankton species growth rates and succession [12–14], selecting species most capable of growth as pH increases [13–16]. Further, the major grazers of phytoplankton, the microzooplankton, can be sensitive to elevated pH [17], just as copepods can show sensitivity to lower pH [8]. Such selective pressures affect the extent of biomass production and also its fate. These pH-sensitive events in future oceans will be affected by the long-term decrease in ocean pH with OA overlain by short-term changes in seawater pH such as basification during phytoplankton blooms. Because of climate change (including changes in terrestrial runoff), human population growth and allied changes in land use, enhanced nutrient release from the land is also expected over the coming decades. There are expected to be particularly severe regional impacts affecting coastal marine ecosystems [18,19], though some areas may see reversal of eutrophication under policies to decrease pollution. Any changes in nutrient loading that affect primary production will also affect the degree of basification.

The implications of the potentially synergistic effects of (de)eutrophication and OA, inducing modifications to phytoplankton growth and species succession, have not previously been explored. With OA, bloom-induced basification will commence at a lower pH. This will fundamentally shift the conditions for competitive interactions between bloom-forming species over a different range of H^+ concentrations, with potentially unanticipated impacts on predator–prey dynamics through to higher trophic levels. Any significant changes to plankton growth in estuarine, coastal and shelf sea systems will have profound implications owing to the substantial contribution these systems make to global productivity [18,19] and marine ecosystem services [20].

This paper explores how these contrasting drivers on seawater pH (pH declining with OA versus basification with inorganic nutrient eutrophication) may affect growth and interactions of different types of phytoplankton. We show, using a combination of experiments and models, that future changes in ocean carbonate chemistry, coupled with changes in the availability of nutrients, will have an important influence on competitive interactions between phytoplankton taxa during bloom development.

9.2 MATERIAL AND METHODS

A brief overview is provided here; detailed methods are described in [21] and in the electronic supplementary material. Three taxonomically contrasting phytoplankton, a weakly calcifying strain of the non-motile prymnesiophyte (*Emiliania huxleyi*), a motile cryptophyte (*Rhodomonas* sp.), and a silicifying diatom (*Thalassiosira weissflogii*), were grown in nitrogen-limiting cultures. Initial pH treatments were set as typical of present day seawater (extant: pH 8.2), future OA (acidic: pH 7.6) or within dense blooms (basic: pH 8.8). pH was then allowed to drift with phytoplankton growth with no CO_2 entry to counter dissolved inorganic carbon (DIC) removal through primary production. (In the oceans, gas exchange to equilibrium is very slow, especially into a mixing depth of many metres.) These pH treatments are henceforth termed ED (start at extant pH with drift), AD (start at acidic pH with drift) and BD (start at basic pH with drift). In addition, comparative treatments were run with pH held fixed at the initial values by addition of HCl or NaOH; these conditions are termed EF (extant pH fixed), AF (acidic pH fixed) and BF (basic pH fixed). Data from experiments (nutrients, biomass C, N and P) were used to configure multi-nutrient, photoacclimative, variable stoichiometry models of phytoplankton growth dynamics, which were coupled to a carbonate chemistry submodel, as described in [22]. These models were then run under different scenarios of nutrient and pCO_2 (OA), and of mixing depth, to simulate growth of single or multi-species phytoplankton communities.

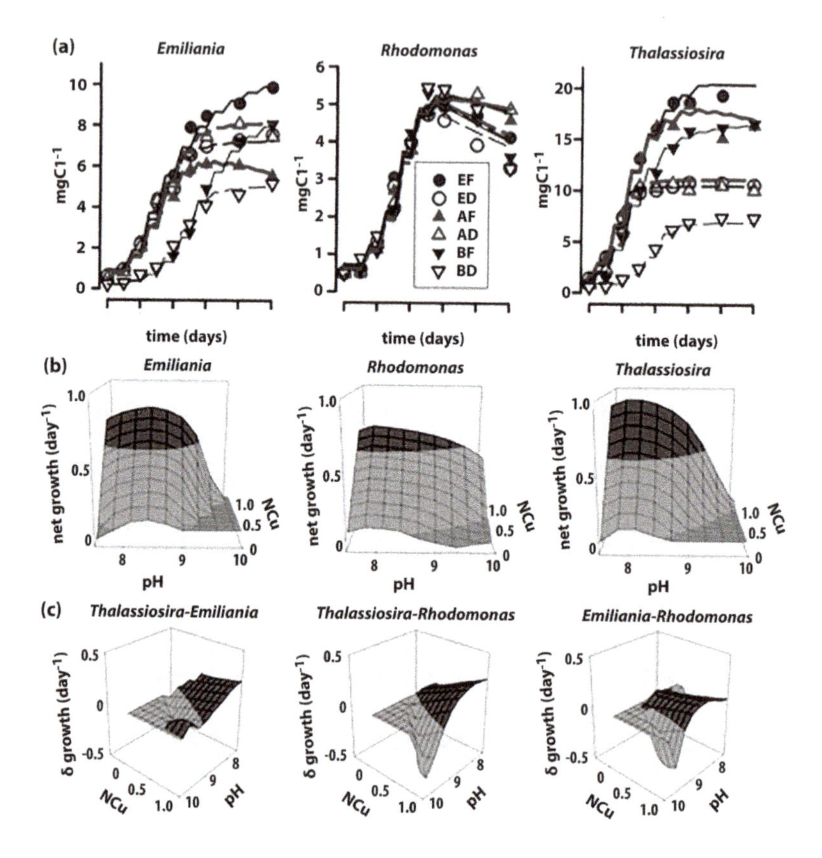

FIGURE 1: Experimental data and model outputs. (a) Experimental data (symbols) for the prymnesiophyte *Emiliania huxleyi*, the cryptophyte *Rhodomonas* sp. and the diatom *Thalassiosira weissflogii* grown under conditions of pH that were extant fixed at pH 8.2 (EF), extant drifting from pH 8.2 (ED), acidic fixed at pH 7.6 (AF), acidic drifting from pH 7.6 (AD), basic fixed at pH 8.8 (BF) or basic drifting from pH 8.8 (BD). Experimental data are averages from duplicate experiments, with the range of those values typically within the symbol size. Lines are model fits to the data. (b) Emergent relationships between growth rate, pH and nutrient status (NCu; where 0 is nutrient-starved and 1 is replete). Measured pH ranged between 7.5 and 10; simulation outputs are shown within these values. Darkest zones (brown in online colour plot) indicate zones with growth rates >0.5 day−1. (c) Differences in net growth potential (δ growth) between pairs of algae, with pH and nutrient status. In each plot, the light zones (blue in online colour plot) indicate where the second named species would outgrow the first named species; a value of δ growth = zero indicates where neither species exhibited positive net growth. (Online version in colour.)

9.3 RESULTS

In comparison with growth in the extant drift (ED) pH systems, growth of Thalassiosira and Emiliania was almost halved in the basic drift (BD) systems, and was similar or slightly enhanced in the acidic drift (AD) systems (figure 1a). There was little difference between *Rhodomonas* grown in drift systems of different initial pH, at least during the nutrient-replete phase. Typically, growth of these single species cultures was greater under fixed-pH rather than under drift-pH conditions (figure 1a), with DIC draw-down continuing until concentrations of substrates for photosynthesis (H_2CO_3 and HCO_3^-) were very low (electronic supplementary material, figures S1, S3–S5). Cessation of growth at high pH in drift systems was thus not a simple consequence of the exhaustion of DIC. Only for Emiliania was growth under AF pH conditions poorer than under AD pH conditions (figure 1a). Within multi-species cultures, *Emiliania* biomass declined following nutrient exhaustion; this could not be explained simply through reference to the results from single species cultures (figure 1 and the electronic supplementary material, figure S1 versus S6), but appears to be related to the extent of growth of total biomass (which was greater in fixed-pH systems), perhaps associated with a lack of DIC and/or an allelopathic (or other cell–cell) interaction.

Fits of the model to experimental data (figures 1a and the electronic supplementary material, figures S1 and S2) were achieved by inclusion of functions relating both phytoplankton growth and death to pH and cellular nutrient status (electronic supplementary material, table S1). The optimal pH, and form of the interaction between pH with nutrient stress, differed between species (figure 1b). When emergent relationships were compared for pairs of species (figure 1c), potential windows of opportunity for each competing species became apparent. Thus, *Emiliania* growth is favoured under extant pH, with more acidic and basic conditions being unfavourable. *Rhodomonas*, which had the lowest maximum growth rate, has its highest competitive scope at elevated pH but only when nutrient-replete; these are conditions likely associated with eutrophic areas during blooms.

Dynamic sensitivity analyses, conducted on simulations for the multi-species cultures under the fixed- and drift-pH regimes (electronic supplementary material, figures S7–S9), indicated that phytoplankton succession is most predictable (i.e. least variable) under extant pH, and least predictable under the acidic (OA) scenarios. This reflects the form of the relationship between pH–nutrient status–growth/death for each species (figure 1b), tracking consumption of DIC and nutrients during growth of the total algal community. Variability in response was further enhanced by the decreased buffering capacity of seawater at lower pH [22], which affects growth under the acidic scenarios. Even within a narrow pH range, the additional impact of nutrient stress can be significant (figure 1c); the prymnesiophyte Emiliania showed greater sensitivity and relatively lower competitive advantage in the acidic fixed-pH (AF) simulations.

Simulations were also made under different conditions of gas exchange equilibration to historic, extant or future atmospheric pCO_2, low medium and high nutrient loadings, and mixing depths (figure 2). These plots should be viewed as being indicative of the potential for different successions, and not necessarily that one or other plankton group (diatom versus cryptophyte versus prymnesiophyte) will dominate in nature under a given set of conditions. Simulated future scenarios generally gave rise to faster and more extensive growth (figure 2), and more extreme elemental stoichiometry (electronic supplementary material, figure S10) when light or Si was not limiting. Simulated bloom composition varied between pCO_2 scenarios, because the deleterious conditions at elevated pH were encountered later during bloom development under future scenarios (the starting pH being lower with OA).

9.4 DISCUSSION

Our study shows species variability within the pH–growth relationship and how this is affected by nutrient stress (figure 1b). Predicted under different OA and (de)eutrophication scenarios (figure 2), the implications of our results for shifting plankton successions in shelf seas are numerous and far reaching. Over time, some waters (e.g. North Sea) will have seen changes from conditions akin to historic pCO_2 with mesotrophy, to extant pCO_2 with eutrophy, to future pCO_2 with mesotrophy, with corresponding scope for significant changes in plankton dominance (figure

2). More broadly, with greater inorganic eutrophy (an increasingly likely event [18,19,23]) coupled with elevated pCO_2, there is potential for an increase in total bloom size and/or of the rate of growth in coastal and shelf seas. The nutrient balance in shallow waters is also affected by basification during blooms leading to the release of bound nutrients [23], and decreased coupling of nitrification–denitrification [24]. With OA, initiation of all these events will occur at a lower average pH, but additional differences in bloom extent and composition as a consequence of other impacts, including temperature changes and pollution control measures may be expected [10,25].

The responses of our coupled experimental–model systems, with respect to growth dynamics and C : N : P, are consistent with empirical studies [1,2,9,26], providing confidence in the behaviour of the models. However, dominance of different taxa in nature will also be affected by other factors, including responses of the zooplankton. Blooms develop only in the absence of effective grazing pressure [27]. The activity of grazers will have several points of interaction. More extreme stoichiometry (electronic supplementary material, figure S10) may be detrimental to grazers [27,28], and also promote toxicity in some phototrophic species [29]. High pH may decrease net growth of grazer populations [17] and hence decrease feeding on phytoplankton, whereas low pH has been seen to affect vertical migration of flagellates [30] and copepod growth [8]. Nonetheless, all these interactions are ultimately secondary to the auto-ecological responses of individual primary producers to nutrients, pH and light.

The fate of this changed primary production, depending on local conditions and bloom composition, when coupled with impacts on grazers, will likely be associated with increased frequency of deleterious events such as harmful and ecosystem-disruptive algal blooms [31,32], and increases in hypoxic and anoxic zones, affecting fisheries [33] and thence food security. For blooms of calcifying phototrophs, and any plankton growing within these blooms, the situation is complicated further as calcification with primary production mitigates pH increases [22]. If coccolithophore growth is adversely affected by lower pH (as seen here), then there is opportunity for enhanced variability in their bloom development; this will depend on how preceding non-coccolithophorid primary production conditions affect the water column (figures 1b,c and 2).

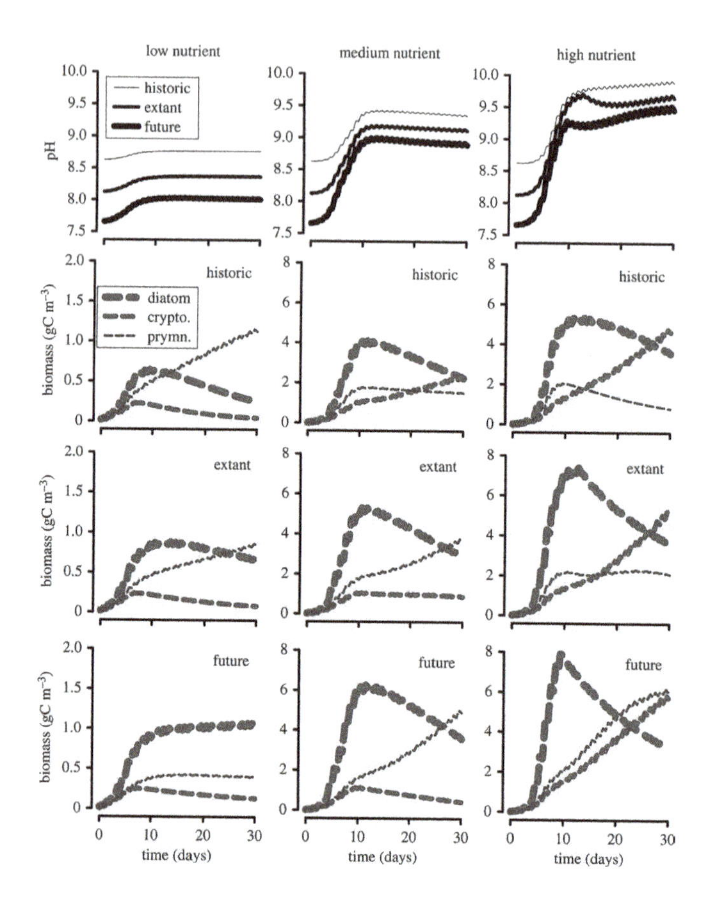

FIGURE 2: Simulations of competitive growth between three phytoplankton types under different pCO_2 scenarios and physico-chemical characteristics. The plankton type models conform to the cryptophyte (crypto.), diatom (diatom) and prymnesiophyte (prymn.) types, as configured against experimental data (figure 1a). Initial algal-C biomass values were the same for each type; biomass has units of gC m^{-3}. Scenarios conformed to historic (preindustrial, pCO_2 280 ppm), extant (pCO_2 390 ppm) and future (prediction for 2100, pCO_2 1000 ppm) conditions. Nutrients (N, P, Si) were supplied at Redfield ratios; the low nutrient regime contained 5 μmol N l^{-1} with mixing depth of 40 m; the medium nutrient regime contained 40 μmol N l^{-1} with mixing depth of 10 m; the high nutrient regime contained 200 μmol N l^{-1} but with only 40 μmol Si l^{-1} with mixing depth of 5 m. In all instances wind speed was set at 10 m s^{-1}, maximum day time surface irradiance at 2000 μmol photons m^{-2} s^{-1} in a 12 : 12 h light : dark cycle, mixing rate between upper and lower layers of 0.05 day^{-1}, temperature of 16°C and salinity of 35. (Online version in colour.)

Our work shows variation in the sensitivity of phytoplankton commencing growth under different initial conditions to changes in pH (figure 2 and the electronic supplementary material, figures S7–S9). While there has always been variability in seawater pH, with OA, the variability in [H+] will be greater, starting from a less buffered state at a higher [H+] [22,34]. Studies on OA, therefore, need to consider the dynamic interplay between primary and secondary producers grown under variable pH, and their physiological mechanisms to cope with these dynamics. While one may expect organisms growing with OA to adapt to lower pH [35], it is uncertain whether such adaptations would then adversely affect scope for growth over the broader pH range with bloom-induced basification. Further, as nutrient stress impacts negatively on the ability to cope with pH-induced stress (figure 1b), the potential for mixotrophs, with their ability to derive nutrients from different sources [36,37], to be favoured under OA scenarios also warrants research. Taken together, we have every reason to expect a more variable and less predictable future for primary production in coastal and shelf seas.

REFERENCES

1. Riebesell U, et al. 2007 Enhanced biological carbon consumption in a high CO2 ocean. Nature 450, 545–548. (doi:10.1038/nature06267)
2. Bellerby RG, Schultz KG, Riebesell U, Neill C, Nondal G, Heegaard E, Johannessen T, Brown KR. 2008 Marine ecosystem community carbon and nutrient uptake stoichiometry under varying ocean acidification during the PeECE III experiment. Biogeoscience 5, 1517–1527. (doi:10.5194/bg-5-1517-2008)
3. Doney SC, Fabry VJ, Feely RA, Kleypas JA. 2009 Ocean acidification: the other CO2 problem. Annu. Rev. Mar. Sci. 1, 169–192. (doi:10.1146/annurev.marine.010908.163834)
4. Artioli Y, Blackford JC, Nondal G, Bellerby RGJ, Wakelin SL, Holt JT, Butenschön M, Allen JI. 2013 Heterogeneity of impacts of high CO2 on the North Western European Shelf. Biogeoscience 10, 9389–9413. (doi:10.5194/bgd-10-9389-2013)
5. Connell SD, Kroeker KJ, Fabricius KE, Kline DI, Russell BD. 2013 The other ocean acidification problem: CO2 as a resource among competitors for ecosystem dominance. Phil. Trans. R. Soc. B 368, 20120442. (doi:10.1098/rstb.2012.0442)
6. Duarte CM, Hendriks IE, Moore TS, Olsen YS, Steckbauer A, Ramajo L, Carstensen J, Trotter JA, McCulloch M. 2013 Is ocean acidification an open-ocean syndrome? Understanding anthropogenic impacts on seawater pH. Estuar. Coast. 36, 221–236. (doi:10.1007/s12237-013-9594-3)

7. Schulz KG, et al. 2013 Temporal biomass dynamics of an Arctic plankton bloom in response to increasing levels of atmospheric carbon dioxide. Biogeoscience 10, 161–180. (doi:10.5194/bg-10-161-2013)
8. Cripps G, Lindeque P, Flynn KJ. 2014 Have we been underestimating the effects of ocean acidification in zooplankton? Glob. Change Biol. 20, 3377–3385. (doi:10.1111/gcb.12582)
9. Schoo KL, Malzahn AM, Krause E, Boersma M. 2013 Increased carbon dioxide availability alters phytoplankton stoichiometry and affects carbon cycling and growth of marine planktonic herbivore. Mar. Biol. 160, 2145–2155. (doi:10.1007/s00227-012-2121-4)
10. Borges AV, Gypens N. 2010 Carbonate chemistry in the coastal zone responds more strongly to eutrophication than to ocean acidification. Limnol. Oceanogr. 55, 346–353. (doi:10.4319/lo.2010.55.1.0346)
11. Cai W-J, et al. 2011 Acidification of subsurface coastal waters enhanced by eutrophication. Nat. Geosci. 4, 766–770. (doi:10.1038/ngeo1297)
12. Macedo MF, Duarte P, Mendes P, Ferreira JG. 2001 Annual variation of environmental variables, phytoplankton species composition and photosynthetic parameters in a coastal lagoon. J. Plankt. Res. 23, 719–732. (doi:10.1093/plankt/23.7.719)
13. Hansen PJ. 2002 The effect of high pH on the growth and survival of marine phytoplankton: implications for species succession. Aquat. Microb. Ecol. 28, 279–288. (doi:10.3354/ame028279)
14. Hinga KR. 2002 Effect of high pH on coastal marine phytoplankton. Mar. Ecol. Prog. Ser. 238, 281–300. (doi:10.3354/meps238281)
15. Lundholm N, Hansen PJ, Kotaki Y. 2004 Effect of pH on growth and domoic acid production by potentially toxic diatoms of the genera Pseudo-nitzschia and Nitzschia. Mar. Ecol. Prog. Ser. 273, 1–15. (doi:10.3354/meps273001)
16. Hansen PJ, Lundholm N, Rost B. 2007 Growth limitation in marine red-tide dinoflagellates: effects of pH versus inorganic carbon availability. Mar. Ecol. Prog. Ser. 334, 63–71. (doi:10.3354/meps334063)
17. Pedersen MF, Hansen PJ. 2003 Effects of high pH on the growth and survival of six marine heterotrophic protists. Mar. Ecol. Prog. Ser. 260, 33–41. (doi:10.3354/meps260033)
18. Howarth RW. 2008 Coastal nitrogen pollution: a review of sources and trends globally and regionally. Harmful Algae 8, 14–20. (doi:10.1016/j.hal.2008.08.015)
19. Doney SC. 2010 The growing human footprint on coastal and open-ocean biogeochemistry. Science 328, 1512–1516. (doi:10.1126/science.1185198)
20. Cooley SR, Kite-Powell HL, Doney SC. 2009 Ocean acidification's potential to alter global marine ecosystem services. Oceanography 22, 172–181. (doi:10.5670/oceanog.2009.106)
21. Clark DR, Flynn KJ, Fabian H. 2014 Variation in elemental stoichiometry of the marine diatom *Thalassiosira weissflogii* (Bacillariophyceae) in response to combined nutrient stress and changes in carbonate chemistry. J. Phycol. 50, 640–651. (doi:10.1111/jpy.12208)

22. Flynn KJ, Blackford JC, Baird ME, Raven JA, Clark DR, Beardall J, Brownlee C, Fabian H, Wheeler GL. 2012 Changes in pH at the exterior surface of plankton with ocean acidification. Nat. Clim. Change 2, 510–513. (doi:10.1038/nclimate1696)
23. Glibert PM, Fullerton D, Burkholder JM, Cornwell J, Kana TM. 2011 Ecological stoichiometry, biogeochemical cycling, invasive species and aquatic food webs: San Francisco Estuary and comparative systems. Rev. Fish. Sci. 19, 358–417. (doi:10.1080/10641262.2011.611916)
24. Gao Y, Cornwell JC, Stoecker DK, Owens MS. 2012 Effects of cyanobacterial-driven pH increases on sediment nutrient fluxes and coupled nitrification-denitrification in a shallow fresh water estuary. Biogeoscience 9, 2697–2710. (doi:10.5194/bg-9-2697-2012)
25. Rabalais NN, Turner RE, Diaz RJ, Justić D. 2009 Global change and eutrophication in coastal waters. ICES J. Mar. Sci. 66, 1528–1537. (doi:10.1093/icesjms/fsp047)
26. Egge JK, Thingstad TF, Engel A, Bellerby RGJ, Riebesell U. 2009 Primary production during nutrient-induced blooms at elevated CO2 concentrations. Biogeoscience 6, 877–885. (doi:10.5194/bg-6-877-2009)
27. Mitra A, Flynn KJ. 2006 Promotion of harmful algal blooms by zooplankton predatory activity. Biol. Lett. 2, 194–197. (doi:10.1098/rsbl.2006.0447)
28. Mitra A, Flynn KJ. 2005 Predator–prey interactions: is 'ecological stoichiometry' sufficient when good food goes bad? J. Plankt. Res. 27, 393–399. (doi:10.1093/plankt/fbi022)
29. Granéli E, Flynn KJ. 2006 Chemical and physical factors influencing toxin production. In Ecology of harmful algae, Ecological studies, vol. 189 (eds Granéli E, Turner JT), pp. 229–241. Berlin, Germany: Springer.
30. Kim H, Spivack AJ, Menden-Deuer S. 2013 pH alters the swimming behaviors of the raphidophyte Heterosigma akashiwo: implications for bloom formation in an acidified ocean. Harmful Algae 26, 1–11. (doi:10.1016/j.hal.2013.03.004)
31. Anderson DA, Glibert PM, Burkholder JM. 2002 Harmful algal blooms and eutrophication: nutrient sources, composition, and consequences. Estuaries 25, 562–584. (doi:10.1007/BF02804901)
32. Heisler J, et al. 2008 Eutrophication and harmful algal blooms: a scientific consensus. Harmful Algae 8, 3–13. (doi:10.1016/j.hal.2008.08.006)
33. Stramma L, Prince ED, Schmidtko S, Luo J, Hoolihan JP, Visbeck M, Wallace DWR, Brandt P, Körtzinger A. 2012 Expansion of oxygen minimum zones may reduce available habitat for tropical pelagic fishes. Nat. Clim. Change 2, 33–37. (doi:10.1038/nclimate1304)
34. Thomas H, et al. 2007 Rapid decline of the CO2 buffering capacity in the North Sea and implications for the North Atlantic Ocean. Glob. Biogeochem. Cycle 21, GB4001. (doi:10.1029/2006GB002825)
35. Lohbeck KT, Riebesell U, Reusch T. 2012 Adaptive evolution of a key phytoplankton species to ocean acidification. Nat. Geosci. 5, 346–351. (doi:10.1038/ngeo1441)
36. Flynn KJ, Stoecker DK, Mitra A, Raven JA, Glibert PM, Hansen PJ, Granéli E, Burkholder JM. 2013 Misuse of the phytoplankton–zooplankton dichotomy: the need to

assign organisms as mixotrophs within plankton functional types. J. Plankt. Res. 35, 3–11. (doi:10.1093/plankt/fbs062)
37. Mitra A, et al. 2014 The role of mixotrophic protists in the biological carbon pump. Biogeoscience 11, 995–1005. (doi:10.5194/bg-11-995-2014)

There are several supplemental files that are not available in this version of the article. To view this additional information, please use the citation on the first page of this chapter.

CHAPTER 10

Coccolithophore Calcification Response to Past Ocean Acidification and Climate Change

SARAH A. O'DEA, SAMANTHA J. GIBBS, PAUL R. BOWN, JEREMY R. YOUNG, ALEX J. POULTON, CHERRY NEWSAM, AND PAUL A. WILSON

10.1 INTRODUCTION

Acidification and reduced carbonate saturation of the oceans are measureable responses to anthropogenic emissions of carbon dioxide into the atmosphere [1]. As major pelagic producers of $CaCO_3$ in the modern ocean, the sensitivity of coccolithophores (single-celled phytoplankton) to changes in surface water chemistry is of particular relevance for ocean biogeochemical cycles and climate feedback systems (for example, ref. 2). Coccolithophores build exoskeletons from individual $CaCO_3$ (calcite) plates—coccoliths—that cover the cell surface and form a protective barrier (the coccosphere). Our current understanding of coccolithophore responses to ocean acidification (OA) is predominantly based on calcification rate experiments (coccolith calcite production per cell, per unit time), which indicate complex, often species- or strain-specific, impacts (for example, refs 3, 4, 5). There is also experimental evidence that coccolithophore species have the ability to evolve and adapt to the acidifiying carbonate chemistry conditions that are projected for the future, over the relatively short timescales of multiple generations (100–1,000 s of generations) [6]. The geological record of fossil coccolithophores is remarkably complete (stratigraphically and taxonomically) throughout the past 220 million years [7], and therefore provides a valuable means by which to test hypotheses of coccolithophore response both to long-term environmental change and abrupt climate perturbations. However, interrogation of the geological record for the specific impact of OA on coccolithophore calcification is challenging, because calcification rate cannot readily be determined in fossil populations, and therefore direct comparison with responses measured in modern culture and field experiments is inhibited.

At a basic level, the fossil remains of coccolithophores provide a means by which to estimate coccosphere calcite quotas (the mass of calcite associated with individual coccospheres or cells), by combining the physical parameters of individual coccoliths (that is, size, thickness and mass) [8, 9, 10] with the cell geometry of intact coccospheres (that is, the size and number of coccoliths per cell) [11]. However, coccosphere calcite quotas cannot be directly used to estimate calcification rate, because high coccosphere calcite quotas can be associated with low calcification rates if the production of coccoliths and/or rate of cell division is low and vice versa. We can overcome this problem by combining measurements of coccosphere calcite quotas (including a newly revised method for calculating coccolith thickness) with estimates of cell division rates, utilizing a recently developed method of coccosphere geometry analysis [11]. This approach enables a closer comparison of fossil and modern species, because it focuses on biomineralization at a cellular level. This technique is underpinned by recent observations of modern *Coccolithus pelagicus* batch culture experiments, supported by field data, which reveal a systematic relationship between coccosphere size (external diameter, Ø), coccolith length (C_L) and the number of coccoliths per coccosphere (C_N) [11]. Importantly, variation in C_N is linked to the growth phase of the cell, such that small cells with few coccoliths are produced during exponential growth phase (normal, rapid division) and larger cells with more coccoliths are produced during early stationary phase (slowed cell division).

Here, we perform a suite of morphometric measurements on fossil coccospheres and their coccoliths to test for skeletal changes across the Paleocene–Eocene Thermal Maximum (PETM), a prominent global carbon cycle perturbation that occurred ~56 million years ago. The PETM is associated with a carbon isotope excursion (CIE) to lower values, global mean surface water warming of ~4–5 °C and widespread deep water acidification [12, 13]. Proxy records of surface water pH change also suggest that variations in surface water chemistry accompanied deep water acidification at the CIE [14]. While the absolute magnitudes of surface water chemistry change remain poorly constrained, modelling studies suggest that the PETM might have been associated with a ~1.5-unit decline in carbonate saturation or a 0.1–0.45-unit decline in pH [15, 16, 17]. In order to enable the study of a greater range of fossil and modern coccolithophore taxa, which can differ greatly in size and crystallography, we have modified existing methods to measure coccolith thickness (for example, refs 8 and 10). The new technique provides an estimate of the amount of calcite per coccolith for ancient taxa, here specifically *C. pelagicus* and *Toweius pertusus*, which were volumetrically and numerically dominant during the late Paleocene and early Eocene (for example, refs 18, 19, 20). In addition, we address the issue of fossil preservation, one of the major problems encountered in geological materials, by utilizing exquisite hemipelagic fossils from palaeo-shelf-sea sediments of New Jersey (Bass River), supplemented by material from California (Lodo Gulch) and Tanzania Drilling Project Site 14 (refs 21, 22), which provide the best-preserved materials available for the PETM. Our morphometric analyses show that calcification rate declined in *T. pertusus* and *C. pelagicus* populations following the CIE onset, and provide evidence for species-specific adaptive response to environmental change during the PETM interval. Furthermore, the thinning of *C. pelagicus* coccoliths immediately before the CIE indicates an additional biomineralization response, likely caused by a different environmental control, possibly OA.

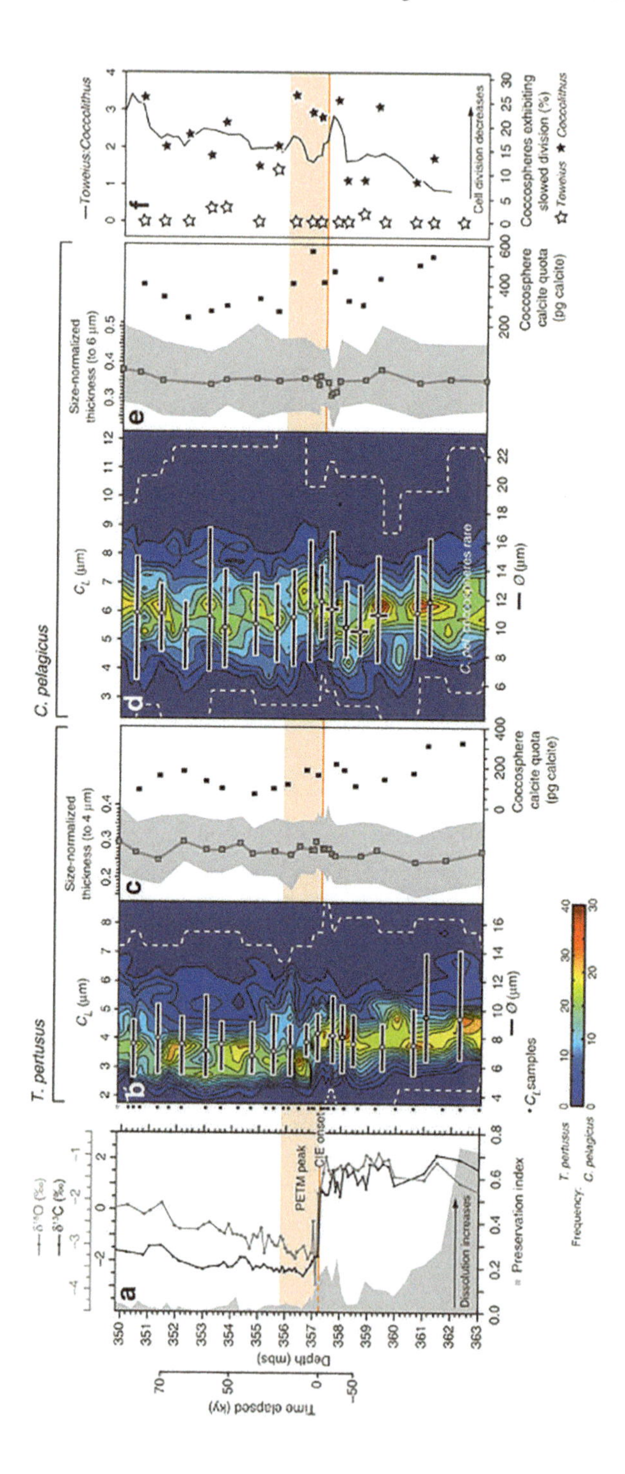
T. pertusus
C. pelagicus
a
b
c
d
e
f
δ18O (‰)
δ13C (‰)
PETM peak
CIE onset
Dissolution increases
Preservation index
Depth (mbs)
Time elapsed (ky)
CL (μm)
Ø (μm)
CL samples
Size-normalized thickness (to 4 μm)
Coccosphere calcite quota (pg calcite)
Size-normalized thickness (to 6 μm)
C. pelagicus coccospheres rare
Frequency
T. pertusus
C. pelagicus
—Toweius:Coccolithus
Cell division decreases
Coccospheres exhibiting slowed division (%)
Toweius ★ Coccolithus

FIGURE 1: Stable isotope, coccolith preservation and morphometric records at Bass River, New Jersey. (a) Bulk carbon (black squares) and bulk oxygen isotopes (grey diamonds; data from John et al.21) with a quantitative nannofossil preservation index (data from Gibbs et al.38; grey shading). Time from the initial carbon isotope excursion (CIE) is indicated (kyrs; following John et al.21), as are depths from which coccolith length (CL) measurements were collected (black squares). (b,d) Frequency data for coccolith lengths (CL; totalling 3,200 and 3,050 measurements for *T. pertusus* and *C. pelagicus*, respectively) are interpolated to equal depth steps of 10 cm, with minimum and maximum size-bins (dashed line). Mean coccosphere diameters (Ø; white squares) with 5th and 95th percentiles of each population (horizontal black bars) and the sampling interval (vertical black bars) are shown, calculated from a total of 507 and 375 coccospheres for Toweius and Coccolithus, respectively. (c,e) Mean coccolith size-normalized thickness for *T. pertusus* and *C. pelagicus* (dark grey squares), with the 5th and 95th percentiles of each sample (grey shading). Uncertainty on mean size-normalized thickness is calculated as two s.d.s across the bootstrap results at the length to which thickness is being normalized and does not exceed ±0.008 μm. Mean coccosphere calcite quotas are shown (black squares). (f) The percentage of each population that exhibits coccosphere geometry typical of slowed cell division (early stationary growth phase, see ref. 11; open stars for *T. pertusus* and closed stars for *C. pelagicus*) and the ratio of *T. pertusus* to *C. pelagicus* coccoliths (black line). The onset of the CIE (orange line) and interval of peak warmth during the PETM (orange shading) are indicated (following John et al.21).

10.2 RESULTS

10.2.1 COCCOLITH MORPHOMETRIC MEASUREMENTS AND COCCOSPHERE GEOMETRY

Our morphometric data reveal a clear, positive relationship between coccolith size (distal shield length, C_L) and coccolith thickness (partial thickness of the proximal shield, C_T; see Methods section) at all sites, which indicates that the thickness of coccoliths typically increases in proportion with their size (Supplementary Figs 1 and 2b,c,g,h), in effect an allometric relationship. To critically assess the variations in C_T that are beyond this relationship, we calculate size-normalized coccolith thickness (see Methods section). Calculating size-normalized thickness provides a quantitative estimate of the divergence between the C_L and C_T records, and therefore describes the variation in C_T that is independent of the 'normal' C_L and C_T relationship (Supplementary Fig. 2d,i). Our high-resolution downcore records at Bass River demonstrate species-specific differences in size-normalized thickness, with a notable minimum in mean size-normalized thickness of *Coccolithus pelagicus* at ~357.56 m below surface (mbs; Fig. 1), immediately before the onset of the CIE. Variations in size-normalized thickness are decoupled from coccolith preservation, with no evidence of coccolith thinning at levels of increased dissolution (Fig. 1), where a secondary modification of C_T would be most likely. This observation is further supported by scanning electron microscope (SEM) images, which show no preservation-related coccolith thinning that could impact our thickness estimates (Fig. 2 and Supplementary Methods).

Modal C_L values indicate a step-shift decrease in *Toweius pertusus* across the peak of the PETM (centred on 356.83 mbs; Fig. 1b) from values typically greater than ~4 µm to values <3.8 µm. For *C. pelagicus*, downcore modal C_L values reveal minor fluctuations and a transient increase during the peak of the PETM, and a longer-term increase in maximum values (Fig. 1d). Coccosphere geometry (the relationship between C_L, Ø and C_N) is more tightly constrained in *T. pertusus* than in *C. pelagicus*, which shows a broader range of Ø and C_L and a highly variable C_N (consistent with ref. 11 and Supplementary Fig. 3). The downcore record of coccosphere Ø broadly tracks variations in modal C_L values for both taxa (Fig. 1b,d), with divergence occurring where C_N varies. Using the proportion of coccosphere population that exhibits stationary phase geometry, with non-dividing coccospheres identified as those that have $C_{N\geq16}$ for *C. pelagicus* and $C_{N\geq12}$ for *T. pertusus* [11], we are able to estimate the general growth phase of our individual populations through time across the PETM interval. *Toweius* populations typically display exponential phase coccosphere geometry (characterized by a low proportion of coccospheres with $C_{N\geq12}$), thereby indicating that high levels of cell division are maintained across the PETM (Fig. 1f). *Coccolithus* coccosphere geometries, however, reveal intervals of slowed division, characterized by an increased proportion of large coccospheres with $C_{N\geq16}$, (and a reduced proportion of post-division coccospheres with $C_{N\leq8}$), particularly across the onset and peak of the PETM (consistent with ref. 11; Fig. 1f). These levels of reduced cell division in *C. pelagicus* are broadly supported by transient increases in the relative numbers of disarticulated *T. pertusus* to *C. pelagicus* coccoliths preserved in the sediments (Fig. 1f).

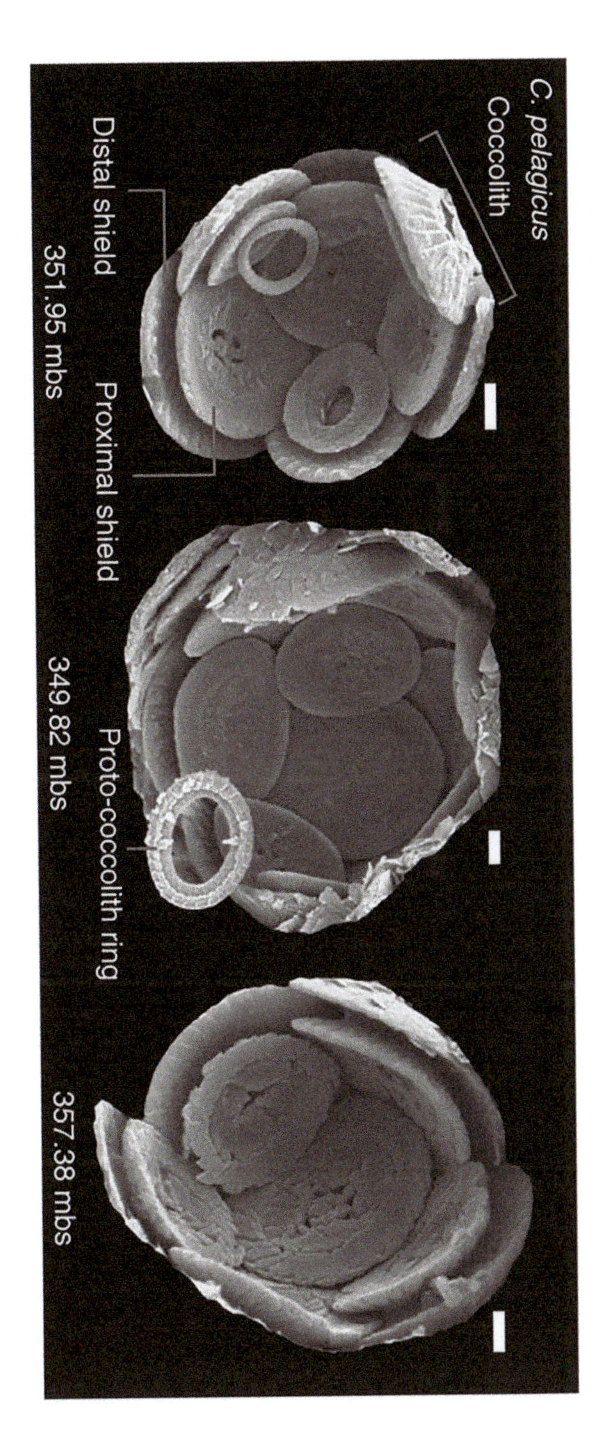

FIGURE 2: Scanning electron micrographs of *C. pelagicus* at Bass River, New Jersey. Sample depths are 351.95, 349.82 and 357.38 mbs as indicated. The minimum in size-normalized thickness of *C. pelagicus* occurs between 357.38 and 357.36 mbs. Individual scale bars indicate 1 μm.

10.2.2 COCCOSPHERE CALCITE QUOTAS AND CALCIFICATION RATES

We use C_L measurements to calculate coccolith mass by applying species-specific shape factors (following ref. 23; see Methods section). We then adjust these shape factor calculations by using the change in our mean size-normalized thickness measurements to provide estimates of variation in coccolith mass that might result directly from the thinning and thickening of coccoliths (see Methods section; Supplementary Fig. 2e,j). We find that the net impact of thickness variation on coccolith mass across the PETM accounts for up to ~5–11% and ~6–16% of the amount of calcite per coccolith for *Toweius pertusus* and *Coccolithus pelagicus*, respectively (see Methods section). This change in calcite mass per coccolith is modest in comparison with variations in coccosphere calcite quotas that result from the observed changes in C_L, Ø and C_N across the PETM, which are up to 500% for *T. pertusus* (that is, up to a fivefold difference in mean cocoshere calcite mass across the record) and 240% for *C. pelagicus* (Fig. 1c,e). Although these variations in coccosphere calcite quota are significant, they do not constitute evidence for a change in the coccolithophore calcification rate during the PETM, because they do not account for potential variations in coccolith production or cell division. Using a similar range of cell division rates to those observed in *C. pelagicus* and *Emiliania huxleyi* culture and field experiments [11] (0.5–1.0 and 0–0.2 divisions per day for exponential and stationary phase, respectively; see Methods section), we conservatively estimate that for *Coccolithus*, the observed shift to early stationary phase cell division could result in at least a near halving of the calcification rate of those populations, from 220–440 to <120 pg calcite per cell per day, despite the increase in maximum coccosphere calcite quotas during the peak of the PETM (Figs 1e and 3). For *Toweius* populations, which maintain high levels of cell division across the PETM, the effect of variation in coccolith production and cell division on calcification rate is less marked. However, unlike *Coccolithus* populations, the step-shift decrease in *Toweius* C_L and Ø values at ~356.83 mbs, and the resultant decrease in coccosphere calcite quota from pre- to post-CIE onset populations (Fig. 1b,c), have a clear influence on calcification rate. Based on the ranges of coccosphere calcite quota for pre- and post-CIE onset populations, and a realistic range of growth rate variation within exponential phase (0.5–1.0 divisions per day), this change in morphology results in a reduction of calcification rate in *Toweius* populations from 140–260 to 70–130 pg calcite per cell per day, again, an approximate halving of calcification rate (Fig. 3; for reference, field populations of modern *E. huxleyi* typically produce calcite at a rate of 25–75 pg calcite per cell per day) [24].

10.3 DISCUSSION

Our records of coccolith size-normalized thickness and coccosphere geometry at Bass River have enabled us to identify three key diagnostic features of coccolithophore calcification across the PETM. First, the minor reduction in *Coccolithus pelagicus* size-normalized thickness immediately before the CIE; second, the significant decrease in calcite production by *Coccolithus* populations during the interval of peak warmth; and third, the decrease in calcite production in *Toweius* populations post-CIE onset (Figs 1 and 3). *Coccolithus* calcite production was likely controlled by changes in growth phase, resulting from the overall environmental changes associated with the PETM [11], with similar growth phase variations also evidenced by low-resolution time series data from other sites including Lodo Gulch (California), the Bay of Biscay and Tanzania [11]. However, the decrease in calcite production in *Toweius* populations reflects a reduction in size of coccospheres and associated coccoliths via a population shift towards smaller cells, likely resulting from either a phenotypic shift or biogeographic introduction, rather than a change in overall cell division. Our combined records of coccosphere calcite quota, C_L, Ø and C_N, therefore indicate that the significant changes in calcification rate of the *Coccolithus* and *Toweius* populations reflect responses that are species specific, involving reduced cell division in *Coccolithus* and a prolonged cell (and coccolith) size decrease in *Toweius*. Importantly, the reduced calcite production is likely linked with the environmental factors (predominantly temperature, nutrient availability and irradiance) that we know have major influence on growth phase, cell size [11, 25, 26] and species biogeography [27] in the natural environment, and is unlikely to predominantly result from OA in either species.

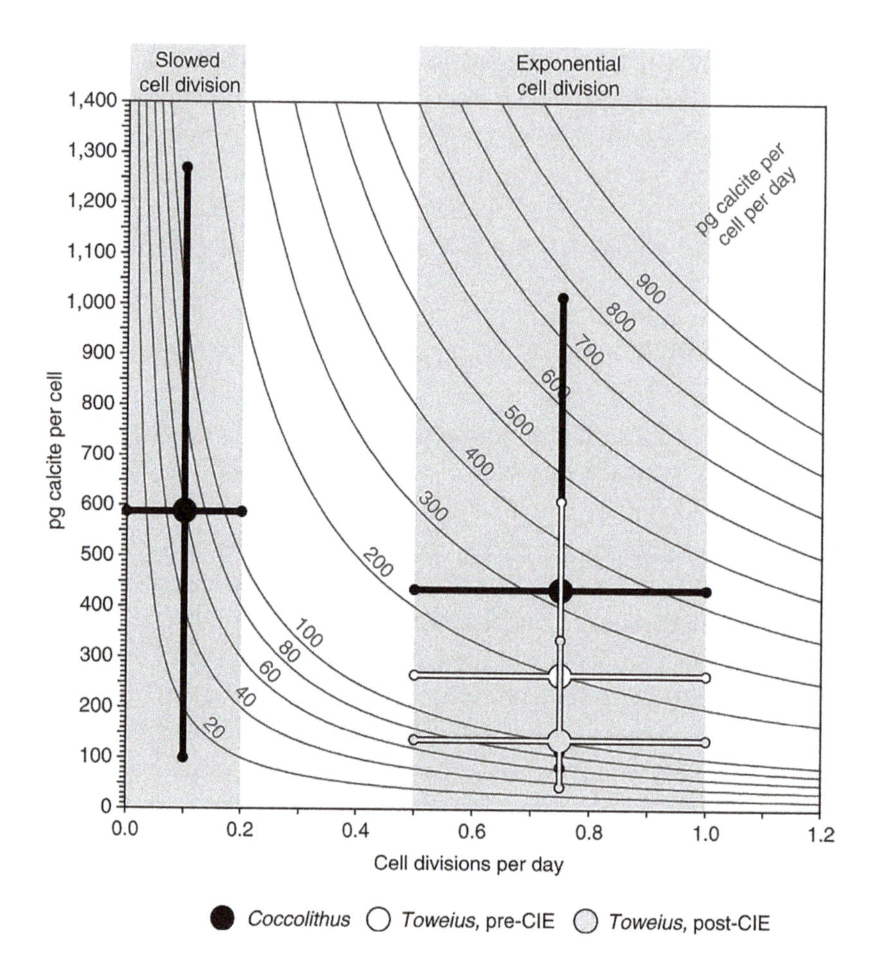

FIGURE 3: Estimating calcification rate for fossil coccolithophore species. Calcification rates (curved red lines) are derived by multiplying the amount of calcite produced per cell by the number of cell divisions per day. Mean (large circles) and the 5th and 95th percentiles (vertical lines) of coccosphere calcite quotas are shown (*Toweius* in white and yellow for pre- and post-CIE onset populations, respectively; *Coccolithus* in black). Populations exhibiting exponential phase coccosphere geometry (normal rates of cell division) are plotted between 0.5 and 1.0 divisions per day and likely vary within this range. *Coccolithus* populations that are exhibiting early stationary phase coccosphere geometry (slowed division; characterized by a high proportion of CN≥16) across the onset and peak of the PETM are plotted between 0 and 0.2 divisions per day. Small circles indicate intersections between calcification rates and either the measured percentile range of coccosphere calcite quotas (on vertical lines) or the inferred range of divisions per day (on horizontal lines).

The reduction in *C. pelagicus* size-normalized thickness differs from the coccosphere geometry data (C_L, Ø and C_N) as it occurred before the onset of the CIE and over a short and discrete time interval. As changes in *C. pelagicus* size-normalized thickness are apparently decoupled from mean cell size and the ecophysiological factors (primarily growth phase) that govern coccosphere geometry [11], the thinning was most likely caused by a different environmental factor, with OA being a primary candidate. Although changing coccolith thickness is the type of biomineralization response we might expect to result from ocean chemistry changes as biomineralization becomes more metabolically expensive, coccolith thinning is still not equivocal evidence for OA and the underlying mechanism for how a thinning response might occur requires cautious consideration. The observed variation in coccolith size-normalized thickness could be owing to, first, a biomineralization response to changes in surface water chemistry with phenotypic plasticity in C_T within successive ancestor-descendent populations, and/or, second, selection of genotypically distinct populations with subtle differences in C_T. A degree of phenotypic plasticity that accounts for the relatively minor changes in mean size-normalized thickness is certainly plausible. However, as internal coccolithogenesis occurs within a pH-regulated vacuole [28], the selection of genotypes with different coccolith thicknesses may represent a more realistic explanation. As a first approximation, mixture analyses (see Methods section) performed on C_L data from all sites support this hypothesis, as they indicate the presence of at least three discrete, size-defined morphotypes of each fossil 'species' of *T. pertusus* and *C. pelagicus*, (Fig. 4a–d and Supplementary Table 1); the rare and smallest morphotype of *T. pertusus* has a slightly different characteristic size-normalized thickness (Fig. 4e and Supplementary Fig. 4). This morphotype diversity is likely a considerable underestimate, because genetic and ecological studies of living coccolithophores show that genotypic diversity within species or subspecies is significantly higher than is suggested by the subtle morphological differences seen in coccoliths and coccospheres (for example, refs 29, 30, 31). In addition, broad genotypic diversity has recently been documented in experiments with modern *Emiliania huxleyi*, revealed as subtle differences in the physiology and ecological niche of individual strains, such that changes in environment result in selection of different genotypes [6].

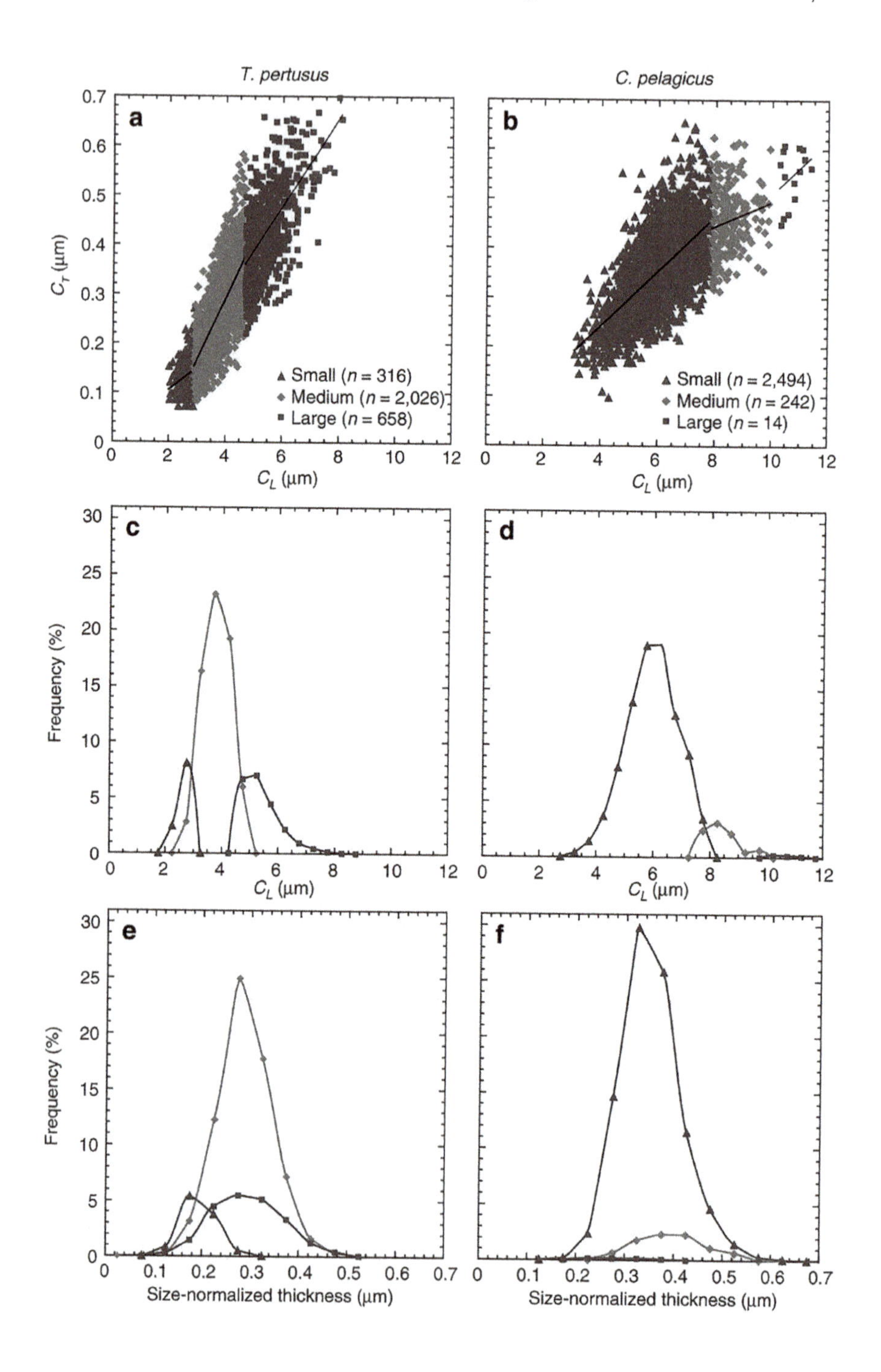
T. pertusus
C. pelagicus
a
b
c
d
e
f
C_T (μm)
C_L (μm)
Frequency (%)
Size-normalized thickness (μm)
▲ Small (n = 316)
◆ Medium (n = 2,026)
■ Large (n = 658)
▲ Small (n = 2,494)
◆ Medium (n = 242)
■ Large (n = 14)

FIGURE 4: Species morphotypes at New Jersey, California and Tanzania. (a,b) Coccolith length (C_L) and thickness (C_T) measurements for size-defined morphotypes of *T. pertusus* and *C. pelagicus*, identified using mixture analyses. Frequency distributions of C_L (c,d) and size-normalized thickness (e,f) of each size-defined morphotype. For *T. pertusus* 'small' is <2.9 μm, 'medium' is >2.9 to <4.7 μm and 'large' is >4.7 μm, whereas for *C. pelagicus*, 'small' is <7.7 μm, 'medium' is >7.7 to <9.9 μm and 'large' is >9.9 μm.

Whether a function of phenotypic plasticity or selection of genotypically distinct populations, if the thinning of *C. pelagicus* coccoliths is linked to changes in surface water carbonate chemistry, this would suggest that OA preceded the onset of the CIE at Bass River by ~3,000–4,000 years (based on the Bass River age model that correlates the CIE to orbitally tuned data from Ocean Drilling Program Site 690 (refs 21, 32)). Although the CIE provides for reliable identification of the PETM in the geological record, a range of other chemical and biotic evidence indicate that significant environmental change preceded the CIE at multiple sites [32]. The thinning of *C. pelagicus* coccoliths is an addition to this existing evidence for pre-CIE changes, which includes TEX86 palaeothermometry data that indicate substantial sea surface temperature increase and an anomalous abundance acme of the subtropical dinoflagellate cyst *Apectodinium*, at least along the New Jersey margin and in the North Sea [32] (Supplementary Fig. 4a). The significant biotic response seen in two key plankton groups, coccolithophores and dinoflagellates, indicates rapid pre-CIE environmental change, because the short generation time of these organisms typically increases their resilience to all but the most abrupt environmental perturbation. While the cause of pre-CIE environmental change remains unclear, evidence for pre-CIE surface water OA indicates carbon cycle involvement. However, any precursor increase in atmospheric carbon dioxide would need to be from a different source to that of the isotopically light carbon, which caused the main CIE, in order to not significantly affect the carbon isotope record [32, 33]. Currently, it is not clear what the source of this carbon might be as, while volcanic outgassing is an obvious candidate, it is currently difficult to reconcile timescales of addition of known carbon sources with the rapid environmental changes necessary to induce the observed range of biotic features [32].

Overall, our PETM data suggest that changes in the environmental factors that affect coccolithophore growth (that is, temperature, nutrient availability and irradiance) had the potential to significantly influence global pelagic carbonate production in the Paleogene, and this is likely to be just as significant today. A change in division rate in modern *Coccolithus* populations could actually have a greater influence on overall calcite production than that observed during the PETM, because the modern *C. pelagicus* subspecies, *pelagicus* and *braarudii*, typically produce larger coccoliths than Paleogene *C. pelagicus* [11]. This is reflected in estimates of mean rate of calcite production during exponential division, which are ~535 and 3,360 pg per cell per day for modern *C. pelagicus* ssp. pelagicus and ssp. braarudii, respectively, compared with ~332 pg per cell per day for Paleogene populations (modern data obtained from ref. 11). However, global calcite production is also dependent on species biogeography and abundance, and while modern *C. pelagicus* is a dominant and large coccolithophore similar to its fossil counterparts, its biogeography is more restricted in the modern ocean (for example, refs 34 and 35). Instead, modern coccolithophore populations are typically dominated by smaller and more lightly calcified taxa, the most abundant species being *E. huxleyi*, a descendent of *Toweius*. Shifts in *E. huxleyi* C_L and Ø, similar to those we have documented in *Toweius* during the PETM, could therefore have significant impact on modern rates of global calcification. Moreover, coccolith thickness variations may play a greater role today than in the past because *E. huxleyi* has an unusually variable, perforate coccolith architecture that results in up to fourfold differences in calcite mass, for very similar sized coccoliths, across its morphotypes [23]. This perforate architecture could potentially accommodate much greater levels of thickness and mass change than we observed in the dominant Paleogene coccoliths, in response to the more rapid OA that is predicted for the coming centuries. Ultimately, though, it is the factors that govern the taxonomic composition of coccolithophore communities, their biogeography, growth rate and adaptive response, which will likely exert the most significant control on overall calcite production by coccolithophores, with comparatively little direct impact on biomineralization from changes in ocean chemistry.

10.4 METHODS

10.4.1 MATERIAL AND SITE DESCRIPTIONS

Our morphometric data are from coccoliths and intact coccospheres from the palaeo-shelf-sea sediments of Bass River, New Jersey (39°36.42 N, 74°26.12 W), supplemented by material from Lodo Gulch, California (36°32.18 N, 120°38.48 W) and Tanzania Drilling Project Site 14 (9°1659.89 S, 39°30 45.04 E). The Paleogene sections at Bass River, Lodo Gulch and Tanzania are all stratigraphically expanded with average sedimentation rates of ~10, ~24.6 and 4 cm per kyr, respectively [21, 20]. The Bass River section is stratigraphically complete for the first ~100 kyr of the 170 kyr CIE [32]. All three sections comprise clay-rich fine sandy and silty sediments deposited above the lysocline in outer shelf to upper slope environments [20, 21, 22], thereby minimizing the effects of secondary dissolution on the preserved coccoliths [21, 20]. Preservation was assessed using SEM images, taken from sediment surfaces following techniques used in ref. 36.

10.4.2 SAMPLING STRATEGY AND MORPHOMETRIC MEASUREMENTS

All measurements were made using standard smear slides, with samples spanning the PETM section at Bass River. The highest resolution sampling of 6 cm intervals was undertaken where carbon isotope values show the greatest stratigraphic variability. We have measured morphometric parameters from statistically significant samples of 100 disarticulated coccoliths per species. Measurements were collected from the first 100 coccoliths of each species identified per slide and include coccolith size (distal shield length, C_L) and coccolith thickness (C_T, to quantify partial thickness of the proximal coccolith shield according to its birefringence under cross-polarized light; see Supplementary Methods). We calculate size-normalized thickness using the mean slope of a bootstrapped linear regression between C_L and C_T for each sample, including 100 measurements that are selected at random from the data set for 100,000 iterations in the bootstrap. To determine the distribution of size-normalized thickness for each sample, we effectively collapse the data to a consistent, species-specific C_L (4 μm for *T. pertusus* and 6 μm for *C. pelagicus*) along the mean slope of the bootstrapped regression (Supplementary Fig. 5), as follows:

$$\text{Size-normalized thickness (to x μm)} = ((\text{x} - \text{length}) \times \text{slope of regression}) + \text{original thickness} \quad (1)$$

In addition, we present the first high-resolution coccosphere size record for the PETM, with measurements of up to 30 fossil coccospheres per species per sample where available, totalling 507 coccospheres for *T. pertusus* and 375 for *C. pelagicus* from 23 and 24 samples, respectively. Morphometric fossil coccosphere data include coccosphere size (external diameter Ø) and the number (C_N) and size (C_L) of intact coccoliths on each coccosphere (following ref. 11). All measurements were collected using Cell^D imaging software, with images taken using a colour DP71 video camera attached to an Olympus BX51 cross-polarizing microscope.

10.4.3 ESTIMATING COCCOSPHERE CALCITE QUOTAS

We estimate coccosphere calcite quotas for intact coccospheres by estimating the mass of each coccolith and multiplying this by C_N. Coccolith mass is estimated as:

$$\text{Density of calcite} \times \text{shape factor} \times C_L^3 \quad (2)$$

following ref. 23, using 2.7 as the density of calcite, 0.060 as a shape factor for *C. pelagicus* and a conservative placolith shape factor of 0.055 for *T. pertusus*. We also include an estimate of C_T variations using our measurements from disarticulated coccoliths, volumetrically modifying the shape factors by adding the difference between mean size-normalized thickness and the expected thickness of the same-sized coccolith, based on its shape factor. We include minimum and maximum values for the estimated net impact of C_T change on coccolith carbonate mass, with minimum values accounting for a change in C_T that is restricted to the proximal shield, while maximum values account for a change in C_T that alters the proximal shield, distal shield and tube cycle.

10.4.4 ESTIMATING CALCIFICATION RATES FOR FOSSIL POPULATIONS

Our estimated fossil coccolithophore calcification rates assume that rates of exponential phase cell division during the Paleogene were similar to those observed in culture and field experiments of modern *C. pelagicus* and *E. huxleyi* (typically between ~0.5 and 1 divisions per day) and that populations displaying early stationary phase (slowed division) coccosphere geometries are dividing slower than minimum exponential rates (<0.2 divisions per day) [11]. These division rates likely represent realistic but conservative rates for Paleogene communities, which lived in warmer waters than modern taxa and may therefore have undergone slightly higher rates of division. Fossil coccolithophore calcification rates are derived by multiplying the amount of calcite produced per cell by the number of cell divisions per day.

10.4.5 USING MIXTURE ANALYSES TO IDENTIFY SIZE-DEFINED MORPHOTYPES

We provide a first order approximation of the presence of different species morphotypes by performing mixture analyses on C_L data from all sites using Palaeontological Statistics. Mixture analyses aim to identify the parameters of normally distributed groups within a pooled sample, and use a maximum likelihood approach to assign each C_L measurement to one group. Here, mixture analyses identify up to three groups, or 'morphotypes', within each species. Results are based on a single morphometric parameter, C_L, and therefore do not provide definitive evidence of full morphotypic variability, but instead indicate a conservative number of possible differences within a species. We calculate the Akaike Information Criterion (AICc) values for each mixture analysis (that is, identifying 1, 2 or 3 morphotypes of each species; Supplementary Table 1) using Palaeontological Statistics, and convert these to Akaike weights between 0 and 1, following ref. 37. Akaike weights can be interpreted as an approximate probability (between 0 and 1) that each of the mixture analysis scenarios tested (that is, 1, 2, 3 or more morphotypes) is the best candidate. The Akaike weights sum to one across the candidate scenarios. For both species, the mixture analysis that defines three morphotypes records the lowest AICc value and the Akaike weight that is nearest to 1 (1.0 and 0.95 for *Toweius* and *Coccolithus*, respectively; Supplementary Table 1).

REFERENCES

1. Feely, R. A. et al. Impact of anthropogenic CO2 on the CaCO3 system in the oceans. Science 305, 362–366 (2004).
2. Doney, S. C. et al. Ocean acidification: the other CO2 problem. Annu. Rev. Marine Sci. 1, 169–192 (2009).
3. Fabry, V. J. Marine calcifiers in a high-CO2 ocean. Science 320, 1020–1022 (2008).
4. Ridgwell, A. et al. From laboratory manipulations to Earth system models: scaling calcification impacts of ocean acidification. Biogeosciences 6, 2611–2623 (2009).
5. Riebesell, U & Tortell, P.D. inOcean Acidification (eds Gattuso J. P., Hansson L. 99–121Oxford Univ. Press (2011).
6. Lohbeck, K. T., Riebesell, U. & Reusch, T. B. H. Adaptive evolution of a key phytoplankton species to ocean acidification. Nat. Geosci. 5, 346–351 (2012).
7. Bown, P. R., Lees, J. A. & Young, J. R. inCoccolithophores: From Molecular Processes to Global Impact (eds Thierstein H. R., Young J. R. Springer (2004).
8. Beaufort, L. Weight estimates of coccoliths using the optical properties (birefringence) of calcite. Micropaleontology 51, 289–297 (2005).
9. Beaufort, L. et al. Sensitivity of coccolithophores to carbonate chemistry and ocean acidification. Nature 476, 80–83 (2011).
10. Cubillos, J., Henderiks, J., Beaufort, L., Howard, W. R. & Hallegraeff, G. M. Reconstructing calcification in ancient coccolithophores: Individual coccolith weight and morphology of *Coccolithus pelagicus* (sensu lato). Marine Micropaleontol. 92-93, 29–39 (2012).
11. Gibbs, S. J. et al. Species-specific growth response of coccolithophores to Palaeocene-Eocene environmental change. Nat. Geosci. 6, 218–222 (2013).
12. Zachos, J. C. et al. Rapid acidification of the ocean during the Paleocene-Eocene thermal maximum. Science 308, 1611–1615 (2005).

13. Dunkley Jones, T. et al. Climate model and proxy data constraints on ocean warming across the Paleocene-Eocene Thermal Maximum. Earth-Sci. Rev. 125, 123–145 (2013).
14. Penman, D. E., Zachos, J. C., Hönisch, B., Eggins, S. & Zeebe, R. inClimate and Biota of the Early Paleogene Conference Program and Abstracts 85, (ed. Egger H. Berichte der Geologischen Bundesanstalt (2011).
15. Ridgwell, A. & Schmidt, D. N. Past constraints on the vulnerability of marine calcifiers to massive carbon dioxide release. Nat. Geosci. 3, 196–200 (2010).
16. Uchikawa, J. & Zeebe, R. E. Examining possible effects of seawater pH decline on foraminiferal stable isotopes during the Paleocene-Eocene Thermal Maximum. Paleoceanography 25, PA2216 (2010).
17. Zeebe, R. History of seawater carbonate chemistry, atmospheric CO2 and ocean acidification. Annu. Rev. Earth Planet. Sci. 40, 141–165 (2012).
18. Bralower, T. J. Evidence of surface water oligotrophy during the Paleocene-Eocene thermal maximum: nannofossil assemblage data from Ocean Drilling Program Site 690, Maud Rise, Weddell Sea. Paleoceanography 17, doi:doi:10.1029/2001PA000662 (2002).
19. Gibbs, S. J., Bralower, T. J, Bown, P. R., Zachos, J. C. & Bybell, L. M. Shelf and open-ocean calcareous phytoplankton assemblages across the Paleocene-Eocene Thermal Maximum: Implications for global productivity gradients. Geology 34, 233–236 (2006).
20. Bown, P. & Pearson, P. Calcareous plankton evolution and the Paleocene/Eocene thermal maximum event: new evidence from Tanzania. Marine Micropaleontol. 71, 60–70 (2009).
21. John, C. M. et al. North American continental margin records of the Paleocene-Eocene thermal maximum: implications for global carbon and hydrological cycling. Paleoceanography 23, PA2217 (2008).
22. Nicholas, C. J. et al. Stratigraphy and sedimentology of the Upper Cretaceous to Paleogene Kilwa Group, southern coastal Tanzania. J. Afr. Earth Sci. 45, 431–466 (2006).
23. Young, J. R. & Ziveri, P. Calculation of coccolith volume and its use in calibration of carbonate flux estimates. Deep-Sea Res. Part II: Topical Stud. Oceanogr. 47, 1679–1700 (2000).
24. Poulton, A. J. et al. Coccolithophore dynamics in non-bloom conditions during late summer in the central Iceland Basin (July—August 2007). Limnol. Oceanogr. 55, 1601–1613 (2010).
25. Paasche, E. A review of the coccolithophorid *Emiliania huxleyi* (Prymnesiophyceae), with particular reference to growth, coccolith formation, and calcification-photosynthesis interactions. Phycologia 40, 503–529 (2002).
26. Müller, M. N., Antia, A. N. & LaRoche, J. Influence of cell cycle phase on calcification in the coccolithophore *Emiliania huxleyi*. Limnol. Oceanogr. 53, 506–512 (2008).
27. Winter, A., Jordan, R. W. & Roth, P. H. inCoccolithophores (eds Winter A., Siesser W. G. Cambridge Univ. Press (1994).

28. Taylor, A. R., Chrachri, A., Wheeler, G., Goddard, H. & Brownlee, C. A voltage-gated H+ channel underlying pH homeostasis in calcifying coccolithophores. Plos Biol. 9, e1001085 (2011).
29. Sáez, A. G. et al. Pseudo-cryptic speciation in coccolithophores. Proc. Natl Acad. Sci. USA 100, 7163–7168 (2003).
30. Quinn, P. S. et al. inCoccolithophores: from Molecular Processes to Global Impact (eds Thierstein H. R., Young J. R. Springer (2004).
31. Iglesias-Rodriguez, M. D., Schofield, O. M., Batley, J., Medlin, L. K. & Hayes, P. K. Intraspecific genetic diversity in the marine coccolithophore Emiliana huxleyi (Prymnesiophyceae): the use of microsatellite analysis in marine phytoplankton population studies. J. Phycol. 42, 526–536 (2006).
32. Sluijs, A. et al. Environmental precursors to rapid light carbon injection at the Palaeocene/Eocene boundary. Nature 450, 1218–1222 (2007).
33. Dickens, G. R., O'Neil, J. R., Rea, D. K. & Owen, R. M. Dissociation of oceanic methane hydrate as a cause of the carbon-isotope excursion at the end of the Paleocene. Paleoceanography 10, 965–971 (1995).
34. Sato, T., Yuguchi, S., Takayama, T. & Kameo, K. Drastic change in the geographical distribution of the cold-water nannofossil *Coccolithus pelagicus* (Wallich) Schiller at 2.74 Ma in the late Pliocene, with special reference to glaciation in the Arctic Ocean. Marine Micropaleontol. 52, 181–193 (2004).
35. Ziveri, P. et al. inCoccolithophores: from Molecular Processes to Global Impact (eds Thierstein H. R., Young J. R. Springer (2004).
36. Bown, P. R. et al. A Paleogene calcareous microfossil Konservat-Lagerstätte from the Kilwa Group of coastal Tanzania. Geol. Soc. Am. Bullet. 120, 3–12 (2008).
37. Anderson, D. R., Burnham, K. P. & Thompson, W. L. Null hypothesis testing: problems, prevalence, and an alternative. J. Wildl. Manage. 64, 912–923 (2000).
38. Gibbs, S. J., Stoll, H. M., Bown, P. R. & Bralower, T. J. Ocean acidification and surface water carbonate production across the Paleocene-Eocene Thermal Maximum. Earth Planet. Sci. Lett. 295, 583–592 (2010).

There are several supplemental files that are not available in this version of the article. To view this additional information, please use the citation on the first page of this chapter.

CHAPTER 11

Near-Shore Antarctic pH Variability has Implications for the Design of Ocean Acidification Experiments

LYDIA KAPSENBERG, AMANDA L. KELLEY, EMILY C. SHAW, TODD R. MARTZ, AND GRETCHEN E. HOFMANN

11.1 INTRODUCTION

The extensive effects of ocean acidification, the systematic reduction of ocean pH due to the absorption of anthropogenic carbon dioxide (CO_2) by surface oceans [1], are predicted to be first observed in high-latitude seas [2]. Cold waters of the Southern Ocean are naturally rich with CO_2, which results in low carbonate (aragonite and calcite) saturation states [2]. As ocean acidification progresses, pH and aragonite saturation state (Ω_{arag}) will decrease and facilitate the dissolution of marine calcium carbonate. From a biological perspective, evolution in the absence of shell-crushing predators in the near-shore Antarctic has left many benthic biogenic calcifiers with relatively brittle shells [3] that may be vulnerable to ocean acidification. Shell dissolution in live Southern Ocean pteropods, *Limacina helicina antarctica*, has already been observed in CO_2-rich upwelled waters ($\Omega_{arag} \approx 1$) [4]. Antarctic marine biota is hypothesized to be highly sensitive to ocean acidification [5], and predicting the impact of this anthropogenic process and the potential for future organismal adaptation is a research priority [6].

To predict how future ocean acidification will affect any marine ecosystem, it is first necessary to understand present-day pH variability. In the Southern Ocean, there are strong seasonal cycles in carbonate chemistry [7,8,9] due to the temporal partitioning of summertime primary production and wintertime heterotrophy [10]. Summertime phytoplankton blooms regularly drive the partial pressure of CO_2 in seawater (pCO_2) well below atmospheric equilibrium and are the primary source for pCO_2 variability in the Southern Ocean [9]. This seasonal carbonate chemistry cycle corresponds to a summertime pH increase of 0.06 units on a regional scale in the Southern Ocean [11] and as much as 0.6 units locally in Prydz Bay [7] and the Ross Sea [12]. The summertime pH increase (e.g. 0.6) can thus exceed the 0.4 pH unit magnitude of ocean acidification predicted for 2100 [13].

Future ocean carbonate chemistry remains challenging to predict due to other environmental processes and biological feedbacks [14]. Southern Ocean aragonite undersaturation (approx. pH ≤ 7.9) is predicted to occur first during the winter season in the next 20 years [11]. However, seasonal ice cover may delay the onset of ocean acidification thresholds by a few decades due to reduced air-sea gas exchange [12]. Likewise, decreasing seasonal ice cover, due to changes in wind and air temperature, are estimated to yield at least a 14% increase in primary production in the Ross Sea by 2100 [15]. This could potentially increase pH and Ω_{arag} in summer. Furthermore, increased stratification in the future may result in phytoplankton community shifts [15]. As an example, diatom communities dominate periods of highly stratified waters in the Ross Sea but drawdown less CO_2 compared to the dominant bloom algae *Phaeocystis antarctica* that proliferate in deeply mixed waters [16]. Thus, seasonal changes in carbonate chemistry (for example, from primary production) may yield alternative scenarios for ocean acidification outcomes [17]. Currently, projections of ocean acidification for near-shore Antarctica are largely based on discrete sampling [11], which may not have detected sub-seasonal (e.g. daily, weekly) pH variability that could be important for biological processes.

Although ocean acidification is generally predicted to be deleterious to marine life, not all taxa and species respond similarly to future conditions [18]. There is emerging evidence that an organism's pH-exposure history can influence its tolerance of ocean acidification. For example, Ref. 19 showed that an Arctic copepod species that experienced varied depth-dependent pH exposure was more tolerant of CO_2-acidified seawater treatments compared to another Arctic copepod species that experiences a smaller range in pH. Comprehensive characterization of the 'pH-seascape' is thus necessary to link CO_2-perturbation experiments with present-day and future organismal performance in the field. Such field time-series are sparse in near-shore Antarctica and are either extremely short [20,21] or low in sampling frequency [7,8,12].

In this study, our main goal was to describe pH variability experienced by organisms in near-shore Antarctica across seasonal transitions in an area with annual sea ice cover. In addition, we use the data to explore how pH variability and changes in seasonal CO_2 drawdown (as a proxy for changes in primary production) may impact future trajectories of ocean acidification in our study region.

11.2 RESULTS

11.2.1 PH DATA

To collect high-resolution pH data, we deployed autonomous SeaFET pH sensors [22] in the austral spring at two sites in near-shore McMurdo Sound, Jetty and Cape Evans, on subtidal moorings in separate years (Fig. 1). Both the Jetty and Cape Evans showed four general sequences of pH variation during the observed period (Fig. 2a, 2b; Table 1). First, pH (reported on the total hydrogen ion scale for all measurements) rapidly increased from approximately 8.0 to 8.3 units in the early austral summer from December to January. Second, monthly pH variability from December through April (s.d. ± 0.03 to 0.08 units) was higher than that observed in November (s.d. ± 0.01 units). The increase in short-term pH variability remained after removing low-frequency pH variability that was inherently included in the monthly standard deviations listed in Table 1. Standard deviation of the 10-day moving average of high-pass filtered pH data was greater during the summer months relative to November, May, and June and peaked in January at both sites (<0.05 pH units, Fig. 3a, 3b). Third, following peak pH in January, pH and short-term pH variability generally declined to the end of April, but remained higher than November and early-December conditions. Fourth, around the onset of 24 h darkness at the end of April and during stabilized temperature pH declined and was followed by lower mean monthly pH and variability in May (s.d. ± 0.02 units) and in June (s.d. ± 0.01 units), relative to summer months. The initial pH increase from fall to peak January conditions corresponded to a decline in the calculated dissolved inorganic carbon (DIC) of 167 and 137 μmol $kgSW^{-1}$ at the Jetty (54 days) and Cape Evans (31 days), respectively.

The seasonal pH range was 0.30 and 0.33 pH units for the Jetty and Cape Evans, respectively (Fig. 1a, 1b), based on a 10-day low pass filter. Short-term pH variability contributed to a total range of observed pH from summer to winter conditions of 0.40 and 0.42 units, at the Jetty and Cape Evans, respectively. Maximum pH was observed in January and minimum pH was observed in May at both sites. Mean pH differed between the Jetty and Cape Evans when comparing pH observations of the same date range (8.15 ± 0.08 at the Jetty; 8.08 ± 0.09 at Cape Evans; Mann-Whitney Wilcoxon test, $p < 0.001$, $W = 3481992$, $n = 2103$). In general, summertime sub-seasonal (Fig. 2a, 2b) and short-term pH variability (Fig. 3a, 3b) was greater at Cape Evans in 2013 compared to the Jetty in 2012. Changes in pH of ± 0.13 units occurred various times over the course of hours to a day at Cape Evans. The largest pH change over a relatively short period was −0.27 units over 5.5 days in March at Cape Evans.

Within the same site, temperature data showed similar patterns in variability as pH: temperature increased from the start of the recording period, peaked in January, after which it declined and stabilized in early April to similar temperatures observed in November and early December (Fig. 2c, 2d). Low-pass filtered data show a seasonal warming of 1.33°C and 1.55°C at the Jetty and Cape Evans, respectively. Like pH, high-pass filtered temperature data showed a seasonal increase in short-term variability from January through April (Fig. 3d, 3d). Absolute seasonal temperature change was 1.8°C and 1.7°C at the Jetty and Cape Evans, respectively.

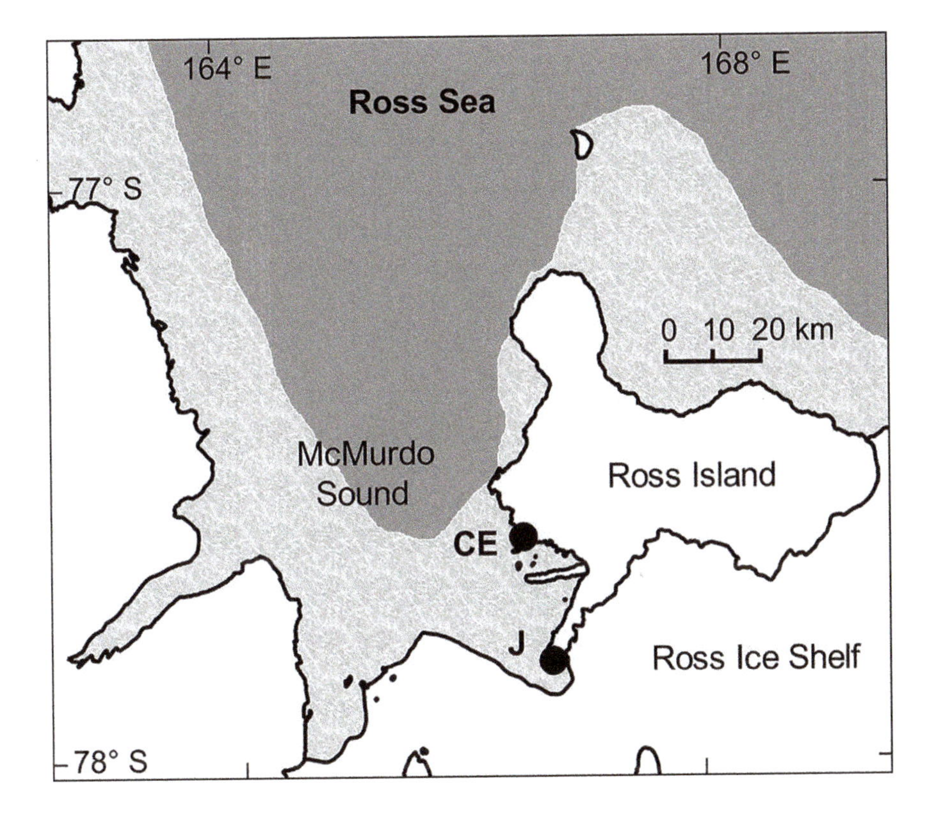

FIGURE 1: Map of pH sensor deployments in McMurdo Sound, Antarctica. Sensors were deployed at the Jetty (J) in 2011 and at Cape Evans (CE) in 2012. Annual sea ice contour (marble color) approximates November conditions for 2011 (RISCO RapidIce Viewer). Mapping data are courtesy of the Scientific Committee on Antarctic Research, Antarctic Digital Database. Map was constructed in QGIS (Version 2.0.1) and sea ice contour was added using GIMP (Version 2.6.11).

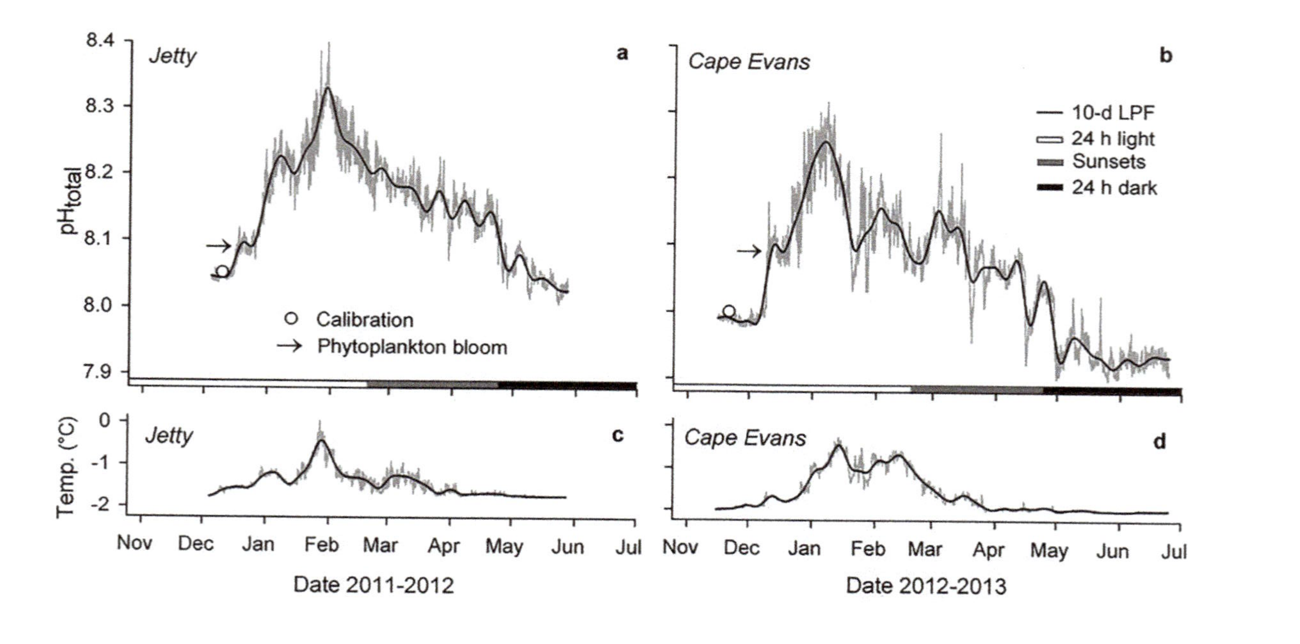

FIGURE 2: pH and temperature cycles in McMurdo Sound, Antarctica. Time-series pH (a, b) and temperature (c, d) at the Jetty and Cape Evans as recorded by SeaFET pH sensors (grey line). A 10-day low-pass filter (10-d LPF) was applied to the pH and temperature observations (blue line). Daylight is noted by colored x-axis bars where 'sunsets' indicates decreasing day length. Arrows indicate anecdotal events of phytoplankton blooms as observed by United States Antarctic Program SCUBA divers. Calibration samples are noted (circle). Ticks on x-axes denote the first day of the month.

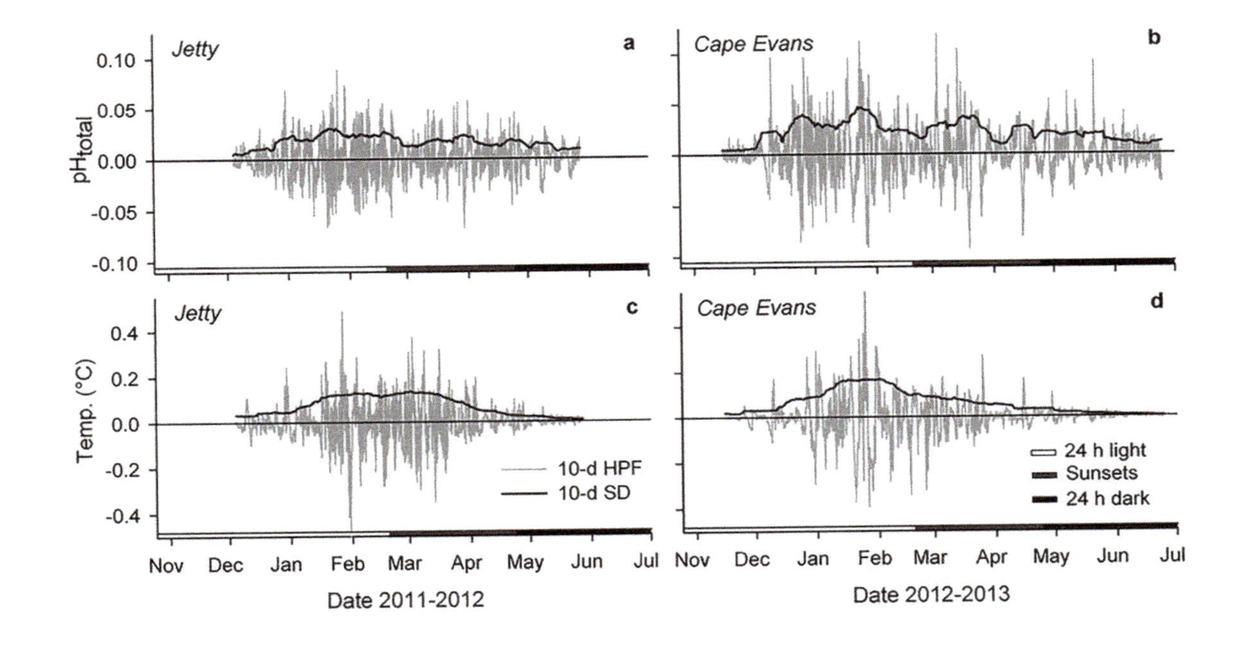

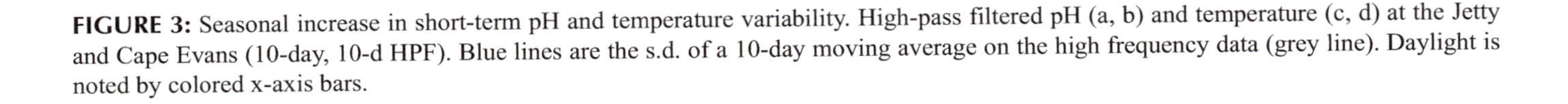
FIGURE 3: Seasonal increase in short-term pH and temperature variability. High-pass filtered pH (a, b) and temperature (c, d) at the Jetty and Cape Evans (10-day, 10-d HPF). Blue lines are the s.d. of a 10-day moving average on the high frequency data (grey line). Daylight is noted by colored x-axis bars.

TABLE 1: Carbonate parameters at two sites in McMurdo Sound, Antarctica

Site, year		pH	T (°C)	DIC (μmol kgSW^{-1})*	pCO_2 (μatm)*	Ω_{arag}*
Jetty	mean	8.15 ± 0.08	−1.45 ± 0.31	2179 ± 37	302 ± 62	1.68 ± 0.29
2011–2012	median	8.16	−1.50	2177	292	1.66
	min	8.01	−1.80	2058	152	1.22
	max	8.40	0.00	2238	428	2.81
	range	0.40	1.80	181	276	1.60
	Dec	8.08 ± 0.04	−1.54 ± 0.14	2216 ± 17	355 ± 36	1.45 ± 0.14
	Jan	8.24 ± 0.05	−1.08 ± 0.36	2142 ± 26	236 ± 31	2.03 ± 0.23
	Feb	8.23 ± 0.04	−1.28 ± 0.25	2138 ± 17	241 ± 23	1.95 ± 0.16
	Mar	8.17 ± 0.03	−1.42 ± 0.19	2170 ± 10	286 ± 19	1.70 ± 0.09
	Apr	8.12 ± 0.04	−1.67 ± 0.05	2191 ± 14	319 ± 31	1.55 ± 0.11
	May	8.05 ± 0.02	−1.73 ± 0.01	2224 ± 7	387 ± 20	1.33 ± 0.06
Cape Evans	mean	8.05 ± 0.10	−1.63 ± 0.47	2218 ± 39	391 ± 92	1.37 ± 0.30
2012–2013	median	8.06	−1.90	2216	372	1.36
	min	7.90	−2.04	2107	192	0.96
	max	8.32	−0.27	2276	559	2.34
	range	0.42	1.77	168	367	1.39
	Nov	7.99 ± 0.01	−1.96 ± 0.0	2256 ± 2	450 ± 6	1.17 ± 0.01
	Dec	8.09 ± 0.08	−1.73 ± 0.2	2212 ± 30	349 ± 67	1.49 ± 0.25
	Jan	8.17 ± 0.07	−0.90 ± 0.3	2170 ± 29	285 ± 53	1.79 ± 0.25
	Feb	8.11 ± 0.04	−0.96 ± 0.3	2183 ± 15	327 ± 31	1.56 ± 0.12
	Mar	8.10 ± 0.05	−1.75 ± 0.1	2198 ± 18	342 ± 42	1.46 ± 0.15
	Apr	8.03 ± 0.04	−1.95 ± 0.0	2225 ± 14	403 ± 44	1.27 ± 0.10
	May	7.94 ± 0.02	−1.99 ± 0.0	2262 ± 7	508 ± 29	1.04 ± 0.05
	Jun	7.93 ± 0.01	−2.00 ± 0.0	2268 ± 3	518 ± 18	1.03 ± 0.02

Error is ± s.d.
**Calculated parameter.*

At both the Jetty and Cape Evans, temperature was significantly and positively correlated with pH over the deployment period ($p < 0.001$; Table 2), opposing the thermodynamic relationship. High-pass filtered temperature was significantly correlated with pH at both sites (Table 2), but the direction of this relationship was different at both sites and explains little of the overall pH variation (< 5%, Table 2).

TABLE 2: Linear regression analysis of pH and temperature

Site	Predictor	Coef	SE Coef	T	p	R^2
Jetty	Temperature	0.20526	0.00358	57.27	<0.001*	0.61
	Temperature 10-d HPF	−0.04749	0.00453	−10.49	<0.001*	0.05
Cape Evans	Temperature	0.14985	0.00275	54.51	<0.001*	0.53
	Temperature 10-d HPF	0.02912	0.0058	5.02	<0.001*	0.01

**Statistically significant.*
10-d HPF, 10-day high-pass filtered data.

When used for carbonate calculations (DIC, pCO_2, Ω_{arag}), pH data indicate that McMurdo Sound is currently supersaturated with respect to aragonite (Table 1). Monthly mean Ω_{arag} in late fall and early winter approached 1. Conditions may have actually reached undersaturation (Ω_{arag} <1) for brief periods at Cape Evans in May and June (minimum of Ω_{arag} 0.96), depending on the error in pH measurements (see Methods).

11.2.2 OCEAN ACIDIFICATION SCENARIOS

McMurdo Sound regional ocean acidification trajectories were made using averaged pH observations from 2011–2013 and forced with the Representative Concentration Pathway 8.5 (RCP8.5) CO_2 emission scenario [23]. Due to the potential offset in pH measurements associated with use of unpurified m-cresol dye (~0.03 pH units, see Methods), our results may slightly overestimate acidification trends. The equilibrium scenario [12], which represents an increase in seawater pCO_2 that tracks atmospheric levels, predicted more extreme acidification than the disequilibrium scenario12, which represents a 65% reduced CO_2 uptake due to seasonal ice cover (Fig. 4, 5).

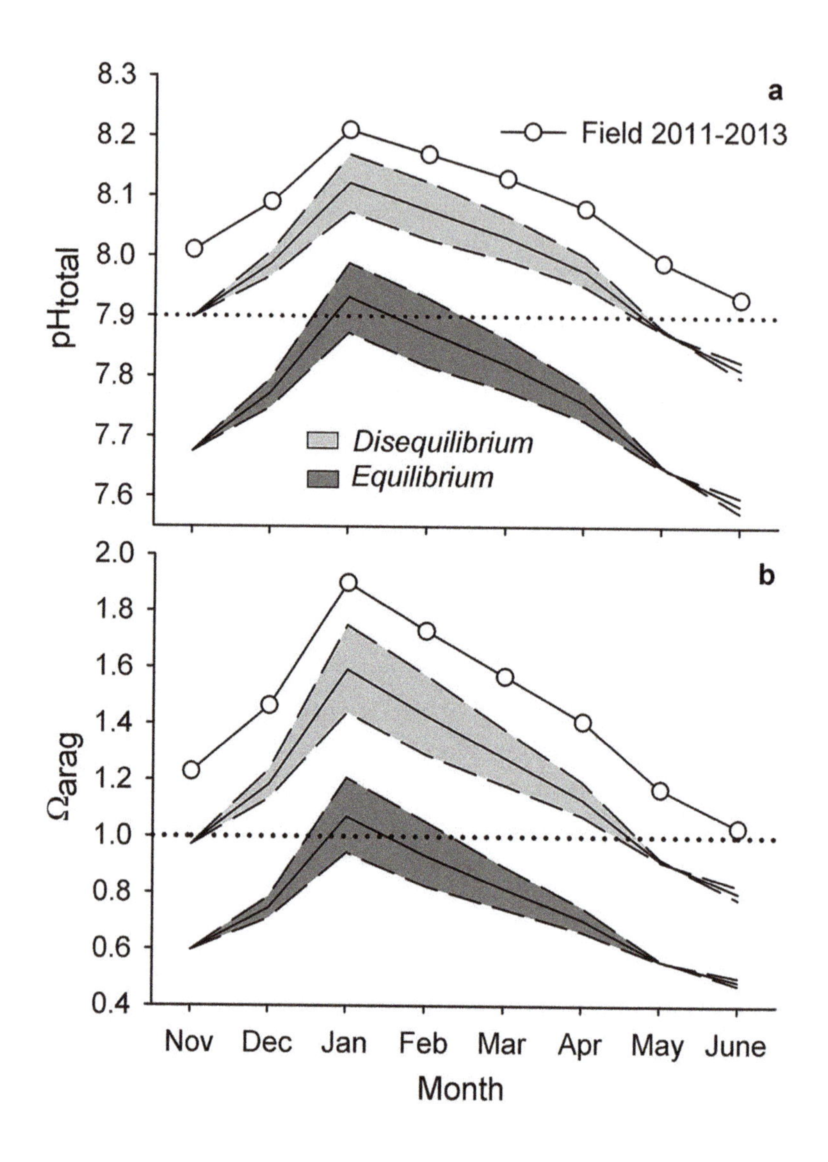

FIGURE 4: Present-day and end-century pH and aragonite saturation state. Present–day (circle) and end-century monthly mean pH (a) and aragonite saturation state, Ω_{arag} (b), in McMurdo Sound, Antarctica, using a disequilibrium and equilibrium scenario (solid line). Within each scenario, a simulated 20% increase (upper dashed lines) and decrease (lower dashed lines) in seasonal DIC amplitude is used to simulated changes in net community production. Dotted lines reference pH 7.9 and Ω_{arag} of 1.

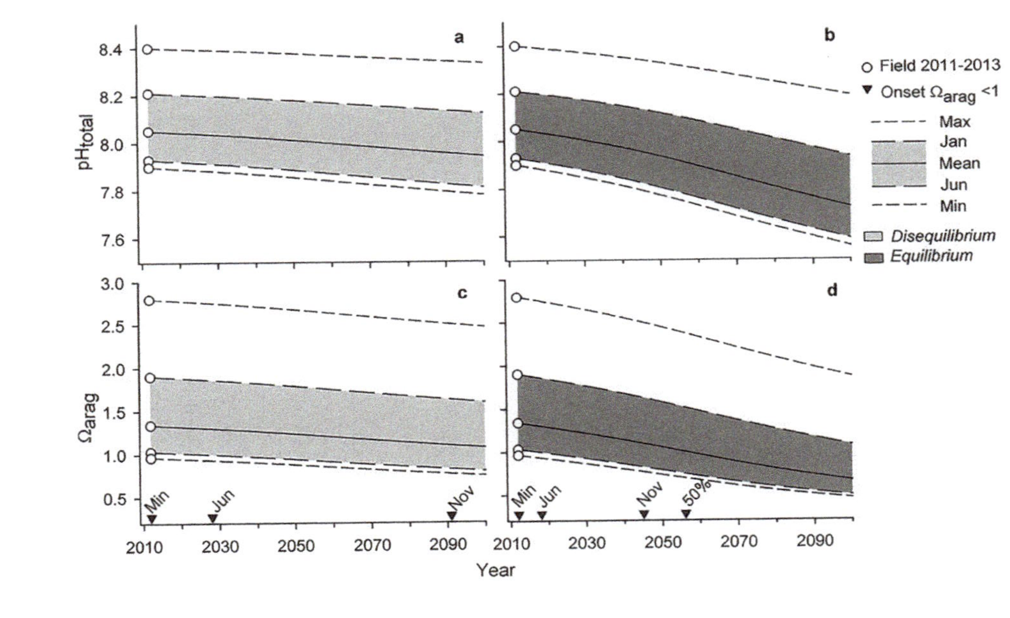

FIGURE 5: Annual changes in pH and aragonite saturation state ranges. Projections of yearly changes in pH and aragonite saturation state, Ω_{arag}, in McMurdo Sound, Antarctica, using a disequilibrium (a, c) and equilibrium (b, d) scenario. Annual range in pH increases and Ω_{arag} decreases with future acidification. End-century maximum pH and Ω_{arag} remain above acidification thresholds of pH 7.9 and Ω_{arag} of 1. Projections are based on field data collected in 2011–2013 (circle). January and June monthly means represent mid-summer and winter conditions, respectively. The overall mean represent mean values from spring into winter conditions. Onset of aragonite undersaturation (triangles) is marked for each parameter and additionally for November monthly mean conditions.

In both scenarios, CO_2 forcing increased the seasonal pH amplitude and reflects the process of reduced ocean buffer capacity as CO_2 is absorbed [24]. For example, present-day range of observed monthly mean pH from January to June was 0.28 units and increased to 0.31 and 0.35 units under the disequilibrium and equilibrium scenario, respectively. For all scenarios, wintertime pH of ~7.9 (approximate aragonite undersaturation) occurred by the end of the century (Fig. 4, 5). Assuming that pH < 7.9 persists for the period that we lack data for (July through October), the disequilibrium and equilibrium models suggest a 7- and 11-month annual duration of pH conditions < 7.9 units and undersaturation by 2100, respectively.

As a proxy for simulating changes in net community production, DIC amplitude was perturbed by ± 20% (Fig. 4). A 20% increase in seasonal DIC amplitude raised pH and Ω_{arag} during the summer and fall but failed to raise pH and Ω_{arag} to present-day levels. For example, under the equilibrium model, a 20% increase in seasonal DIC amplitude marginally extended end-century duration of summertime pH > 7.9 from January (pH 7.93, Ω_{arag} 1.07) to January (pH 7.99, Ω_{arag} 1.21) and February (pH 7.93, Ω_{arag} 1.05).

Any reduction in the amplitude of seasonal DIC will exacerbate the effects of ocean acidification. For example, during the month of peak pH, mean January pH remained above 7.9 units in all scenarios, except under the equilibrium scenario with a simulated 20% reduction in seasonal DIC amplitude (January pH 7.87, Fig. 4). This latter scenario was the only scenario that exhibited permanent aragonite undersaturation in McMurdo Sound by 2100.

Due to the increase in pH variability observed during summer months (Fig. 3), organisms at our study sites will likely still periodically experience pH > 7.9 and Ω_{arag} > 1 by 2100 (Fig. 5). For instance, under the equilibrium scenario, maximum pH was pH 8.19, and 0.47 units above mean January conditions (pH 7.72). Acidification thresholds (pH ~7.9 and Ω_{arag} < 1) were crossed earlier under the equilibrium model compared to the disequilibrium model (Fig. 5). Here, onset of June (i.e. winter) undersaturation was projected to occur by 2018, a decade earlier than under the disequilibrium scenario. November (i.e. spring) aragonite undersaturation was predicted to first occur by 2045 in the equilibrium model, 46 years earlier than predicted by the disequilibrium model. Timing of the threshold crossings may be delayed given the potential offset in Ω_{arag} associated with the pH measurement error.

11.3 DISCUSSION

The observed pH regime in McMurdo Sound can be grouped into two seasonal patterns: (1) stable pH with low variability during the winter and spring, and (2) elevated pH with high variability during the summer and fall. While our pH sensors did not record data from July through October, previous studies of pH (in October [21]) and temperature [25] in this region support our hypothesis of low environmental variability during the winter. Note, observations from Prydz Bay [8] (68°S) suggest that pH may decline slightly (<0.1 units) from June to September.

The amplitude of summertime pH elevation (0.3–0.4 units) observed in McMurdo Sound is among one of the greatest observed in the ocean and matches pH cycles at a northern coastal site in Prydz Bay, Antarctica [7,8]. In McMurdo Sound, the intense summertime DIC drawdown started in December and matched the timing of the annually recurring Phaeocystis sp. phytoplankton blooms, which are well-described and typically centered on 10 December (R. Robbins pers. comm.) [26]. The initial pH increase at Cape Evans (pH 8.01 to 8.12) occurred within 24 h on 9 December 2012 during which SCUBA divers noted sudden increase in phytoplankton presence in McMurdo Sound (R. Robbins pers. comm.).

Given that (1) the sudden increase in pH at our study sites followed a period of extremely stable pH conditions [20,21], (2) maximum observed pH corresponded to pCO_2 ~200 µatm below atmospheric equilibrium, and (3) productive waters from the Ross Sea are advected south into east McMurdo Sound [10], the initial rapid pH increase in December is likely the signature of phytoplankton blooms that originated in the Ross Sea and reached our coastal sites. Calculated DIC drawdown from fall to summer at the Jetty and Cape Evans (167 and 137 µmol $kgSW^{-1}$ DIC) matches the timing and magnitude of CO_2 cycles observed at similar depths in the Ross Sea [27] and Prydz Bay (~135–200 µatm $kgSW^{-1}$ DIC) [7,8].

Following the peak pH in January, pH steadily declined to pre-summer conditions by the end of April. A recent study of autonomous pCO_2 measurements on incoming seawater at Palmer Station from Arthur Harbor (64°S) observed a summertime increase in primary production, starting in November [28]. Here, a phytoplankton bloom was captured with peak production corresponding to an observation of 50 μatm pCO_2. Contrary to the slow return of carbonate chemistry to pre-summer conditions observed in McMurdo Sound over 4–5 months, pCO_2 at Arthur Harbor rapidly returned to atmospheric equilibrium in December and persisted to the end of the study in March. The authors attributed the crash of the bloom to physical mixing and zooplankton grazing, which would control phytoplankton density and contribute respiratory CO_2. Depending on the year-to-year pH variability on the Antarctic Peninsula, the season of high pH in Arthur Harbor may potentially be much shorter compared to that in McMurdo Sound. For example, interannual carbonate chemistry variability in the Weddell Sea is linked to the timing of sea-ice melt and phytoplankton productivity in the mixed layer [29]. The decline in pH observed at McMurdo is likely a combination of reduced primary production, increased heterotrophy and deepening of the mixed layer, as has been suggested to occur in Prydz Bay [7] and observed in other notable bloom regions such as the North Atlantic [30].

Calculated pCO_2 at Cape Evans in April (403 ± 44 μatm) nears observations from the Ross Sea made in April 1997 (320–400 μatm) [9]. A stabilization of pH in May and June at Cape Evans corresponded to ~500 μatm pCO_2. We were unable to collect validation samples during this period, however, biofouling was not an issue at our sites and SeaFET pH sensors have been shown to maintain stability over >9 months [31]. Similar observations have been made elsewhere in near-shore Antarctica. For example, high pCO_2 (~490 μatm) was observed near the Dotson Ice Shelf in the Amundsen Sea Polynya in summer and was correlated with the deepening of the mixed layer relative to the surrounding area [32]. In addition, the range of Ω_{arag} from mean summer (January) to winter (May) conditions was 0.70 and 0.75 at the Jetty and Cape Evans, respectively, matching the latest observations from Prydz Bay (0.738) and the Weddell Sea (0.7729). The low pCO_2 recorded in May and June at Cape Evans may thus be a combination of water column mixing and heterotrophy, as well as a potential 37 μatm pCO_2 overestimation associated with the offset of our pH measurement.

The observed ~0.3 unit summertime increase in pH in McMurdo Sound is much larger than that of northern high-latitudes [33]. While primary productivity in Antarctic waters is comparable to that of the high-latitude North Atlantic and Pacific [9], the observed < 2°C annual temperature variation is typical of McMurdo Sound [25] and plays almost no role in the seasonal amplitude of pH (1.8°C warming corresponds to a pH decrease of 0.03 units). In contrast, at locations such as the North Pacific the temperature cycle can be ~5 times greater than the observed range of temperatures in this study [9]. At our sites, the seasonal temperature forcing on pH counteracts seasonal forcing by primary production. As a result, the absence of a significant temperature forcing in near-shore Antarctica leads to a more pronounced seasonal pH cycle with greater amplitude compared to other bloom regions in the world [33].

As captured in our dataset, the summer season in McMurdo Sound is marked by an increase in sub-seasonal and short-term pH variability from December through April. In terms of s.d. of unfiltered (monthly s.d.) and high-pass filtered (10-day s.d.) pH, pH variability in McMurdo Sound is of similar magnitude to that observed in temperate kelp forests (e.g. ± 0.043 – 0.111) and tropical coral reefs (e.g. ± 0.022) over 30 days [34]. This is surprising due to absence of large temperature forcing and structural macrophytes and holobionts, which induce diurnal pH cycles at lower latitudes. On a Hawaiian reef, variability in pH was correlated with environmental parameters such as wave and height, wind speed, and solar radiation [35], suggesting a combination of influential abiotic and biotic drivers on coastal seawater pH variability.

We did not directly measure abiotic and biotic factors that influence carbonate chemistry in our study region and more measurements would be needed to quantify the sources of variability over different frequencies. For instance, air-sea gas exchange contributes to pH on a seasonal time-frame, where summertime CO_2 uptake by the ocean during ice-free periods masks the total contribution of net community production to DIC drawdown [7]. Likewise, summer meltwater dilutes DIC and AT [7,8] and may contribute to short-term pH variability in summer. The timing of sea ice melt onset may impact the duration and magnitude of carbonate chemistry seasonality where early melting enhances phytoplankton production under optimal mixed layer depths, as has been observed in the Weddell Sea [29]. Small pH variability (8.009 ± 0.015) observed from late October through November in McMurdo Sound may be explained by algal photosynthesis, although tides may play a small role as well [36]. Tidal exchanges of shallow and deeper water masses could play a larger role in summer pH variability, compared to spring [36], when the water column is highly stratified [37]. Low pH variability observed in winter and spring could also stem from a decrease in respiratory CO_2 contributions to DIC due to metabolic depression during periods of low food availablity, as has been observed to occur in pteropods [38]. In contrast, increased pH variability during the summer and fall is potentially influenced by the dominant biological forcing on the carbonate system in the Ross Sea at that time [9]. Such phytoplankton blooms create large spatial differences in pCO_2 [32] that could lead to sub-seasonal and short-term pH variability through bloom patchiness across water mass movement. Quantification of abiotic and biotic parameters described above would improve estimations for future ocean acidification when incorporated into sensitivity models [17].

We explored how seasonal pH variability may influence future ocean acidification in our study region in order to provide guidelines for biological experiments assessing future species' and ecosystem responses. The equilibrium and disequilibrium models provide boundaries for potential worst- and best-case acidification under a CO_2 emission scenario that does not account for climate mitigation efforts [23]. Within all model parameters we employed, marine biota at our study sites are anticipated to experience changes beyond the envelope of current conditions, as has been predicted for lower latitude marine ecosystems as well [17]. As atmospheric CO_2 continues to increase, (1) pH and duration of summertime high pH (> 7.9) will decrease and (2) the magnitude of seasonal and short-term pH variability may increase.

Previous studies of ocean acidification in the Southern Ocean and the Ross Sea identify the importance of seasonality and predict onset of wintertime aragonite undersaturation (Ω_{arag} < 1) between 2030 and 2050 under Intergovernmental Panel on Climate Change emissions scenario IS92a [2,11,12]. Our calculations of Ω_{arag} show wintertime undersaturation in McMurdo Sound occurring within this same timeframe, despite the higher CO_2 emission scenario and high-resolution data used in our study, and potential over estimation of acidification trends associated with the offset in pH measurements. Given that pH and Ω_{arag} may decrease slightly from June through September [8] and the lack of pH observations during these months, it is possible that periodic aragonite undersaturation may occur sooner than our predictions based on June observations. For context, the consequences of such periodic undersaturation could lead to calcium carbonate dissolution of live animals, as was observed for *L. helicina antarctica* at $\Omega_{arag} \approx 14$. Likewise, studies on Antarctic sea urchin, *Sterechinus neumayeri*, early development conducted during the period of stable spring pH and urchin spawning in McMurdo Sound, suggest that persisting conditions of pH < 7.9 (approximate aragonite undersaturation) may to impair larval growth [39] and calcification (G. E. Hofmann and P. C. Yu, unpubl.). Such conditions could occur in the latter half of this century during the sea urchin spawning season. Future carbonate chemistry conditions will ultimately depend on the rate at which anthropogenic CO_2 is released to the atmosphere and any future changes in local physical and biological processes that our model does not account for (e.g. changes in temperature, meltwater, wind, mixing and stratification, upwelling, gas-exchange, and phytoplankton blooms).

Despite the dominant biological footprint in pH seasonality in the Southern Ocean, a 20% increase in seasonal DIC amplitude (simulating an increase in net community production) failed to raise pH to present-day levels at our study site. This suggests that relatively large changes in seasonal primary productivity may have a small effect on the pH exposure of coastal organisms relative to the changes induced by ocean acidification. Phytoplankton blooms, as a food source however, may impact species responses to ocean acidification. For example, a study of *L. helicina* antarctica collected in McMurdo Sound found that (1) feeding history (e.g. weeks, months, seasons) impacted oxygen consumption rates and (2) metabolic suppression due to low pH exposure was a masked during periods of food limitation [38]. This study highlights the importance of incorporating environmental history when interpreting experimental results. As the feeding history is likely correlated with pH exposure in the bloom, parsing out the effects of pH history and food availability will present a challenge for Antarctic physiology.

In Antarctic ocean acidification biology, 'control' conditions used in experiments are often ~pH 8.0 (e.g. Ref. 39, 40) and represent current spring conditions in McMurdo Sound. Based on our future projections, this 'control' treatment will only occur during summer months if at all. Regardless of the exact rate of ocean acidification, the seasonal window of pH > 7.9 and $\Omega_{arag} > 1$ will likely shorten in the future. This shrinking and seasonally shifting window of high pH may lead to unpredictable ecological consequences through changes in physiological and seasonally dependent biological processes (e.g. sea urchin larval development). It remains largely unknown how summertime pH levels currently contribute to animal physiology and whether or not a reduction in future peak pH and duration of high pH exposure influences physiological recovery following 7–11 months unprecedented low pH conditions. As an example, oxygen consumption and gene expression of heat shock protein 70 in the Antarctic bivalve Laternula elliptica increased when adults were exposed to experimental conditions near the habitat maxima (pH 8.32, categorized as 'glacial levels' by the authors) and below their current pH exposure (pH 7.77), relative to performance at ~pH 8.040. These results suggest that summer exposures may induce stress similar to conditions predicted with ocean acidification. Understanding how organisms are adapted to their present-day exposures will help elucidate how they will respond to future conditions.

As the exposure period of pH > 7.9 shrinks under simulated ocean acidification, the magnitude of annual pH variability increases. These changes suggest that calcifying marine biota of Antarctic coastal regions will experience larger seasonal pH cycles in addition to exposure to lower environmental pH. Due to the reduced buffer capacity of the ocean under high CO_2, it is likely that the short-term pH variability in McMurdo Sound will be amplified in the future as well [24]. This has been predicted for coral reefs under ocean acidification scenarios [17] and shown experimentally in pelagic field mesocosms [41] where primary production drives diurnal pH cycles.

Our results provide guidance for the design of biological experiments aimed to address the potential for Antarctic species to adapt to a seasonally shrinking window of future high pH conditions. Although ocean acidification is likely to create an unprecedented marine environment, the existing presence of high pH variability in near-shore Antarctica may have beneficial implications for biological tolerance of ocean acidification. The distinct summertime increase in pH and pH variability in near-shore McMurdo Sound suggests that marine biota here have some capacity to deal with large fluctuations in the carbonate system, as has also been suggested by Ref. 42 in relation to the seasonal pH cycle. Unlike temperate upwelling regions where pH variability frequently drops below pH 8.0 [43], elevation of summer pH in McMurdo Sound opposes the direction of future ocean acidification. Future studies are necessary to describe how this pH-seascape may select for physiological tolerances of ocean acidification. For example, are natural positive (e.g. near-shore Antarctica) or negative deviations (e.g. temperate upwelling systems [43]) from pH 8.0 important for tolerance of future acidification? Will high summertime pH prepare organisms for low pH conditions in the winter? What frequency of pH variability promotes acidification tolerance?

A few recent studies have tackled such questions in temperate regions with mixed results. For example, Ref. 44 found that larval growth of mussel Mytilus galloprovincialis veligers was reduced under low static pH but recovered under similar conditions of low mean pH when semi-diurnal pH variability was introduced. However, congener *M. californianus* did not exhibit this 'rescued' response with diurnal cycles [44]. Although the Southern Ocean does not experience year-round diurnal photoperiods, a similar experimental approach can be used to guide studies on the impact of pH seasonality on ocean acidification tolerance [45], and ultimately, adaptation.

We highlight a coupled oceanography and biology research strategy for studying ocean acidification biology in the Southern Ocean. Studying physiological tolerance and local adaptation to variable seawater chemistry ideally requires large differences in spatial and temporal pH variability [34,43]. If patterns of pH variability differ spatially around the Antarctic continent (e.g. McMurdo Sound vs. Arthur Harbor [28]), we can begin to investigate possible levels of adaptation to local pH regimes as a proxy for evolutionary adaptions to future conditions [46]. In other words, evidence of adaptation in space suggests that animals may be able to adapt in time, as the capacity to do so is linked directly to standing genetic diversity in populations [47]. As illustrated in the Southern Ocean, population level differences (e.g. Ross Sea vs. Western Antarctic Peninsula biota) and local adaptation in tolerance of future anthropogenic stressors may be possible due to different rates in regional warming [48]. Some studies have shown genetic structure across the biogeographic boundary of the Drake Passage (reviewed by Ref. 49). Studies regarding population differences in pH tolerances and exposures in circum-Antarctic species can be accomplished with strategic placement of oceanographic sensors and design of biological experiments with environmentally relevant pH treatments [43,50]. In addition, use of autonomous pH sensors would address the need for pH observations at high-latitudes [5,38].

11.4 METHODS

11.4.1 STUDY SITES AND DEPLOYMENT

Autonomous SeaFET pH sensors containing Honeywell DuraFET® electrodes [22] were deployed in the austral spring at two sites in separate years on subtidal moorings in near-shore east McMurdo Sound (Fig. 1). Two SeaFETs were deployed side-by-side in December 2011 at a site near McMurdo Station (the Jetty, -77.85115, 166.66425), and one SeaFET was deployed during November 2012 at Cape Evans (-77.634617, 166.4159). Cape Evans is located 25 km north of the Jetty and is a highly productive site with an abundance of fish, macrophytes and marine invertebrates, including the sea urchin *S. neumayeri*. This site has previously been important for ocean acidification biology [20,39]. Subtidal moorings were anchored at approximately 27 m with sensor depth of 18 m. SeaFETs sampled on a two-hour frequency.

11.4.2 CALIBRATION

All reported pH is on a total hydrogen ion scale and listed as ‘pH’. Raw voltage recorded by the SeaFETs was converted to pH using one discrete seawater sample per sensor deployment following methods from Ref. 31. Calibration samples were collected via SCUBA following sensor conditioning to seawater within the first two weeks of each deployment, using a 5 L GO-FLO sampling bottle. Ideally, additional validation samples are collected throughout a sensor deployment. However, the remoteness of our sites restricted this work to one discrete sample per sensor deployment.

Calibration samples were preserved with saturated mercuric chloride according to Standard Operating Procedure (SOP) 1 [51]. Spectrophotometric pH was determined at 25°C following SOP 6b [51] using m-cresol purple from Sigma-Aldrich®. Total alkalinity (A_T) was measured via open-cell titration with a Mettler-Toledo T50 (SOP 3b [51]). Salinity was measured using a calibrated YSI 3100 Conductivity Instrument. Certified Reference Materials of seawater (CRMs) and acid titrant were supplied by Dr. Andrew G. Dickson (University of California San Diego, Scripps Institution of Oceanography). pH at in situ temperature, as recorded by SeaFETs, was calculated from spectrophotometric measurements of $pH_{25°C}$ and AT and salinity on the bottle sample using the program CO2Calc [Version 1.0.1, 2010, U.S. Geological Survey] with CO_2 constants from Ref. 52 refit by Ref. 53. All reported carbonate system calculation were conducted according to these constants.

11.4.3 DATA PROCESSING AND ANALYSIS

Raw data from the SeaFETs were cropped based on battery exhaustion, which occurred before sensor recovery. One of the two sensors deployed at the Jetty failed quality control analyses, and data from this instrument are not reported. Inspections of raw voltages recorded by the functional SeaFETs confirmed that the calibration samples were collected after the period of sensor conditioning to seawater. In the absence of biofouling (as was the case for our sensors), sensor stability has been demonstrated over similar deployment times [31] thereby generating high-quality pH datasets. A comparison of pH from each site was conducted using a Mann-Whitney Wilcoxon test as pH values were not normally distributed (Minitab® 16, Kolmogorov-Smirnov test, $p < 0.10$, for each site). All time is reported as UTC.

Time-series carbonate parameters were calculated from pH measurements using CO2calc for a depth of 18 m. Monthly mean salinity data was used from prior measurements in McMurdo Sound [54] (Table 3). A_T was calculated from the empirical relationship between sea surface salinity (SSS) and sea surface temperature (SST, as measured by SeaFETs) for the Southern Ocean as reported by Ref. 55:

$$A_T = 2305 + 52.48 \times (SSS - 35) + 2.85 \times (SSS - 35)^2 - 0.49 \times (SST - 20) + 0.086 \times (SST - 20)^2 \quad (1)$$

TABLE 3: Model inputs for seasonally variable parameters. See Methods for details.

Month	pH	Temp. (°C)	Salinity [54]	Total Alkalinity (µmol $kgSW^{-1}$)∞	Total PO_4 (µmol $kgSW^{-1}$)	Total Si (µmol $kgSW^{-1}$)
Nov*	8.01	−1.9	34.82	2348	1.9 [10,37,57,58]	65.1 [10,57]
Dec	8.09	−1.6	34.76	2343	1.3 [10,57,59]	62.1 [10,57,59]
Jan	8.21	−1.0	34.65	2335	0.9 [10,57,59]	50.7 [10,57,59]
Feb	8.17	−1.1	34.37	2322	1.3 [59]	54.5 [59]
Mar	8.13	−1.6	34.44	2327	1.5 [57]	70.5 [57]
Apr	8.08	−1.8	34.50	2331	2.1 [12]	78.9 [12]
May	7.99	−1.9	34.62	2337	2.1 [12]	78.9 [12]
Jun*	7.93	−1.9	34.69	2341	2.1 [12]	78.9 [12]
mean*	8.05	−1.6	34.65	2335	1.6	67.5
min*	7.90	−1.9	34.62	2337	2.1	78.9
max§	8.40	−0.4	34.61	2336	0.9	50.7

**pH and temperature data from Cape Evans only, based on data collected in this study and in Refs. 20,34,36.*
§pH and temperature data from the Jetty only.
∞Calculated from salinity and temperature [55].

A_T measurements on SeaFET calibration samples matched the calculated AT within the accuracy of titrator (AT and salinity were 2342 µmol $kgSW^{-1}$ and 34.3 for the Jetty; 2351 µmol $kgSW^{-1}$ and 34.6 for Cape Evans, respectively). Monthly mean nutrient concentrations were estimated from the literature for McMurdo Sound and various Ross Sea stations in close proximity following the directions of ocean currents (max measurements $month^{-1} = 4$). Due to the lack of published phosphate measurements for this region, the Redfield ratio was applied to estimate phosphate from nitrate and silicic acid concentrations, in some cases (W. O. Smith, Jr. pers. comm.).

Summertime decrease in DIC was calculated for both sites from stable fall mean DIC conditions to minimum DIC observed in summer. Temperature and pH data were analyzed for event-scale to seasonal (10-day low-pass filter) and short-term (10-day high-pass filter) trends. Standard deviation of a 10-day moving average window on high-pass filtered data was calculated to describe seasonal changes in short-term pH and temperature variability. Unfiltered and 10-day high-pass filtered pH and temperature data from the duration of the entire deployment was investigated for each site using a linear correlation analysis (Matlab R2012b, Minitab® 16).

11.4.4 ERROR ESTIMATES

SeaFET thermistors were not individually calibrated resulting in a maximum estimated temperature error of ~0.3°C. The estimate of the combined standard uncertainty associated with the pH measurement of the calibration samples is ± 0.026 pH units (quadratic sum of partial uncertainties). The quantified sources in pH error are: use of unpurified m-cresol dye (0.02 [56]), spatio-temporal mismatch of the calibration sample (± 0.015 [31]), user differences (± 0.006), and calibration of the SeaFET thermistor (± 0.005). Measurements of spectrophotometric pH on CRMs, although not specified by the SOP, suggest that our benchtop methods may underestimate $pH_{25°C}$ by 0.032 (± 0.006, n = 18, across different users and days) relative to theoretical CRM pH calculated from DIC, A_T, and salinity. It is hoped that, in the future, purified indicator dye will become widely available to the oceanography community in order to improve accuracy of pH measurements. The estimated uncertainty for the pH of calibration samples does not impact the relative changes in pH recorded by the SeaFET on hourly to monthly time scales, which in the absence of biofouling can be resolved to better than 0.001. Thermistors provide a stable temperature reading with resolution of better than 0.01°C. Based on replicate analyses of CRMs, the precision of the titration system used for calibration samples is ± ≤10 μmol $kgSW^{-1}$ and did not impact the pH calculation of our calibration samples at in situ temperatures. Errors in salinity were not quantified. Instead, calculations of DIC, pCO_2, and Ω_{arag} from the pH time-series were conducted using monthly estimates of AT and salinity (Table 3). For reference, a +0.026 pH error corresponds to errors under November (January) conditions of -9 (-11) μmol kgSW−1 DIC, -27 (-17) μatm pCO_2, and +0.07 (+0.10) Ω_{arag}.

11.4.5 OCEAN ACIDIFICATION SCENARIOS

RCP8.5, which predicts atmospheric CO_2 to reach 935.87 ppm by 2100 [23], was used to generate four ocean acidification scenarios. The equilibrium scenario assumes an increase in DIC at the same rate as would be expected if seawater pCO_2 tracks the atmospheric value (~100 μmol $kgSW^{-1}$ increase in DIC by 2100) and (2) the disequilibrium scenario assumes a DIC increase at a 65% slower rate due to seasonal ice cover [12]. Secondary simulations of a ± 20% change in the observed seasonal amplitude of DIC are included along with the CO_2 forcing scenarios. The disequilibrium model likely overestimates pH and Ω_{arag} as horizontal advection of northern ice-free water masses with longer surface residence times was not accounted for Ref. 12.

First, November was used as a baseline for CO_2 forcing scenarios because it is a period of stable pH and has been measured for three consecutive years at Cape Evans [20,34,36]. Based on these prior studies and data collected in November 2012 during this study, mean November pH from 2010–2012 was pH 8.01. Calculated mean November seawater pCO_2 was then forced with pCO_2 from the RCP8.5 emission scenario assuming air-sea equilibrium, and annual changes in pCO_2 were used to calculate annual changes in November DIC up to 2100.

Second, monthly mean pH and temperature observations from the Jetty and Cape Evans from 2011–2013 were averaged to calculate a partial (8-month), present-day, regional DIC climatology. Calculations were performed in CO2calc following methods listed above, with the exception that monthly mean temperature in June at Cape Evans was corrected from −2.0°C up to −1.9°C to match previous long-term observations [25]. Input variables are listed in Table 3. Starting from the November baseline, present-day changes in DIC where calculated by month (December – June) and for the maximum and minimum observed DIC and overall mean. Monthly changes in DIC were assumed constant for future projections and were applied to end-century November DIC to generate a DIC climatology for 2100. Annual DIC trajectories were modeled for observed minimum and maximum DIC, November, January, June, and overall mean. For simulations of ± 20% change in seasonal DIC amplitude, monthly changes in DIC were increased or decreased by 20%. Owing to the lack of projections of future warming for coastal Antarctica, the effects of future temperature change were not included in our simulation.

REFERENCES

1. Doney, S. C., Fabry, V. J., Feely, R. A. & Kleypas, J. A. Ocean acidification: The other CO2 problem. Ann. Rev. Mar. Sci. 1, 169–192, 10.1146/annurev.marine.010908.163834 (2009).
2. Orr, J. C. et al. Anthropogenic ocean acidification over the twenty-first century and its impact on calcifying organisms. Nature 437, 681–686, 10.1038/nature04095 (2005).
3. Aronson, R. B. et al. Climate change and invasibility of the Antarctic benthos. Annu. Rev. Ecol. Evol. Syst. 38, 129–154, 10.1146/annurev.ecolsys.38.091206.095525 (2007).
4. Bednaršek, N. et al. Extensive dissolution of live pteropods in the Southern Ocean. Nat. Geosci. 5, 881–885 (2012).
5. Fabry, V. J., McClintock, J. B., Mathis, J. T. & Grebmeier, J. M. Ocean acidification at high latitudes: The bellweather. Oceanography 22, 160–171 (2009).
6. Kennicutt, M. C. et al. Six priorities for Antarctic science. Nature 512, 23–25 (2014).
7. Gibson, J. A. & Trull, T. W. Annual cycle of fCO2 under sea-ice and in open water in Prydz Bay, East Antarctica. Mar. Chem. 66, 187–200 (1999).
8. Roden, N. P., Shadwick, E. H., Tilbrook, B. & Trull, T. W. Annual cycle of carbonate chemistry and decadal change in coastal Prydz Bay, East Antarctica. Mar. Chem. 155, 135–147 (2013).
9. Takahashi, T. et al. Global sea-air CO2 flux based on climatological surface ocean pCO2, and seasonal biological and temperature effects. Deep-Sea Res. II 49, 1601–1622 (2002).
10. Rivkin, R. B. Seasonal patterns of planktonic production in McMurdo Sound, Antarctica. Am. Zool. 31, 5–16 (1991).
11. McNeil, B. I. & Matear, R. J. Southern Ocean acidification: A tipping point at 450-ppm atmospheric CO2. Proc. Natl. Acad. Sci. 105, 18860–18864, 10.1073/pnas.0806318105 (2008).

12. McNeil, B. I., Tagliabue, A. & Sweeney, C. A multi-decadal delay in the onset of corrosive 'acidified' waters in the Ross Sea of Antarctica due to strong air-sea CO2 disequilibrium. Geophys. Res. Lett. 37, L19607, 10.1029/2010gl044597 (2010).
13. IPCC. . Climate Change 2013: The Physical Science Basis. Contribution of Working Group I to the Fifth Assessment Report of the Intergovernmental Panel on Climate Change. [Stocker T. F., , D. Qin G.-K., Plattner M., Tignor S. K., Allen J., Boschung A., Nauels Y., Xia V., Bex, and Midgley P. M., eds. (eds.)]. Cambridge University Press, Cambridge, United Kingdom and New York, NY, USA, 1535 pp. (2013).
14. Riebesell, U., Körtzinger, A. & Oschlies, A. Sensitivities of marine carbon fluxes to ocean change. Proc. Natl. Acad. Sci. 106, 20602–20609 (2009).
15. Smith, W. O. J., Dinniman, M. S., Hofmann, E. E. & Klinck, J. M. The effects of changing winds and temperatures on the oceanography of the Ross Sea in the 21st century. Geophys. Res. Lett. 41, 1624–1631 (2014).
16. Arrigo, K. R. et al. Phytoplankton community structure and the drawdown of nutrients and CO2 in the Southern Ocean. Science 283, 365–367 (1999).
17. Shaw, E. C., McNeil, B. I., Tilbrook, B., Matear, R. & Bates, M. L. Anthropogenic changes to seawater buffer capacity combined with natural reef metabolism induce extreme future coral reef CO2 conditions. Global Change Biol. 19, 1632–1641 (2013).
18. Kroeker, K. J., Kordas, R. L., Crim, R. N. & Singh, G. G. Meta-analysis reveals negative yet variable effects of ocean acidification on marine organisms. Ecol. Lett. 13, 1419–1434, 10.1111/j.1461-0248.2010.01518.x (2010).
19. Lewis, C. N., Brown, K. A., Edwards, L. A., Cooper, G. & Findlay, H. S. Sensitivity to ocean acidification parallels natural pCO2 gradients experienced by Arctic copepods under winter sea ice. Proc. Natl. Acad. Sci. 110, E4960–E4967 (2013).
20. Kapsenberg, L. & Hofmann, G. E. Signals of resilience to ocean change: high thermal tolerance of early stage Antarctic sea urchins (Sterechinus neumayeri) reared under present-day and future pCO2 and temperature. Polar Biol. 37, 967–980, 10.1007/s00300-014-1494-x (2014).
21. Matson, P. G., Martz, T. R. & Hofmann, G. E. High-frequency observations of pH under Antarctic sea ice in the southern Ross Sea. Antarct. Sci. 23, 607–613, 10.1017/s0954102011000551 (2011).
22. Martz, T. R., Connery, J. G. & Johnson, K. S. Testing the Honeywell Durafet® for seawater pH applications. Limnol. Oceanogr. Methods 8, 172–184 (2010).
23. Riahi, K., Grübler, A. & Nakicenovic, N. Scenarios of long-term socio-economic and environmental development under climate stabilization. Technol. Forecast. Soc. Change 74, 887–935 (2007).
24. Egleston, E. S., Sabine, C. L. & Morel, F. M. Revelle revisited: Buffer factors that quantify the response of ocean chemistry to changes in DIC and alkalinity. Global Biogeochem. Cycles 24, GB1002 (2010).
25. Cziko, P. A., DeVries, A. L., Evans, C. W. & Cheng, C.-H. C. Antifreeze protein-induced superheating of ice inside Antarctic notothenioid fishes inhibits melting during summer warming. Proc. Natl. Acad. Sci. 111, 14583–14588 (2014).
26. Putt, M., Miceli, G. & Stoecker, D. K. Association of bacteria with Phaeocystis sp. in McMurdo Sound, Antarctica. Mar. Ecol. Prog. Ser. 105, 179–189 (1994).

27. Sweeney, C. The annual cycle of surface water CO2 and O2 in the Ross Sea: A model for gas exchange on the continental shelves of Antarctica. Antarct. Res. Ser. 78, 295–312 (2003).
28. Tortell, P. D. et al. Metabolic balance of coastal Antarctic waters revealed by autonomous pCO2 and ΔO2/Ar measurements. Geophys. Res. Lett. 41, 6803–6810 (2014).
29. Weeber, A., Swart, S. & Monteiro, P. Seasonality of sea ice controls interannual variability of summertime ΩA at the ice shelf in the Eastern Weddell Sea – an ocean acidification sensitivity study. Biogeosci. Disc. 12, 1653–1687, 10.5194/bgd-12-1653-2015 (2015).
30. Körtzinger, A. et al. The seasonal pCO2 cycle at 49°N/16.5°W in the northeastern Atlantic Ocean and what it tells us about biological productivity. J. Geophys. Res. 113, C04020 (2008).
31. Bresnahan, P. J. J., Martz, T. R., Takeshita, Y., Johnson, K. S. & LaShomb, M. Best practices for autonomous measurement of seawater pH with the Honeywell Durafet. Methods Oceangr. 9, 44–60 (2014).
32. Mu, L., Stammerjohn, S., Lowry, K. & Yager, P. Spatial variability of surface pCO2 and air-sea CO2 flux in the Amundsen Sea Polynya, Antarctica. Elementa: Sci. Anthrop. 2, 000036 (2014).
33. Shadwick, E., Trull, T., Thomas, H. & Gibson, J. Vulnerability of polar oceans to anthropogenic acidification: comparison of Arctic and Antarctic seasonal cycles. Sci. Rep. 3, 2339 (2013).
34. Hofmann, G. E. et al. High-frequency dynamics of ocean pH: a multi-ecosystem comparison. Plos One 6, e28983, 10.1371/journal.pone.0028983 (2011).
35. Lantz, C., Atkinson, M., Winn, C. & Kahng, S. Dissolved inorganic carbon and total alkalinity of a Hawaiian fringing reef: chemical techniques for monitoring the effects of ocean acidification on coral reefs. Coral Reefs 33, 105–115, 10.1007/s00338-013-1082-5 (2014).
36. Matson, P. G., Washburn, L., Martz, T. R. & Hofmann, G. E. Abiotic versus biotic drivers of ocean pH variation under fast sea ice in McMurdo Sound, Antarctica. Plos One 9, e107239 (2014).
37. Barry, J. Hydrographic patterns in McMurdo Sound, Antarctica and their relationship to local benthic communities. Polar Biol. 8, 377–391 (1988).
38. Seibel, B. A., Maas, A. E. & Dierssen, H. M. Energetic plasticity underlies a variable response to ocean acidification in the pteropod, Limacina helicina antarctica. Plos One 7, e30464 (2012).
39. Yu, P. C. et al. Growth attenuation with developmental schedule progression in embryos and early larvae of Sterechinus neumayeri raised under elevated CO2. Plos One 8, e52448 (2013).
40. Cummings, V. et al. Ocean acidification at high latitudes: potential effects on functioning of the Antarctic bivalve Laternula elliptica. Plos One 6, e16069, 10.1371/journal.pone.0016069 (2011).
41. Schulz, K. G. & Riebesell, U. Diurnal changes in seawater carbonate chemistry speciation at increasing atmospheric carbon dioxide. Mar. Biol. 160, 1889–1899 (2013).
42. McNeil, B. I., Sweeney, C. & Gibson, J. A. E. Short Note Natural seasonal variability of aragonite saturation state within two Antarctic coastal ocean sites. Antarct. Sci. 23, 411–412 (2011).

43. Hofmann, G. E. et al. Exploring local adaptation and the ocean acidification seascape – studies in the California Current Large Marine Ecosystem. Biogeosciences 11, 1053–1064 (2014).
44. Frieder, C. A., Gonzalez, J. P., Bockmon, E. E., Navarro, M. O. & Levin, L. A. Can variable pH and low oxygen moderate ocean acidification outcomes for mussel larvae? Global Change Biol. 20, 754–764, 10.1111/gcb.12485 (2014).
45. Murray, C. S., Malvezzi, A., Gobler, C. J. & Baumann, H. Offspring sensitivity to ocean acidification changes seasonally in a coastal marine fish. Mar. Ecol. Prog. Ser. 504, 1–11, 10.3354/meps10791 (2014).
46. Sanford, E. & Kelly, M. W. Local adaptation in marine invertebrates. Ann. Rev. Mar. Sci. 3, 509–535, 10.1146/annurev-marine-120709-142756 (2011).
47. Sunday, J. M. et al. Evolution in an acidifying ocean. Trends Ecol. Evol. 29, 117–125 (2014).
48. Steig, E. J. et al. Warming of the Antarctic ice-sheet surface since the 1957 International Geophysical Year. Nature 457, 459–462 (2009).
49. Kaiser, S. et al. Patterns, processes and vulnerability of Southern Ocean benthos: a decadal leap in knowledge and understanding. Mar. Biol. 160, 2295–2317 (2013).
50. McElhany, P. & Busch, D. S. Appropriate pCO2 treatments in ocean acidification experiments. Mar. Biol. 160, 1807–1812 (2013).
51. Dickson, A. G., Sabine, C. L. & Christian, J. R. Guide to best practices for ocean CO2 measurements. PICES Special Publication 3, 191 pp. (2007).
52. Mehrbach, C., Culberso, C. H., Hawley, J. E. & Pytkowic, R. M. Measurement of apparent dissociation constants of carbonic acid in seawater at atmospheric pressure. Limnol. Oceanogr. 18, 897–907 (1973).
53. Dickson, A. G. & Millero, F. J. A comparison of the equilibrium constants for the dissociation of carbonic acid in seawater media. Deep-Sea Res. I 34, 1733–1743, 10.1016/0198-0149(87)90021-5 (1987).
54. Littlepage, J. L. Oceanographic investigations in McMurdo sound, Antarctica. Antarct. Res. Ser. 5, 1–37 (1965).
55. Lee, K. et al. Global relationships of total alkalinity with salinity and temperature in surface waters of the world's oceans. Geophys. Res. Lett. 33, L19605 (2006).
56. Liu, X., Patsavas, M. C. & Byrne, R. H. Purification and characterization of meta-cresol purple for spectrophotometric seawater pH measurements. Environ. Sci. Technol. 45, 4862–4868 (2011).
57. Gordon, L. et al. Seasonal evolution of hydrographic properties in the Ross Sea, Antarctica, 1996–1997. Deep-Sea Res. II 47, 3095–3117 (2000).
58. Noble, A. E., Moran, D. M., Allen, A. & Saito, M. A. Dissolved and particulate trace metal micronutrients under the McMurdo Sound seasonal sea ice: basal sea ice communities as a capacitor for iron. Front. Chem. 1, 25 (2013).
59. Smith, W. O. J., Dinniman, M. S., Klinck, J. M. & Hofmann, E. E. Biogeochemical climatologies in the Ross Sea, Antarctica: seasonal patterns of nutrients and biomass. Deep-Sea Res. II 50, 3083–3101 (2003).

Author Notes

CHAPTER 1

Conflict of Interest

The author declares that the research was conducted in the absence of any commercial or financial relationships that could be construed as a potential conflict of interest.

Acknowledgments

I acknowledge funding by the UK Natural Environment Research Council, and the German Research Foundation Collaborative Research Center 754 (DFG SFB754) Programme on Climate-Biogeochemistry Interactions in the Tropical Ocean.

CHAPTER 2

Acknowledgments

The Surface Ocean CO_2 Atlas (SOCAT) is an international effort, supported by the International Ocean Carbon Coordination Project (IOCCP), the Surface Ocean Lower Atmosphere Study (SOLAS), and the Integrated Marine Biogeochemistry and Ecosystem Research program (IMBER), to deliver a uniformly quality-controlled surface ocean CO_2 database. The many researchers and funding agencies responsible for the collection of data and quality control are thanked for their contributions to SOCAT. Special thanks go to Dorothee Bakker (University of East Anglia, UK) and Ute Schuster (University of Exeter, UK) for their advice regarding the use of the SOCAT database. The calculations presented in this paper were carried out on the High Performance Computing Cluster supported by the Research Computing Service at the University of East Anglia. S. Jones was supported by a tied NERC Studentship, reference NE/F005733/1. A. Olsen was supported by the Centre for Climate Dynamics at the Bjerknes

Centre, and appreciates additional support from the Norwegian Research Council project SNACS (229752), and the EU H2020 project AtlantOS. S. Jones, C. Le Quéré and A. Manning were supported through the EU FP7 projects CARBOCHANGE "Changes in carbon uptake and emissions by oceans in a changing climate" (grant agreement 264879) and GEO-CARBON (agreement 283080). The code and input data used to calculate the gap-filled fCO_2, with instructions for use, are published at Pangaea. Alongside the code is the calculated gap-filled fCO_2 data set from SO-CAT v2 that is presented in this study. Both the code and output are available from Pangaea, doi 10.1594/PANGAEA.849262 [Jones et al., 2015]. We thank the reviewers for their thoughtful and helpful comments, which have proved invaluable.

CHAPTER 3

Acknowledgments

We acknowledge Martin Frank for re-estimating the rain rates according to the revised age model. We are grateful to James Collins for his assistance to revise our English. We also thank Ute Bock, Ulrike Böttjer, Ruth Cordelair and Birgit Glückselig for technical support. Financial support for this work was provided by the Deutsche Forschungsgemeinschaft (DFG), AWI Helmholtz-Zentrum für Polar- und Meeresforschung and MARUM Center of Marine Environmental Sciences, University of Bremen. Funding by the 'Helmholtz Climate Initiative REKLIM' (Regional Climate Change), a joint research project of the Helmholtz Association of German Research Centres (HGF), is gratefully acknowledged (G.K.).

CHAPTER 4

Conflict of Interest

The authors have not declared any conflict of interest.

Acknowledgments

The EuroSITES Project data was used for this research, contributions of the principal investigator and other scientists involved in the PAP project are acknowledged. The first author is particularly thankful to Professors

G. A. McKinley and Arne Körtzinger for technical guidance. The fellowship opportunity provided by the Fulbright Scholarship Program to the Department of Atmospheric, Space and Ocean Sciences, University of Wisconsin, Madison is acknowledged. The authors would like to thank anonymous reviewers for their comments and suggestions made to improve the original manuscript.

CHAPTER 5

Acknowledgments

S. Saux-Picard and T. Kristiansen are acknowledged for their help in preparing Figure 1. E. Boss and N. Briggs are thanked for discussions on spikes in optical measurements. L. Polimene, D. Raitsos, J. Bishop, and, especially, an anonymous reviewer are thanked for their comments on an earlier version of this paper. These data were collected and made freely available by the International Argo Program and the national programs that contribute to it (http://www.argo.ucsd.edu,http://argo.jcommops.org). The Argo Program is part of the Global Ocean Observing System. G.D.O. acknowledges funding from the UK National Centre for Earth Observation and Marie Curie FP7-PIRG08-GA-2010-276812. The Editor thanks James Bishop and an anonymous reviewer for their assistance in evaluating this paper.

CHAPTER 6

Acknowledgments

The data reported are available from the Malaspina digital repository. This research was funded by the Spanish Ministry of Economy and Competitiveness through the Malaspina 2010 Circumnavigation Expedition project (Consolider-Ingenio 2010, CSD2008-00077) and a JGI-Community Sequencing Programme Grant to S.A., to whom we are indebted. We thank M. Delgado for examining surface phytoplankton samples, R. Gutierrez, P. Puerta and J. Boras for help during sampling, M. Gonzalez-Calleja for help with figures drawing and M. Thums for comments. We thank the captain and crew of RV Hespérides for support throughout the cruise.

CHAPTER 7

Acknowledgments

Katsiaryna Pabortsava, Richard Sanders, Jeff Benson, Andy Milton, Sinhue Torres-Valdes, Polly Hill (NOC), Frances Hopkins, John Stephens (PML), Mark Moore (University of Southampton), Eric Achterberg (GEOMAR Kiel/University of Southampton), and Robert Thomas (BODC) are acknowledged for help and advice. This research cruise could not have been undertaken without the efforts of a large number of people located in several organizations within the UK. We are grateful for the help and assistance of all those involved including: the Master Graham Chapman, officers and crew of the James Clark Ross; the scientists and technical support staff of cruise JR271; the staff of the British Antarctic Survey, Cambridge, National Marine Facilities, Southampton, and the Scottish Association for Marine Science, Oban. Raymond Leakey is warmly acknowledged for leading cruise JR 271 and scientific advice. We are also grateful to the UK Natural Environment Research Council (NERC), the UK Department of Environment, Food, and Rural Affairs (Defra), and the UK Department of Energy and Climate Change (DECC) for funding the research cruise via the UK Ocean Acidification research programme (NERC grant NE/H017097/1), and to the Danish, Icelandic, and Norwegian diplomatic authorities for granting permission to travel and work in Greenland, Iceland, and Svalbard coastal and offshore waters. We would like to thank two anonymous reviewers for providing constructive comments. This research was also funded by the NERC SeasFX project (grant NE/J004383/1). Data are held at the British Oceanographic Data Centre (http://bodc.ac.uk/).

CHAPTER 8

Author Contributions

Conceived and designed the experiments: CSC JMA WCC. Performed the experiments: CSC JMA ETC EF. Analyzed the data: CSC EYC WCC. Contributed reagents/materials/analysis tools: CSC JMA EYC EF. Wrote the paper: CSC JMA EYC EF WCC.

CHAPTER 9

Data Accessibility

Raw experimental data used in this work are available from the corresponding author; these are graphically displayed in this work and its allied electronic supplementary material. Model code is available in, and is described in, the electronic supplementary material; code may also be obtained in the form used for this work (i.e. as a Powersim Constructor file) from the corresponding author.

Funding Statement

This work was supported by UK funders NERC and Defra through grants nos NE/F003455/1, NE/H01750X/1 and NE/J021954/1 to K.J.F., D.R.C., J.C.B., C.B. and G.L.W.; A.M. was supported by project EURO-BASIN (ref. 264933, 7FP, European Union).

CHAPTER 10

Acknowledgments

Samples used in this research were provided for Bass River by the Ocean Drilling Program for Lodo Gulch courtesy of J. Zachos (University of California, Santa Cruz) and colleagues, and for Tanzania by the Tanzania Drilling Project (TDP), courtesy of P. Pearson (TDP project leader, Cardiff University). S.A.O. was funded by a Natural Environment Research Council (NERC) studentship (reference NE/H524922/1) and S.J.G. was funded by a Royal Society University research fellowship. S.J.G., P.R.B., A.J.P. and J.R.Y. acknowledge grant support from the United Kingdom Ocean Acidification (UKOA) project, in particular via an added value award (NE/H017356/1). C.N. was funded by a NERC research experience placement. We thank L. Beaufort, A. Purvis and A. Martin for their advice regarding measuring coccolith size-normalized thickness, and UKOA participants for their helpful discussions. Hardware and software support was provided by Olympus Keymed.

Contributions

S.J.G. conceived the project. The research was designed by S.A.O and S.J.G., with advice from P.R.B. and J.R.Y. S.A.O. collected the data, with contributions from S.J.G. and C.N. P.R.B. performed SEM imaging. S.A.O., S.J.G. and P.R.B. interpreted findings and wrote the paper, with advice from J.R.Y., A.J.P. and P.A.W.

CHAPTER 11

Acknowledgments

We thank Dr. Paul G. Matson for sensor preparation in 2011 and United States Antarctic Program staff members, Rob Robbins and Steven Rupp, for SCUBA diving support. We thank Dr. Craig A. Carlson for insightful discussions. This research was supported by U.S. National Science Foundation (NSF) grants ANT-0944201 and PLR-1246202 to GEH. LK was supported by a NSF Graduate Research Fellowship, and ALK was supported by the NSF Postdoctoral Fellowship in Polar Regions Research, award number ANT-1204181.

Contributions

G.E.H. and L.K. conceived deployment strategy. L.K. conducted sensor deployment in 2012 and all data processing. L.K. and T.R.M. determined data quality. A.L.K. assisted with sensor recovery in 2013 and conducted regression analysis. L.K. and E.C.S. conducted model projections. All authors reviewed and contributed to writing the manuscript.

Index

A

abundance, xx, 85, 126, 129–130, 136, 138–140, 143, 155, 161–162, 175, 177, 179–181, 231–232, 258
accumulation, 4, 92, 111–113, 115
acidification, xvii, xxi–xxiii, 5, 7–8, 10, 97, 155, 189, 191–192, 201–203, 207–208, 219–221, 239–241, 247, 249–250, 254–258, 262, 270–271
Actinomma antarctica, 68, 89–90
aerosol, 7–9
Akaike Information Criterion (AICc), 236
algae, 131, 210, 240
anthropogenic, xviii, xxi–xxii, 3–4, 7, 17, 97, 125, 219, 239–240, 255, 258
Arctic, xx, 5–6, 149–151, 157, 160, 165–166, 169, 177–179, 181–183, 241
atmosphere, xvii, 3–4, 17, 36, 64, 67, 77, 83–84, 87, 97, 125, 219, 255, 267
atmosphere-ocean general circulation model (AOGCM), 67, 77, 80
autocorrelation, 20, 23–24, 26, 29, 31, 33, 193
autocorrelation functions (ACFs), 23

B

backscattering, xix, 108, 110, 112–113, 117–118
basification, xxi–xxii, 207–209, 213, 215
bathymetry, 108
bias, 46, 74, 88
biodiversity, 202
biogeochemistry, xvii–xviii, 1, 3–10, 18, 22, 33, 98–99, 108, 119, 219, 267
biomass, xx–xxi, 80, 153, 155, 157, 161–162, 164, 169–172, 180–181, 208–209, 211, 214
biomineralization, xxii, 220–221, 229, 232
bloom cycle, 170
Bottle-Net, 126–127, 132, 135–139

C

calcification, xxii, 213, 219–221, 226–228, 232, 235, 255
calcite ($CaCO_3$), xxii, 8, 117, 136, 168, 219–221, 223, 226–228, 232, 234–235, 239, 255
calcium (Ca^{2+}),117, 192, 198–200, 239, 255
carbon, xvii–xxii, 3–8, 10, 15, 17, 52, 54, 56, 58, 60, 63–64, 68, 80, 84–85, 97–98, 101, 107–108, 112–113, 116–120, 123, 125, 130, 136–137, 149–151, 155–156, 158, 160–162, 164, 166–172, 174, 176–184, 186, 191–193, 198, 201–203, 209, 219, 221, 223, 231–233, 239, 242, 264, 267–268
  biological carbon pump (BCP), xix, 8, 118, 149–151, 168, 181, 202
  carbon budgets, 117
  carbon cycle, xix, xxi, 6–7, 10, 17, 52, 54, 56, 58, 60, 64, 84,

118–120, 123, 158, 160, 162, 164, 166, 168, 170, 172, 174, 176, 178, 180, 182, 184, 186, 201–203, 221, 231, 264
carbon export, xix, 107–108, 116, 118, 149–151, 160, 168–169, 172
carbon flow, 191
carbon flux, 7, 117–118, 136, 161, 191–192, 198, 202–203
carbon isotope excursion (CIE), 221, 223–224, 226–229, 231, 233
carbon pump, xix, 8, 118, 149, 202
carbon sequestration, xix, 7, 64, 118, 171
carbon sink, xix, 63, 68, 80, 85, 183, 202
carbon storehouse, 98
carbon supply, 5, 202
carbon uptake, 107, 268
dissolved inorganic carbon (DIC), xxi, 101, 156, 209, 211–212, 242, 247–248, 250–251, 253–255, 261–263
particulate organic carbon (POC), xx, 108, 111–115, 117–118, 150–151, 155, 159, 161, 164–169, 172, 177, 183
carbon dioxide (CO_2), xvii–xix, xxi–xxii, 4–8, 17–18, 22, 26, 29, 36, 39, 44, 46–47, 54–55, 57, 63–64, 77, 83–84, 86, 92, 97–100, 103, 107, 125, 151, 177, 203, 207–209, 219, 231, 239–241, 247, 250–256, 259, 262, 267
CO_2 variability, 18, 63
cell division, 220–221, 223–224, 226–228, 235
chlorophyll, 6–7, 18, 39, 108, 140, 143, 152, 154–155, 157, 163
chlorophyll a (Chla), 7, 39, 43
chlorophyll fluorescence (chl), 108, 112–113, 155, 157, 163–164, 170–171, 173, 180
subsurface chlorophyll maxima (SCM), 154, 157, 163, 165
climate, xvii–xix, xxii, 3–7, 18, 52, 54, 56, 58, 60, 63–64, 67, 77, 84, 86, 98, 107, 118, 120, 149, 151, 158, 160, 162, 164, 166, 168, 170, 172, 174, 176, 178, 180, 182–184, 186, 192–193, 197, 201–203, 207–208, 219–220, 254, 264, 267–268, 270
climate change, xvii–xviii, xxii, 3, 5–7, 18, 52, 54, 56, 58, 60, 118, 120, 149, 158, 160, 162, 164, 166, 168, 170, 172, 174, 176, 178, 180, 182–184, 186, 197, 202, 207–208, 219, 254, 264, 268, 270
climate cycles, 63
climatology, 5, 49, 51–53, 262–263
coccolith, xxii, 220–221, 223–224, 226–227, 229, 231–235, 271
coccolithophore, xxii, 151, 164–165, 171–172, 175, 181, 213, 219–221, 226–229, 231–232, 235
Coccolithus, xxii, 220, 223–224, 226–228, 232, 236
Coccolithus pelagicus, xxii, 220–221, 223–227, 229, 231–232, 234–235
current, 4, 6, 8, 23, 39, 51–52, 99, 108, 117–118, 125, 178–179, 199, 220, 254, 256, 260
Azores Current (AC), 99, 270
North Atlantic Current (NAC), 99
curve fitting, 18, 21–22, 25–26, 29–30, 43, 56
cyanobacteria, 128–129, 131–132, 143
Cycladophora davisiana, 85, 90
cycling, xvii–xviii, xxii, 3–4, 9, 64, 84, 191, 198, 202–203

D

decay, xx, 133–135, 140, 208

depth, xx, 10, 18, 67, 69, 71, 75, 78, 86, 90, 99, 108, 110–113, 115–118, 125–126, 130–131, 133–138, 140–141, 143, 150–152, 156–157, 159–161, 163–165, 171, 177–179, 181, 202, 209, 214, 223, 241, 258–259
diagenesis, 88
diatom, xviii, xx, 64, 67–71, 73–75, 77–78, 80, 83–89, 91–92, 125, 127–129, 131–133, 135–136, 151, 164–165, 170–172, 175, 179–181, 209–210, 212, 214, 240
dinitrogen, 7–9
dinoflagellate, 127–129, 131–133, 140, 231
displacement, 81–82, 84
dissolution, xxi, 88, 224, 233, 239–240, 255
dissolved organic matter (DOM), xxi, 116, 191–192, 194, 196–199, 201–203
DNA, 128, 141–143
drift, 209, 211
  acidic drift (AD), 209–211
  basic drift (BD), 159, 209–211
  extant drift (ED), 209–211

E

ecosystem, xvii, xxiii, 3, 6–7, 119, 150, 169, 192, 208–209, 213, 240, 254, 267
efficiency, xx, 108, 113–116, 138, 149–151, 160, 168–169, 171, 181
  export efficiency, xx, 149–150, 160, 168–169, 171
El Niño, 39, 52
Emiliania, xxii, 154, 164, 209–212, 226, 229
  Emiliania huxleyi, xxii, 154, 164, 170, 209–210, 226–227, 229, 232, 235
energy, xvii, 3, 88, 270
Ephemera, 164
equilibrium, 69, 164–165, 194, 196–201, 209, 240, 247–252, 254, 262
  equilibrium model, 250
error, 30, 32, 35, 38–39, 42–43, 45–48, 50–52, 67, 74, 77, 92, 116, 162, 172, 246–247, 251, 261–262
Eucampia antarctica, 68, 85, 88–89
euphotic, 64, 84, 102, 108, 112, 150, 154, 156, 169, 172, 202–203
  euphotic zone, 64, 84, 102, 112, 150, 154, 156, 169, 172, 202–203
eutrophication, xxi–xxii, 7, 207–209, 212
eutrophy, 212–213

F

feedback, 219
filter, 112, 139, 141–142, 155–157, 160–161, 192, 197, 242, 244, 261
  filtering, 56, 136, 138, 141, 143
  filtration 138, 156, 167, 192
float, xix, 6, 43, 108–112, 113–114, 116–119
flow cytometry, 131, 143, 159
flux, xix–xxi, 7, 17, 81, 83, 92, 108, 111–118, 135–136, 150–151, 159–161, 164–168, 170, 172, 177, 191–192, 198, 202–203
food, xvii, 3, 213, 254–255, 270
fossil, xxii, 220–221, 228–229, 232, 234–235
fractionation, 67, 69–71, 75, 82
Fragilariopsis kerguelensis, 80–81
fugacity, xviii, 17

G

GEOTRACES, 9–10
greenhouse gases (GHGs), 4–5, 97

growth rate, xxii, 210–211, 227, 232

H

habitat, 68, 70, 75, 78, 86, 90, 162, 256
heterotrophic, 129, 130–131, 151, 177, 179, 181
Holocene, 70–71, 73–75, 77, 82–85, 90
homogeneity, 112, 117–118
hydrophobicity, 192, 199–202
  hydrophobic binding, 198–199

I

ice, xviii–xx, xxii, 5–6, 8, 63–65, 67, 71, 73–75, 77–81, 83–86, 89, 92, 149–150, 152, 156, 162–168, 170–173, 175, 178–182, 240–241, 243, 247, 252–253, 262
  ice cover, xx, 8, 78, 149, 173, 182, 240–241, 247, 262
  iceberg, xviii, 73–74
interpolation, xviii, 18, 21–22, 24–27, 29, 31–32, 38, 42, 46–47, 50–52, 56
irradiance, 156–157, 214, 227, 232
isotope, xviii, 9,64, 67–69, 71, 74–75, 77, 85–90, 130, 221, 223, 231, 233

L

land use, 208
LDEO, 47, 50–51
life cycle, 80
light, 8, 35, 65, 73, 111, 132, 138–140, 151, 155, 157, 182, 210, 212–214, 231, 233

M

mapping, 18, 82, 243
melting, 73–75, 78, 80, 84, 92, 254
mesopelagic, xx, 107–108, 112–113, 115–118, 129, 150–151, 171, 177–183
mesotrophy, 212
metabolism, 179, 202
methane (CH_4), 4–5, 98
migration, 9, 82, 213
mixed layer depth (MLD), 18, 75, 77–81, 83–84, 92, 110, 112, 163, 167, 178
mixing, xix, 6, 68, 78, 81, 84, 99, 101–104, 118, 178, 209, 212, 214, 252, 255
mortality, 80, 134, 140

N

neural networks, 22
nitrification, 8, 213
nitrogen (N_2), xviii, 3–4, 7–9, 64, 131, 141, 143, 209
nutrient, xvii–xxii, 3, 5, 7–9, 64, 68, 81–85, 98, 101–102, 107, 113, 151, 182, 202, 207–215, 227, 232, 260
  macronutrients, 8–9
  nutrient availability, 207, 227, 232
  nutrient cycling, xvii, 3, 84
  nutrient exchange, 64
  nutrient injection, 81
  nutrient limitation, 8

O

ocean
  deep ocean, xviii, 64, 80, 107, 126–130, 132–134, 136–137, 143, 150, 198
  ocean acidification (OA), xvii, xxi–xxiii, 8, 10, 97, 155, 189, 202–203, 207–209, 212–213, 215, 219–221, 227, 229, 231–232,

239–241, 247, 250, 254–258, 262, 270–271
ocean currents, 23, 52, 260
ocean oxygen, 10
ocean ventilation, 64, 83
Southern Ocean, xviii, xxii, 6, 33, 38–39, 47, 50–52, 56, 63, 65, 79, 171, 181, 239–240, 254–255, 257, 260
surface ocean, xviii, 7, 9, 17–18, 22, 29, 47, 56, 64, 75, 81, 84, 97, 104, 130, 181, 198, 202–203, 267
oligotrophic, xx, 8, 130, 135, 137, 208
opal, xviii–xix, 64, 67–69, 77, 80–84, 86, 88, 136, 168
operational taxonomic units (OTUs), 142–143
oxygen (O_2), xviii, xx, 8, 10, 64, 67, 69, 71, 75, 77, 85, 87–88, 151, 155, 161–162, 174, 176–179, 223, 255–256
dissolved oxygen, 155, 174, 177
oxygen demand, xx, 151
oxygen saturation, xx, 151, 177
oxygenation, 5, 7–8

P

paraformaldehyde (PFA), 143, 159–160
partial pressure, xix, 17, 240
particulate, xx–xxi, 7, 108, 110, 112–113, 135, 150, 155, 161, 191
permeability, 132–133, 138
pH, xxi–xxiii, 97, 192–193, 195–199, 201–203, 207–213, 215, 221, 229, 239–262
pH amplitude, 250
Phaeocystis, xx, 151, 164–165, 167, 171–172, 175, 180–181, 240, 251
Phaeocystis antarctica, 181 240
Phaeocystis pouchetii, xx, 164, 167, 170–172, 175, 179–181
phosphorus, xviii, 3–4
photosynthesis, xxi, 112, 177, 202, 211, 254
plankton, xvii, xxi, 89, 125, 129–130, 137, 139, 141, 150, 164, 207–208, 212–214, 231
phytoplankton, xviii–xxii, 5, 8, 123, 125–141, 149–151, 153–155, 157, 163–165, 168–170, 172, 180–182, 202, 207–209, 211–215, 219, 240, 244, 251–252, 254–255, 269
protozooplankton, 64
zooplankton, xx, 151, 161–162, 176–177, 179–181, 213, 252
plasticity, 229, 231
polar, 5, 65, 67, 82, 84, 150, 162, 171, 177–178, 181, 268, 272
pollution, xxii, 7, 208, 213
population, 80, 134, 140, 208, 223–224, 227, 257–258
pressure, xix, 17, 80, 213, 240
partial pressure, xix, 17, 240
productivity, 5–9, 64, 80, 84–85, 98, 101, 209, 252–253, 255
pump, xix, 8, 116, 118, 125, 137, 141, 149, 167, 198, 202
biological pump, 125, 137, 202
purity, 74, 88

R

radiolarian, xviii, 64, 67–71, 73–75, 77–78, 81–90
radionuclides, 10
reliability, 70
remineralization, 116, 150–151, 164, 171, 177, 179–182
remineralized, xx, 151, 171
respiratio,n 151, 161–162, 176–181, 201, 208
Rhizosolenia, 89, 131

Rhodomonas, 209–211
Richelia, 128, 131
run off, 5, 208

S

salinity, 18, 74–75, 78–79, 83, 92, 160, 165, 178–179, 214, 259–262
saturation, xx, 151, 174, 177–179, 219, 221, 239, 248–249
  undersaturation, 103, 240, 247, 249–250, 254–255
scanning electron microscope (SEM), 224, 233, 272
sea level, 6, 92
sea ice, xx, xviii–xix, xxii, xix, 5–6, 63–65, 67, 71, 74–75, 77–79, 80–81, 83–85, 89, 92, 149, 152, 162, 166, 241, 243, 252–253
  sea-ice zone (SIZ), xix, 63–64, 68, 71, 77, 80–81, 83, 85
seasonality, 51, 150, 254–255, 257
sediment, 65, 67–69, 71, 73, 77–78, 80, 83–84, 87, 91–92, 115–117, 167–168, 221, 224, 233
  sediment trap, 68, 92, 116–117, 167–168
sensitivity, 64, 77, 84, 92, 202, 208, 211–212, 215, 219, 254
shelf seas, xxii, 7, 212–213, 215
shoaling, 113, 115–116
silicon (Si), xviii, 9, 64, 67, 69–71, 73–75, 77, 80–83, 85, 87–88, 90, 212, 214
  silicon inventory, 64
small particles, xix, 107, 115–117
SOCAT, xviii, 18–19, 22–23, 26, 33–35, 44–50, 53–54, 56, 267–268
solubility, 9, 98
spectrometry, 88
Spongotrochus glacialis, 68, 89–90
standard deviation, 23, 116, 172, 192, 241, 261
stratification, xviii–xix, 5, 8, 64, 68, 75, 78, 80–81, 98, 101–104, 116, 182, 240, 255
stratigraphy, 77, 85
supersaturation, 104, 178–179
Synechococcus, 131

T

temperature, xvii, xix, xxi, 7–8, 18, 26, 39, 64, 69, 71, 73, 82, 85, 97–104, 108, 133, 140, 143, 151, 161–162, 178–179, 192–203, 207, 213–214, 227, 231–232, 240, 242, 244–247, 251, 253, 255, 259–263
  sea surface temperature (SST) xix, 18, 39–41 43, 73, 85–86, 97–104, 161, 231, 260
  temperature dependency, 198
  temperature variation, 98, 194, 253
tephram 74, 88
Thalassiosira 68, 164, 209–211
  Thalassiosira lentiginosa, 68, 88–89
  Thalassiosira weissflogii, 209–210
thermodynamic, xix, 92, 98, 101, 103–104, 246
time, xxii, 5, 18, 21–22, 24–29, 31–33, 35, 38–39, 44, 46, 71, 73, 75, 78, 83–84, 86, 98, 100–101, 108, 110, 112–113, 115, 134, 140, 142, 160, 167, 170, 173, 177, 180, 193, 196–198, 212, 214, 220, 223–224, 227, 229, 231, 241, 244, 254, 257, 259, 261–262
temporal, xviii, xxii–xxiii, 10, 18, 22, 24–25, 29, 33, 46, 50, 52, 57, 98, 173, 240, 257, 261
Toweius, 221, 223–224, 226–228, 232, 236

Toweius pertusus, 221, 223–224, 226, 229, 231, 234
trace elements, xviii, xxi, 3–4, 7, 9–10, 202
transport, xvii, xix–xx, 3, 9, 51, 64, 98, 102, 125, 130–131, 136, 139, 202
Trichodesmium, 128, 131–132
troposphere, 4
turbulence, 103

U

uncertainties, xviii, xix, 5–6, 18, 22, 29–32, 35–39, 44–57, 70, 82, 85, 108, 113, 117–118, 181, 223, 261
  root mean squared (RMS), 29, 31, 39, 44, 48, 51–52

V

variability, xviii–xx, xxii–xxiii, 4, 18, 20, 23–24, 29, 32–33, 38–39, 43, 47, 50–52, 55–57, 63–64, 69, 71, 77–78, 83–84, 86, 99–104, 150–151, 157, 166, 168, 212–213, 215, 233, 236, 239–242, 245, 250–254, 256–257, 261
  interannual variability (IAV), xviii, 18, 32, 39, 43, 47, 52, 55–57, 99
  seasonal variability, 84, 100
  spatial variability, xx, 20, 23–24, 29, 33, 39, 50, 57, 84
vegetation, 77
viability, 134, 138–139

W

warming, xxi–xxii, 5–6, 8, 75, 84–85, 191–192, 197, 201–203, 221, 242, 253, 258, 263
water, xvii–xix, xxi, 5, 7–9, 17, 39, 64–65, 67–71, 73, 75, 79, 81–84, 87, 89–90, 97–104, 108, 111–113, 116, 118, 133, 135–136, 138–142, 150–151, 155–157, 160, 162, 165, 177–181, 192–193, 201, 207, 213, 219, 221, 229, 231, 252, 254, 262
  Circumpolar Deep Water (CDW), 6, 70–71, 73, 81–82
  freshwater, 5, 73–75
  seawater, xxi–xxii, 67, 69–70, 73, 98, 102–103, 138, 141, 143, 156, 159, 161, 192–193, 197, 207–209, 212, 215, 240–241, 247, 252–253, 257–259, 262
  surface water, xix, 7, 9, 17, 64, 68, 73, 75, 81–83, 97–98, 100–104, 219, 221, 229, 231
  water column, 5, 8, 64–65, 68–71, 75, 82, 90, 108, 111–113, 116, 118, 133, 135, 139, 150–151, 157, 162, 165, 177, 180–181, 201, 213, 252, 254
  water depth, 67, 69, 71, 75
wind, 5–6, 64, 77, 81, 83–84, 92, 214, 240, 253, 255